A MACHINE TO MOVE OCEAN AND EARTH

A MACHINE TO MOVE OCEAN AND EARTH

The Making of the Port of Los Angeles—and America

JAMES TEJANI

W. W. NORTON & COMPANY
Independent Publishers Since 1923

For information about permission to reproduce selections
from this book, write to Permissions, W. W. Norton & Company, Inc.,
500 Fifth Avenue, New York, NY 10110

For information about special discounts for bulk purchases,
please contact W. W. Norton Special Sales at
specialsales@wwnorton.com or 800-233-4830

Manufacturing by Lakeside Book Company
Book design by Lovedog Studio
Production manager: Anna Oler

ISBN: 978-1-324-09355-8

W. W. Norton & Company, Inc.
500 Fifth Avenue, New York, N.Y. 10110
www.wwnorton.com

W. W. Norton & Company Ltd.
15 Carlisle Street, London W1D 3BS

1 2 3 4 5 6 7 8 9 0

CONTENTS

Part Three: REBELLIONS

Part Four: CAPITAL

Part Five: EMPIRE

A MACHINE TO MOVE OCEAN
AND EARTH

Introduction

EXCAVATING
THE LOST COAST

I T WAS NO PLACE FOR A HARBOR. HE SAID SO UNTIL THEY stopped asking. Now the Port of Los Angeles was under construction. Promoters hailed its artificiality. They predicted it would be, upon the opening of the Panama Canal, the premier harbor of the Pacific coast. The port, they said, signaled the United States' rise to global prominence. Yet, for him, the port marked an ending.

George Davidson, famed geodesist and the West Coast's only resident member of the National Academy of Sciences, had lived to the year 1911, long enough to see numerous marvels of history. He was eighty-six years old, frail, and nearly blind from a career that depended on his eyes above all else. Over six decades, he had traveled four hundred thousand miles—equivalent to sixteen times around the Earth—in pursuit of knowledge. Visitors to Davidson's home near Lafayette Park in San Francisco noted how his memory remained sharp as a blade. That gave him comfort. But there were unknowns, irksome things beyond his view.[1]

Despite Davidson's accolades, he felt estranged from the world of his old age. History seemed to grow distant, like a steamship heading off to the horizon, while he watched it move on without him. Things had been different in 1850, when a ship first brought him to California. He was young then, sent out by the government to survey an unknown coast. His work would bind the recently conquered territory to the United States, fulfilling the nation's expansionist designs. It would require decades and measurements by the tens of thousands. Through his efforts, pieces became one. Some were displaced, however, pushed to the margins or lost. Davidson still could see them in his mind. He remembered the region's Native Americans. Many were survivors of the Franciscan missions. Others, farther north, nearly killed him to protect their homelands. He recalled Mexican ranchers, some old enough to have sworn allegiance to the Spanish crown. They felt sufficiently secure, even in the wake of the US conquest, to threaten him off their property. Davidson thought of others still: government officers, soldiers, political luminaries, fellow scientists, and business proprietors. He had worked alongside them. Few remained alive in 1911.

Most of all, Davidson remembered the wild coast. He had learned its contours like a loved one's face. He had calculated its expanse. He had recorded its indigenous and Hispanic place-names so they would not be forgotten. At times, the Pacific coast left him awestruck and afraid. It certainly left his body broken. He once knew its secrets better than anyone, yet his task was to divulge them so that settlers and commerce might overrun the region. He hoped science would steer all toward a greater good. Instead, fracture and conflict recurred. The nation endured a bloody civil war and the pains of industrialization. Powerful corporate masters arose from the West to bend local and national governments to their whim. They diminished science, at least the public-minded kind Davidson practiced. The country itself, its people also, had changed. Meanwhile, the wilderness of his youth disappeared. Property and capital were everywhere. Machines tore apart the shore or buried it under boulders and concrete. Animal life vanished. Nature's beauty remained, but it was no longer so fearsome.

All of this was, on one level, the achievement of Davidson's life. By the twentieth century, his maps had made the world smaller and more comprehensible. Ocean and earth moved, joining places and people in ways impossible before. But the world had grown complex also, too vast and faceted for Davidson to take in. That was his life's dilemma. Parts became whole. And as they did, their workings became harder to discern. Nowhere more so, perhaps, than at San Pedro Bay, near Los Angeles. It was a site he knew well. Decades earlier, he had written that a harbor of scientific merit could never be built there. Its boosters proved irrepressible no less. Their success told him that his time had passed, and that stirred memories all the more. Somewhere among them were the answers. What had happened?

THIS BOOK IS the first to answer. It explains how people and events transformed San Pedro Bay by the start of the twentieth century from estuary into the nation's global gateway. The process culminated in the construction of the Port of Los Angeles, which today operates as a central engine of the US and world economy. The port is all around us. Objects we use on a daily basis pass through it or through supply chains for which it is an essential node. This is especially true for Americans, with their appetite for East Asian imports. Furniture, apparel, plastics, footwear, electronics, and automobiles top the list. In a recent year, the harbor transshipped $311 billion worth of cargo, weighing 219 million metric revenue tons. San Pedro Bay's port complex claimed one-third of all US ocean shipping and a link to one in sixty (or two and a half million) US jobs. It is the busiest container port in the Western Hemisphere. Hence, it is, along with the Panama Canal, a key component of a machine circulating goods and services, culture and power, across the Pacific and beyond. When cargoes bypass San Pedro Bay, they arrive at places for which it is the archetype and just a few steps removed on the chain. Even in our digital age, we all have a port somewhere.[2]

Despite its significance, the port remains invisible and taken for granted. That is because of its success and our dependence upon it. Without the items that pass through its associated networks, our world is simply unimaginable. Yet the port's absence—out of mind and out of sight—is a result, too, of history. Astonishingly, San Pedro Bay has been left out of American historical narratives. Its story was lost and forgotten, and perhaps never fully understood even by participants. Some disowned it, as reputation and politics required. Davidson kept—or was kept—quiet, as time proved him wrong. Silences allowed a local mythology, of modern Los Angeles and its enterprising founders, to hold sway. Lost is the reality that the making of the Port of LA spanned from the estuary to mountains and deserts; across the continent to the Apacheria, to Washington, to New York, and to Civil War Richmond; overseas to Europe, Hawaii, and Asia; and back again. Those places, seemingly peripheral to the bay, often were the cause for its development. And the bay, seemingly peripheral to national and international events, was in fact anything but.

This is a story of the United States' formative yet troubled expansion to the Pacific, of its relationship to an envisioned ocean world, and of Americans' attempts to master this world—a story that led early on to San Pedro Bay and continues still. By the mid-nineteenth century, Americans identified the western shore as the republic's grand object, a place of commercial opportunity as well as a marker of the nation's maturity among nations. They saw this as a unifying project, a way to join the country's fragmented pieces across vast territory and the divides of party and section. Likewise, they saw reaching the Pacific as a way to evade social inequalities that beset humanity elsewhere. This vision was never quite realized. Instead, Americans' attempts at it tore open gaping divides, over slavery and railroads most of all. Even so, the attempts produced maps, rails, machine technology, capital, and stronger government institutions. Together, they shifted the nation westward. And together, they focused US expansion and engagement with the Pacific upon Southern California's coast.

The result was the Port of Los Angeles, an engineering marvel that broke the limits of nature and distance, along with a new metropolis made in part by its claims to an artificial harbor. Both have transformed the United States, providing a foundation for the way Americans live today. But this is a global story, too, and one with great relevance. The port continues to shape a troubled relationship between the US and the Pacific. Knowing its origins may guide us through the troubles of our own time.

Long before I knew this story, San Pedro Bay took hold of my imagination. A traffic jam of idled ships, as the port struggled to keep pace with globalization, provided a backdrop to my hours in the surf at the bay's southern limit. From a car window, I saw the port's industrial grandeur: giant cranes, colorful containers stacked high or carried on railways, and lots with hundreds of identical unpainted cars. The clean angles of slips and channels stood out while cresting the suspension bridge over the harbor. Refinery plumes and chemical tanks appeared endless. These visible signs are only a fraction of the bay's tale, however. I set out to understand its unseen connections, to uncover its buried secrets, including the great estuary that once was.

Today's port, entrance to the inner harbor, and former estuary

That search led me to a set of fragments, of oddly connected persons and scattered archival sources: letters, maps, reports, land patents, court papers, memoirs, and newspapers. Each document revealed little—a meaningful phrase, perhaps, mixed among the chatter of long-gone daily affairs. But notes and paper accumulated through years of research. Names, relationships, and life paths drew out in webs. Dates fell into sequence. When aligned, the fragments revealed a previously hidden story of the bay, from the 1850s to century's end and beyond. Like my hours immersed in the sources, this story is one of pieces and the human need to unite them in meaning. But it is also one of incapacity, the blindness that prevents us from seeing our historical moment in full and from steering the outcome as we otherwise might. The book's characters continuously found themselves caught between intentions and consequence. We still do.

THE CHAPTERS THAT FOLLOW detail how the Port of Los Angeles came to be, from a bay of mud and salt marsh. They tell of an unlikely set of individuals brought together by an estuary, and of contests to control it. The narrative begins in the wake of the US-Mexico War, when the United States first established a claim to the Pacific coast and interior. That claim quickly slid into chaos, as the California Gold Rush brought settlement, violence, and unfettered development. The US Coast Survey (predecessor to today's National Oceanic and Atmospheric Administration) sent out scientists like Davidson with the hope that its precision maps might bring order to diverse spaces and people. The effort drew Davidson and the Survey to San Pedro Bay, and, in turn, into the republic's sectional crisis. The southwest desert—Southern California included—offered superior access to the Pacific, making it the place where politicians competed to locate a transcontinental railroad and thereby shape Western empire and federal union. Despite the Coast Survey's hope to keep science pure from politics, its leaders became trapped in the schemes of others.

Yet the Southwest was a region where various Americans had to negotiate conquest. The desert represented a borderlands. This key term in scholarship on US history views the Southwest as a place beyond the control of governments at Mexico City and Washington, one characterized by indigenous resistance and local identities that spanned culture and nationality.[3] San Pedro Bay marked its far western edge. Here, Manuel Domínguez and his family controlled a massive Spanish-era land grant. US expansion, science, and sectionalism came to them. Nevertheless, Domínguez embraced the national project, as his family sought to transform their grant into private property. That won them allies among Anglo settlers and real estate developers, but it also put the Domínguez family into contention against imperialist slaveholders like Jefferson Davis and land speculators like Edward Ord—the family's friend, Coast Survey officer, and secret informant to Davis. The political contest over rails and slavery would fracture the republic and bring on the Civil War. In Southern California, the contest had similar results while also confounding our assumptions. Expansion and profit made partnerships regardless of the slavery issue, pairing Mexican ranchers with Southern immigrants and a future hero of the Union Army with the Confederate president-to-be.

During the Civil War, San Pedro's estuary served as a harbor for the first time, enabling the US government to fight secessionist rebels and to subjugate the indigenous peoples of the desert. That forged a new partnership at the bay, one of speculative business and federal power, that edged out many of the prior players. Local freighter and developer Phineas Banning gained lucrative contracts to supply a Union Army commanded by James H. Carleton. Their joint efforts weakened the resistance of the Mojave, Apache, and Navajo and sped the influx of soldiers and settlers. At the same time, the partnership set up future conflict, as Unionist entrepreneurs like Banning coalesced with corporations such as Leland Stanford and Collis P. Huntington's Southern Pacific Railroad. The company arrived at San Pedro Bay in the 1870s to take control of shipping.

It soon monopolized much of California, the coast, and the desert, turning the region into its business empire. Army and government engineers—the victors of the Civil War—tried to harness the power of capital for the nation's purposes. Some already looked to the Pacific. However, whether they were returning scientists like Davidson or triumphant generals like Ord, they were no match for wealthy railroad titans whose influence corrupted public institutions and locked up harbor access and waterfront lands.

The reversal of corporate power at San Pedro Bay had an unexpected start in the borderlands property disputes left over from conquest. In 1886, Manuel Domínguez's daughters sued to regain estuary lands that the US had removed from the family's grant decades before. Their suit, decided by the Supreme Court and Justice John Marshall Harlan, enhanced the authority of government at the very moment the United States extended its reach to seize Pacific island colonies and Asian commerce. As Harlan recognized, imperialism abroad required seaports. And that required public control over the places where ocean and continent meet. Yet harbor development would take much more: brilliant new technology to tear apart nature and reconstitute its elements. Army engineers like Amos Fries and bureaucrats like William Howard Taft, both veterans of the Philippine War and occupation, would wield immense dredging machines to carve out the estuary's shallows. That gave opportunity to a new speculator. Los Angeles enlarged its city limits to take charge of the bay and hitched its prosperity to an emerging US Pacific in the twentieth century.

The Port of Los Angeles is the artifact of this history. As it finally took shape, the actions and motives of individuals had less effect than the port's overall, machine-like operation. Cargoes moved by an integrated system of ship, rail, and truck. Mechanization replaced tasks previously done by workers. Large institutions, staffed by interchangeable hires, directed the port. Market forces of supply and demand did, too. For this reason, this book does not tell the port's story beyond the first decades of the twentieth century. It stops at the estuary's destruction, the moment at which humans severed San Pedro Bay from its natural and indigenous past to make it serve the needs of modern mass society. Stories of lost ecology and of the bay's First People, the Gabrieleño-Tongva, remain threads throughout. They provide context to the port's making, a counternarrative no less important, although limited by available historical sources.

Despite the intricate logistics and highly managed waterfront that constitute the port today, its history is one of remarkable conflict and adverse consequences. That is the story's great surprise, among many I found. The port emerged against doubts and long odds from a cacophony of interests and events. Most of all, it formed because people aspired to connect the fragments of their world and to gain the material blessings that might bring. Their efforts exacted a price on nature and persons caught in the way. But it often did as well on persons who dared to partake of ambition and risk. Like the people in this book, today's Americans (along with a large share of the world's population) are engaged in similar efforts, even if they are unaware of it. San Pedro Bay is evidence of the human capacity to construct a deeply intertwined existence. Its history should caution us, too, that the world we make ultimately may not be the world we want.

So it was in 1871, during the era of nascent American and global connection, when Walt Whitman wrote to celebrate the distances "brought near" and lands and peoples "welded together." Amid his exhilaration about railroads, steamships, and canals, Whitman pondered the reckless soul of ambition, with its unknown paths and ending. In the intrepidness to sail forth, he noted, "we will risk the ship, ourselves and all."[4] The present is not so different. Ships still press off to the horizon and back again. As people living between past and future, we, too, strain our eyes to see clearly.

Part One
BORDERLANDS

The Invaders

ＴHEY CAME BY SEA.

Fur trappers, cavalry, and covered wagons provide the lasting images of US westward expansion. But outsiders first came to California by way of the coast. It was the unexpected arrival of a supply ship that saved the 1769 "Sacred Expedition" from failure. The Franciscan missionary Junípero Serra pronounced it a miracle and proceeded to set up the first Spanish colony in Alta California. The Yankee vessel *Lelia Byrd* trespassed through San Diego and San Pedro Bays in 1803, one year before Lewis and Clark's Corps of Discovery started up the Missouri River. Forty-three years later, hundreds of US troops landed ashore as navy commodore Robert Stockton declared that California was no longer Mexican territory. The soldiers displaced a smaller force led by John Frémont that had trekked across the continent. These Americans of 1846 were forerunners. The discovery of gold brought invaders by the hundreds of thousands. The earliest, during 1848 and 1849, arrived by ship. They took first sight of California from the vantage point of Pacific waters, not the other way around.

This was true as well for George Davidson. On a June morning in 1850, he disembarked the overcrowded Pacific Mail steamer *Tennessee* and walked a makeshift gangplank to the shores of San Francisco. He did not look like an invader. The twenty-five-year-old from Philadelphia wore civilian clothes. He carried the instruments, notebooks, and meager salary of the United States Coast Survey. But

"You know me too well to think that I desire a life of indolence. . . . What is life without ambition?" George Davidson wrote in 1853 to his future mother-in-law.

his bookish appearance masked a taste for adventure and profanity. Like many Americans of his time, he was familiar with outdoor life and adept with a gun.[1]

Davidson would need all of his qualities. He came to research in the uncharted wilderness of the Pacific coast. Most Gold Rush migrants arrived to seize land and resources for themselves. They pushed prior occupants aside and justified this as progress, civilization, and racial superiority. Davidson shared those assumptions. But they were not the larger part of him. Instead, California offered the opportunity to study and banish the unknown. Science was his life's object. He pursued it not merely for himself, but for the US republic. Davidson was a *government* scientist. That proved irreconcilable at times. It drove him to overexertion, suffering, and acclaim. His steps held many dangers. They led both north and south and soon to the estuary at San Pedro Bay.

There, Davidson's path would intersect with that of Manuel Domínguez. He, too, was an invader, but without the need to voyage anywhere. Domínguez was a "native" Californian (as Americans of the era called the Spanish-speaking upper class) and the mixed-race descendant, two generations removed, of a soldier who led the

1769 Spanish conquest. He saw Southern California as his birthright. And he wanted more. Domínguez spent his adulthood consolidating a massive land grant—the legacy of Spain's and Mexico's expropriation of Native peoples—located on the shores of San Pedro Bay. Using politics and law, he drove off challengers and guarded his family's Rancho San Pedro. But he also grew frustrated at Mexican authorities. US conquest seemed to improve his prospects. Domínguez would succeed. But he would fail, too. Both results were set in motion by the young, arriving scientist and the instruments he carried.

That an estuary brought Davidson and Domínguez together speaks to the complex story of conquest on the Pacific coast. Invasion was most visible in military action and the flood of settlers. Yet its enduring force came by less visible things like law and bureaucracy, the building blocks of order in society. Federal agents like Davidson came to study and catalog. Claimants like Domínguez appealed to a once-foreign government to perpetuate their property and privilege. Both looked to piece together California's past into a future they desired. But the more Davidson or Domínguez acted to realize their ambitions, the more they found the outcome beyond control. That caused their paths to encircle, along with many others in far-off places. The story began decades before, with a flawed though brilliant idea to fix the American nation.

Chapter 1

EUROPEAN MEASURES,
WESTWARD AMBITION

THE WORLD OF 1800 WAS FLAT AND FRAGMENTED, AT least on paper. People of the post-Enlightenment age knew the Earth to be a vast globe. But they could neither quantify nor visualize it. Maps of the time falsified space. Worse still, they did so in proportion to their reach. Only the most local of maps—those least in the spirit of discovery—had accuracy. This problem preoccupied European science and its patron-monarchs. For the American republic, the same posed an existential dilemma. The young United States looked west with democratic spirit and desire to overspread the continent. Expansion of that scale might occur. But growth also promised distention, dissolution, and chaos. Its violence might never be calmed. Conceived as a hopeful experiment and imperialist practice, the American nation contained a destructive paradox.

In the years before George Davidson's birth, the United States Coast Survey (USCS) set out to solve this problem. Congress established the office in 1807 to study shorelines, harbors, and maritime hazards. The nation was smaller then. Its boundaries ran along the Atlantic Ocean and Gulf of Mexico, with the exception of Florida (Spanish territory until 1819) and Texas. The office worked under the Treasury Department, as its charts would assist in tariff collection. During its first decades, the Coast Survey looked to the Atlantic world and rarely strayed beyond the realm of existing commerce and settlement. It took no role in frontier expansion.

That set the agency and its science on a collision with the rising democratic age.

Head officer Ferdinand Hassler personified the Survey's aspirations and inborn difficulties. He was a refugee from Europe who sought order and rationality in the New World. As a boy in the canton of Bern, Hassler studied land surveying, and later mathematics and astronomy in Paris. But the French Revolution, Reign of Terror, and Napoleon's armies intervened. War engulfed the continent. In 1805, Hassler fled to the United States as part of a Swiss farm colony. He might have vanished into obscurity had an agent not embezzled the colonists' funds. Hassler contacted the American Philosophical Society in Philadelphia, hoping to sell off his instruments and sizable library. In turn, the society alerted Treasury Secretary Albert Gallatin and President Thomas Jefferson. The pair looked to build an enlightened republic by importing scientific knowledge and practitioners. Hassler seemed to wash up to their good fortune. Through the president's influence, he received a professorship at the US Military Academy at West Point. When the federal government solicited proposals for a coastal mapping office, Gallatin and a peer jury selected Hassler's from the pile.[1]

Over the next three decades, Hassler would model the Coast Survey on the great geographic societies of Britain and Europe. By that same model, he adopted the cutting-edge practice of geodetic triangulation. Geodesy offered a solution to a problem of the age, one especially troublesome to expanding empires, kingdoms, and the US republic. Each claimed territory by hopelessly wrong methods. Their maps treated the surface of the Earth as a perfect plane rather than the imperfectly shaped sphere (or oblate spheroid) that it is. Because of the Earth's true shape, parallel north–south lines converge as they move from the equator to the poles. In addition, any straight line will bend along the planet's curved surface to become an arc of added length (eight inches per mile, approximately). Put simply, there are no straight paths across the Earth. By ignoring this fact, flat-Earth surveys produced imprecision. Errors became noticeable at a scale longer than five miles. At a span of one hundred miles north–south, the error grew to nearly four percent. Nevertheless, for the sake of simplicity, governments clung to their faulty methods and uncertain domains.[2]

Hassler chose otherwise. Rather than mapping on an imaginary plane, geodesy employs advanced mathematics to locate on a three-dimensional sphere. Surveyors set up a triangle of points on the landscape. Using astronomical observation and precision instruments, they measure an arcing baseline (or base) for the triangle's first side. From the two endpoints of the base, surveyors then run imaginary courses to a third point, or vertex, in the distance. Then, by calculating the baseline's angles and length, surveyors employ Euclidean geometry to solve for the triangle's two unknown sides. To account for the Earth's curvature, surveyors observe from each vertex the latitude, longitude, and azimuth. The latter is the spherical angle (on a 360-degree field) between north and an astronomical reference— the moon or a prominent star, for example—as it appears above the horizon. When combined, the data transforms the triangle initially set forth into a three-dimensional shape that fits in true position on the planet's surface. Surveyors repeat the process with any of the known sides serving as the baseline for an adjacent triangle. Because geodetic data interconnects and compounds, repetition corrects the minutest of errors toward ever-greater precision.[3]

Geodetic triangulation promised to transform the United States and the way it was understood. Hassler's staff began to assemble a cohesive map of the republic, one greater than the sum of its parts. The map would bring regions, landscapes, and many peoples into one. Mathematical triangles made for a more perfect union, which, in itself, was a statement to the Old World about social organization and free government, the vitality of American independence. It was a statement, as well, on the potential for republican science. North America, with its expanse and wilderness, offered an untrammeled laboratory for both politics and geodesy. A precise grand-scale map to encompass several hundred or several thousand miles would be the envy of Europe, where complex borders, warring nations, and compact kingdoms obstructed such an endeavor. Hassler's geodetic framework lives on today as the National Spatial Reference System, managed by the Coast Survey's successor, the National Oceanic and Atmospheric Administration.[4]

At the time, however, critics were unimpressed. The Coast Survey faced constant doubts from Congress, which controlled its funding. Conventional maps, legislators said, showed far more progress per dollar spent. Errors, although regrettable, seemed not to be of much consequence. They starved the Survey of money, downplayed its accomplishments, and gave away resources to rival agencies. The US Land Office expanded its surveying throughout the Mississippi River Valley as it followed Indian removal and land preemption policies. The navy increased its mapping. Meanwhile, the Army Corps of Topographical Engineers—established in 1838 and led by a Hassler student—emerged as the premier federal science bureau. Topographical engineers improved inland roads and waterways (including the Great Lakes), explored the West, and designed forts and lighthouses. The Coast Survey's scope, by contrast, grew relatively smaller. Its triangulation covered only the seaboard between Narragansett Bay and Cape Henlopen (roughly due east of Washington, DC). Given the labor involved, that was no small achievement. But congressional representatives from outside the North Atlantic region saw little benefit.[5]

Hassler was ill equipped to counter the skeptics. As the temper of Washington turned more democratic, he grew unpopular for his European leanings—for example, his ornate horse carriage (maintained by government stipend) and his efforts to promote the metric system. Foes attacked his salary, which ballooned to $6,000 and compared to that of cabinet ministers. Treasury Secretary Levi Woodbury made it his personal crusade to force a pay cut. Hassler took his case straight to President Andrew Jackson, declaring in his German accent that politicians of the age could—and apparently did—make a treasury secretary out of just about anyone. But, he added, the president could *never* make another Ferdinand Hassler. Jackson famously championed the common man against the few. He was, however, a nationalist and military general who knew the value of reliable maps. Hassler kept his salary. Critics simply waited. In 1842, Congress convened a special committee to eliminate the Coast Survey altogether.[6]

Hassler faced a predicament. He had set up the agency to pursue elite science and long-range objectives. Yet, its lifeblood depended on democratic politics and immediacy. He planned for geodetic triangles to meticulously walk the nation's shoreline from start to finish. Still, this left the larger portion uncharted. And the project meant little to voters, especially the growing number of inland settlers. Republican science had become a contradiction, one that Hassler could not resolve. As a final act, he gathered his few allies—navy commanders and army topographical engineers—to defend the USCS and its mission. They proposed reforms, which Congress adopted into law in 1843. Hassler died that year, at age seventy-three, his legacy secured but left for a successor to realize.

THE REFORMS OF 1843 inaugurated the Coast Survey's golden era, which would last through the Civil War. The bureau emerged with enhanced status and guarantees. Its chief, the Survey superintendent, would report to Congress, which could no longer withdraw funding arbitrarily. The superintendent could request navy officers to command ships and West Point–trained army officers to lead fieldwork on land. That made the Coast Survey the indispensable cartographic service. It would serve as a bridge between the army and navy and between civilian and military science.[7]

At the same time, the Survey gained a leader able to triumph over opponents, rivals, and obstacles. Alexander Dallas Bache was one of the United States' preeminent scientists. He was also a great-grandson of Benjamin Franklin, the republic's first internationally acclaimed polymath. Bache burnished the Survey's academic reputation, and he would use the title of "professor" throughout his twenty-four-year term. He had a sterling military record as well, having graduated at first rank in the West Point class of 1825. The feat qualified him for just about any position within the army's top echelons. Instead, Bache left for the University of Pennsylvania, whose faculty he joined in 1828, at age twenty-two. His army credentials continued to command respect, even awe. Aspiring officers eagerly applied to the Coast Survey to train under its famed former cadet.[8]

If that were not enough, the superintendent hailed from a family steeped in federal politics. In addition to having a constitutional framer and revolutionary hero as his forebear, Bache could claim one grandfather who served as James Madison's secretary of the treasury, (acting) secretary of state, *and* (acting) secretary of war; another grandfather who served as postmaster general of the Continental Congress; a paternal uncle (by marriage) who served as Jackson's treasury secretary; a maternal uncle who, between 1845 and 1849, held the office of vice president; and a brother-in-law (via sister Mary Bache Walker) appointed as treasury secretary during the same years. Bache was reared in the power networks of Philadelphia

Revered by politicians, military leaders, and scientists, Alexander Dallas Bache had no equal.

and Washington City. He had many friends, few enemies, and keen political instincts. As one of his staff wrote, Bache's "skill consists in his perfect directness, truthfulness, disinterestedness, and good temper."[9] "Such men," the observer marveled, "have most influence with all men of sense, whether members of Congress, or men in official station, or men in private life. This is the secret of his getting along so well."

As a star of equal brilliance in academics, military affairs, and politics, Bache raised the Coast Survey's profile and disarmed its critics. Yet his most notable exploit was how he accommodated the agency to the democratic ethos without sacrificing scientific rigor. Bache understood that much of the Survey's prior difficulty stemmed from its slowness and neglect of large sections of coastline. Hence, he reorganized fieldwork to proceed concurrently in eight zones (increased to eleven by the mid-1850s). This untethered the Survey from the Atlantic Northeast and freed it to move apace with the nation's borders: around Florida, across the Gulf, and beyond.[10]

Bache soon enlisted the Survey in territorial expansion. In 1847, during the US-Mexico War, he lobbied Congress to extend his purview to the Pacific. This was a strategic step. Enough work remained in the East to keep the agency busy for decades. The annexation of Texas had recently added 3,360 miles of true shoreline (the aggregate length of every bend, nook, inlet, and island). But as the war wrapped up, dozens of capable army and navy officers sat idle at Monterey and San Francisco. Bache could not allow them to take up surveying. Instead, he offered his staff's geodetic services. Congress declined at first but reconsidered once the Gold Rush caused a boom in ship traffic to California. In 1849, legislators appropriated funds for a Pacific Coast Survey.[11]

Bache possessed an intuition about American impulses that Hassler never had. In the span of several years, he repositioned the USCS in service to frontier and empire and to the popular passions beneath. By the same stroke, the Survey would harmonize sections increasingly at odds over slavery and the profits of Western development. Maps promised to contain the divide within coastal borders and shared ocean commerce. The agency benefited politically on both accounts. It leaped ahead as the strongest and best-funded scientific bureau. Research entered new areas like weather and oceanography (starting with study of the Gulf Stream) and investigation of the Earth's magnetic field. The Survey emerged as a point of consensus in Washington. The superintendent took that as a sign that science, his priority and passion, was a matter of agreement, too.

WASTING NO TIME, Bache dispatched the first cohort of staff to California. Their mission was to bring order to a territory that seemed the heart of chaos. Migrants flooded California by the tens of thousands. Gold of similar value in dollars flowed outbound. Shipwrecks occurred regularly, jeopardizing lives, investments, and treasury. The Coast Survey would need to chart shorelines and select sites for lighthouses or warning signals. At the same time, the Survey had a more abstract task. San Francisco, Sacramento, and the mining camps were a place apart from the republic's genteel communities. Unrestrained growth made California lawless and competitive. Crime was rampant. Morality was loose. Living conditions were rough and primitive. Society lacked libraries, theaters, churches, and schools. The saloon served as pastime and town hall. But Coast Survey charts would change this. They would give an outlet to commerce and the volatile energy of settlers. Like written law, printed maps were thought to tame the wilderness and civilize human existence.

They would enable government, as well. Since the seventeenth century, Europeans had produced rudimentary charts of the Pacific coast. These proved fanciful, for a time depicting a Northwest Passage, or California as an island. They were dangerous, too. One Coast Survey employee recalled charts with errors of *five to nineteen miles* east to west—more than enough to send a ship and its passengers smashing into rocks. Inaccuracy revealed the limits of statecraft. The western coast was a borderlands. The United States had yet to understand the territory, much less control it. That fact grew starker each day. Gold drew migrants from a world unsettled by capital and revolutionary conflict: from Mexico, Latin America, the British Isles, non-English-speaking Europe, Australia, Hawaii, and China in addition to the US. Indigenous people were pushed to the mines, as the intruders disrupted traditional ways of life. California contained a diverse swath of humanity, beyond any place on Earth. But diversity exacerbated tensions, spiraling toward greater fracture.[12]

Duty on the Pacific was not for the meek. Assistant James S. Williams and Sub-Assistant Joseph Ruth arrived during the summer of 1849 to measure latitude and longitude near San Francisco. For riskier exploration, Bache sent William Pope McArthur, a thirty-five-year-old navy officer. He would command the topsail schooner *Ewing* in the urgent work of hydrography (soundings of depth and the sea bottom). McArthur was harsh and hardened (with a lead musket ball lodged in his leg). He had to be. Ship crews consisted of unsavory toughs and cutthroats, some impressed against their will. A successful officer had to be domineering and profuse with corporal punishment.[13]

Still, California tested McArthur's mettle. He traveled separately from the *Ewing*, intending to arrive early. The journey took six months. He might sooner have walked to California from his native Missouri. McArthur's troubles started in Panama, which was bottlenecked with gold seekers. At the Caribbean-side town of Chagres, crowds waited to cross to the Pacific by foot or mule while opportunists preyed upon them. Twenty-seven-year-old Collis Huntington of Oneonta, New York, arrived with a load of whiskey, sugar, dried beef, and flour. He sold the goods at inflated prices, then parlayed his winnings into a train of burros and a river schooner, which he used to profit further. For the same reasons, Chagres attracted criminals and pirates. They were less tolerated. McArthur declared martial law and deputized a committee to drive out the brigands. He found a different problem on the Pacific side. There were too few ships to give outbound passage. McArthur caught a break when a group of three hundred emigrants (including Huntington) purchased the merchant bark *Alexander von Humboldt* and hired him as captain. After one hundred days, with food and water severely rationed, the vessel reached San Francisco. The *Ewing*, meanwhile, voyaged ten thousand additional miles around Cape Horn, weathering storms of the Southern Hemisphere. By the time the Coast Survey crew reunited in September 1849, they were battered and exhausted.[14]

They found San Francisco's waterfront in a worse state. Armed gangs ran amok. Facilities were insufficient. Migrants—already twenty thousand in number—lived in hovels and tents that readily caught fire. Abandoned, derelict ships filled the adjacent coves and beaches. Some were scuttled or stripped of wooden boards, metal hardware, and rigging. Some were unmoored, floating free with the currents. Crews had jumped these ghost ships to try their luck in the mines. The disorder so frightened the *Ewing*'s officer in charge that, while awaiting McArthur's arrival, he retreated the schooner and its sailors to desolate Tomales Bay, thirty miles north.

San Francisco as a frontier outpost—growing quickly and dangerously, and crowded with ships.

The Coast Survey could avoid the mayhem no longer. At San Francisco, on the night of September 13, five *Ewing* sailors mutinied in hopes of reaching the gold country. McArthur was absent and escaped harm. But the conspirators nearly drowned the navy midshipman left in charge. Following their capture and military trial, two leaders of the *Ewing* mutiny were hanged. Ship captains chained their crews in irons to prevent similar plots. The event dashed any hope that surveys would commence before year's end. McArthur fired the remaining malcontents. Local labor demand and high wages made it impossible to replace them. Instead, he sailed to Hawaii, where he hired recruits from King Kamehameha III.[15]

The *Ewing* finally began work in April 1850. McArthur and crew explored north of San Francisco. They made a reconnaissance of the Farallon Islands, a treacherous set of rock mounts and shoals thirty miles offshore from the Golden Gate. They sounded Humboldt Bay and major inlets: the Eel, Klamath, Rogue, Coquille, Umpqua, and Yaquina. The Humboldt-to-Oregon coast bustled with new timber operations and mines. That made it a flashpoint for war between colonists and Native peoples. At the Columbia River, McArthur updated latitude and longitude, and he sounded a seven-mile channel upriver to Astoria. Regardless, his charts lacked geodetic measurement. And they failed to keep pace with settlement and shipping.[16]

It did not help that McArthur grew tempted by the fortune-seeking all around. He speculated in real estate in Oregon's Willamette Valley. He bought land at Mare Island, at the juncture of California's San Pablo and Suisun Bays, which he happened to recommend for a military installation. These distractions kept the *Ewing* from charting the high-traffic navigation lane south of San Francisco. McArthur's mix of duty and interest was common among US agents. It provided reason to serve in a distant, unpleasant land. It supplemented federal wages that lagged below local rates. Military officers felt particularly entitled to speculate because they took a pay cut while working for the Coast Survey. Superintendent Bache had little patience for this. He closely watched staff and warned them against profiteering. But McArthur was too independent and too savvy to listen.

Aside from the social chaos and monetary spoils, surveyors faced the dangers of an uncharted coast. The Pacific would surprise the most experienced among them, at times with tragic result. While the *Ewing* began its work, a parallel US Navy reconnaissance followed, led by Richard Bache III, youngest brother of the superintendent. In March 1850, the party's landing boat capsized in heavy surf while investigating anchorages at Point St. George, near the California-Oregon boundary. Richard Bache and several others drowned. The *Ewing*'s crew recovered their remains. As the year closed, McArthur left by steamer to report in Washington, DC. The officer, who suffered from recurring dysentery, fell gravely ill. He died at sea on December 23.[17]

MCARTHUR'S DEATH separated what came before from what followed. George Davidson was not Bache's first choice to send to California. But that was because of his value. His arrival reinvigorated the Coast Survey's frontier venture and would guide it for the next fifty years. Born in 1825 in Nottingham, England, Davidson emigrated to the United States at age seven. His parents came from Scottish textile-manufacturing families, but had fallen on hard times. They settled near Philadelphia, a city renowned for science and education. The entire family exemplified this spirit. Mother Janet Drummond was an enthusiast of steam-powered machines and taught their principles to her young children.

Davidson became Bache's student while attending Philadelphia's Central High School. Bache served as principal, but held a joint appointment at the nearby private academy, the Girard College for Orphans. He referred to Central—one of the earliest public high schools—as the "People's College" and recruited its best students to assist at Girard's well-endowed observatory.[18] Merit rather than social class, Bache believed, would elevate American science.

Davidson stood first among the crop. He excelled at mathematics and worked as a "computer," back when the term applied to a person of high skill. Under Bache's watch, Davidson learned geodesy and how to use its key instruments: the chronometer and theodolite. He became accustomed to sleeplessness, as most measures were read by night.

After Bache took charge of the Coast Survey in late 1843, he hired several of his Philadelphia protégés. Davidson became his clerk. Their apprentice–master relationship continued, with the superintendent addressing orders to "my young friend."[19] Davidson sometimes chafed at this sort of "paternal interest." He also grew bored of Washington, DC, which he dubbed "Dreary City," and longed for adventure. Following the US annexation of Texas, Bache sent him to the Gulf Coast to be mentored by Robert Henry Fauntleroy. The young man honed his field skills until Fauntleroy's death from cholera in December 1849.

As Bache steered his agency to a greater role in US expansion, Davidson followed. In 1850, the professor chose his star pupil to make the first geodetic surveys on the Pacific shore. He would toil for an $800 salary on a coastline as remote as the edge of the Earth. Bache promoted Davidson to the rank of assistant and allowed him three aides. James Lawson (b. 1828) and Alexander M. Harrison (b. 1829) were alumni of Central High. John Rockwell (b. 1829) was a graduate of Yale College. Bache pledged their well-being to worried mothers and fathers. Yet, because he valued them so, he sent them into the breach. "I have been advised to give up the work on the Western Coast," he wrote Davidson on the eve of departure, "but wish you to understand that I am resolutely bent on prosecuting it. . . . You have put your hand to the plough & I beg you do not look back."[20]

The four men—scarcely whiskered and beyond their teens—entered a world of bandits, gamblers, and scoundrels. San Francisco's newspapers were filled with stories of robbery, duels, lynching, suicide, riots, and massacres. Assignments would take them into racial and ethnic battlegrounds and close to some of the worst carnage suffered by indigenous peoples. They were unmarried and hoped to find domestic comforts. But in California, men outnumbered women in the extreme. The *Tennessee*, which ferried the group to San Francisco, was so overbooked that passengers slept on deck, even in lifeboats. Yet of 487 persons on board, only twenty-six were women.[21]

Among the group, Davidson shouldered the heaviest burden. Geodetic coordinates were necessary—in fact, preliminary—to order. The *Ewing* could identify landmarks or hazards. But without a grid of verified numbers to "square off" and know its exact location, nautical charts represented nothing more than an informed guess. Davidson's job was to fix this by long-distance, or "primary," triangulation. Small triangles, with sides of six to forty miles, would aggregate into larger ones with sides of hundreds of miles. Shapes would span the coast from north to south. They would reveal where the sea met the Mexican boundary, as well as that with what would become British Columbia and every significant feature in between. (International joint commissions had begun to survey the borders inland.) Triangles would root out the unknown, correct falsity, and dispel fates threatened by the untamed Pacific.

Yet Davidson had a greater ambition to link his triangles to the United States' East. Parts and pieces would form a machine, a clockwork as stunning as the instruments he used. The assemblage would map a transcontinental republic, sea to sea, an empire of liberty. It was a heavy responsibility. All other Coast Survey projects depended on him. So, too, did government maps everywhere. He knew the stakes, as evinced by his long workdays and breakdowns in body and health.[22]

Because Davidson's assignment involved building a coordinate system—a "survey network" in geodetic parlance—Bache allowed him to select the initial point of study. Davidson looked to Southern California. The region was a backwater of settlement and commerce. But all shipping had to pass through it to reach centers farther north. In order to avoid prevailing winds, northbound vessels sailed close to shore through a twisting maze of headlands, cliffs, kelp beds, rogue waves, and submerged rock, sandbar, and reef. Mapping the stretch was critical.[23]

Davidson chose Point Conception, forty-five miles west of the village of Santa Barbara. This was the most prominent headland on the California coast and—with nearby Point Arguello—its most abrupt shift (roughly ninety degrees) from an east–west orientation to a north–south one. Point Conception also was the gateway to the San Pedro–Santa Barbara Channel, a 150-mile-long water corridor sheltered by islands from the open ocean. As ships turned north, they faced unpredictable wind, swell, and weather. Fog blanketed the area. Sailors likened the danger to Cape Horn or Cape Hatteras. It was the place that prior maps got most wrong. Here, Davidson would begin his transcontinental patchwork.

That was not all to begin, however. In Southern California, unlike San Francisco or the gold country, US conquest had brought few changes. For two years, maps and international treaty declared this to be United States territory. Yet, on the ground it was still Mexican California. This was Manuel Domínguez's world. And within a few years, Davidson's steps would lead him to an unanticipated site on Domínguez land overlooking San Pedro Bay.

Chapter 2

AN HEIR OF INVASION
WELCOMES THE NEXT

W HEN THE UNITED STATES ANNEXED SOUTHERN
California in 1848, it was a frontier unlike those Americans
had experienced elsewhere. Stage dramas and literature pitted wes-
tering adventurers against deadly beasts, imposing wilderness, and
fearsome Indians. But the plains around Los Angeles had none of
this. Instead, settlers found a land of cattle and sheep, watched over
by a Hispanic ranching elite and their laborers. Ranchers' property
became the great foil to annexation. While the story would be one
of loss for many Mexican-era landholders, for others it offered resil-
ience. Manuel Domínguez saw this and set out to navigate the uncer-
tainties of US conquest to his own advantage.

His object had taken shape one century earlier. At that time, the
country looked more as Americans might have expected. Domes-
ticated animals and livestock were absent. Instead, deer, ante-
lope, grizzly bear, coyote, cougars, and other large animals moved
through a landscape of grass, chaparral, and oak. To Native occupants, this was no wilderness, but rather a world rich with meaning.
The People, whose descendants today call themselves Gabrieleño-
Tongva, controlled a territory framed by mountains later named
Santa Monica to the northwest, San Joaquin and Santa Ana to the
southeast, and San Gabriel due north. Villages along the watershed
and coast centered an indigenous world. Archaeologists estimate
that prior to European settlement, the Gabrieleño-Tongva numbered five thousand persons—quite dense for Native America.

Lands adjacent to San Pedro Bay had the greatest concentration of place-names. Perhaps nine historic villages existed here. The People harvested fish, plants, shellfish, birds, and mammals including whale and sea otter. Long before Europeans arrived, the bay served as a site for commerce, its raw materials traded to the offshore islands and to inland mountains and deserts. One early Spanish visitor marveled at how these people of the coast "live by buying, selling, and bartering."[1]

Gabrieleño-Tongva on the island of Pimu (today's Catalina) made first contact with Europeans in October 1542 during the exploratory voyage of Juan Rodríguez Cabrillo. Sixty years later, Sebastián Vizcaíno's fleet passed through. Gaspar de Portolá's Spanish soldiers did, too, during the summer of 1769. Colonization, begun two years later, altered the indigenous world. Franciscans established Mission San Gabriel along one of the region's rivers and christened the people for the patron saint-archangel. They hoped to save souls. Instead, their so-called converts perished from disease. When the United States seized the region seventy-five years later, survivors lived at a rancheria near Los Angeles. Others labored for the church or for settlers. The village, Suanga, persisted on the bluffs by San Pedro's estuary.[2]

As the missions displaced indigenous tribes, royal governors granted the Franciscans vast acreage. This started with the first post at San Diego and continued with San Gabriel, the fourth mission, established in 1771. Officials granted smaller lands near military forts (*presidios*) for agriculture, livestock, and settlement. Soon they chartered three civilian towns (*pueblos*). The Pueblo of Los Angeles, established in 1781 at a village site the Gabrieleño-Tongva called Yaagna, was the second. By the year 1800, it had a population of three hundred persons.

To extend settlement into the countryside, Spanish governors next offered large concessions called *ranchos*, usually in reward for military or political service. Among the oldest and largest near Los Angeles was the Rancho San Pedro. In 1784, Governor Pedro Fages awarded the estate to Juan José Domínguez, a prominent army officer. The grant sprawled over seventy-five thousand acres (180 square miles) and included one-third of the coastal plain south of Los Angeles. On today's map this would span from San Pedro Bay, west around the Palos Verdes Peninsula, then north to Los Angeles International Airport, and east to the Los Angeles River and Long Beach.[3]

Despite the incredible size of ranchos, settlers had little interest in them. As of the mid-1790s, only five existed in the Los Angeles area. By the end of Spanish rule in 1821, there were fourteen locally and twenty-seven throughout Alta California. One reason for the small number was that regulations kept ranchos from functioning as property. Grantees gained right of occupancy, which they could pass on to descendants or heirs, but they could neither sell the land nor their rights. Land had little value anyhow. It was the livestock feeding upon it that was the greater, movable asset. Ranchos also came with heavy obligations, which if neglected could result in forfeiture and disgrace. Grantees had to maintain at least one thousand cattle—a significant expense. They were obligated to construct a house and improvements. For many, it was easier to graze cattle on the commons or the land of others.

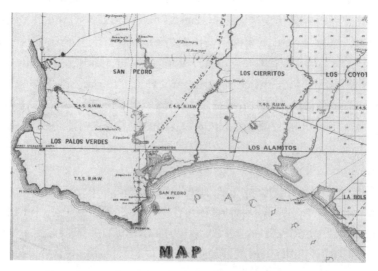

Ranchos and estuaries of San Pedro Bay, late 1860s

But Mexican independence changed things. Liberal reforms made the ranchos more like property, and demand multiplied accordingly. Mexico's federal government promoted foreign commerce and free trade (both illegal under Spanish rule), as well as individual rights, including ownership of real estate. California began to export cattle hides via visiting British and US ships. The proceeds made ranching lucrative and a vehicle for social mobility. By the 1830s, Mexico used land grants to entice immigration—especially to remote territories like Alta California and Texas. Governors readily gave them out in the decade prior to American conquest.[4]

Provincial authorities hoped the grants would stabilize Mexico's northern frontier. Yet they had the opposite effect. An enlarged class of landed elites, the Californios, emerged. They formed rival coalitions that chased influence, commercial advantage, and more lands. A small number of immigrants from the United States entered the rancho class by the 1840s. Abel Stearns and Benjamin D. Wilson married into landowning families. Other US visitors, like Richard Henry Dana, briefly saw the cattle economy, then departed. Dana would write of Mexican California's dysfunction and underdevelopment. But as Stearns and Wilson proved by staying, the profits were real. And that caused further problems.[5]

Given ranchers' wealth and rising expectations, land grants did not liberalize to their satisfaction. Claimants gained some property rights, but counterclaims strengthened, too. Prior to the 1830s, when the cattle trade became profitable and prestigious, most grant holders were absentees. They lived at the pueblo and attended to church, politics, and society. Day-to-day management of estates fell to foremen, employees, and relatives—all of whom gained tenant status. Hispanic legal tradition favored community and family rights over those of the individual. Once a counterclaim took root, it gained legal standing, and the grant holder was obliged to respect it. Whenever conflicts erupted, Mexican officials chose arbitration. Their agreements did not last long. Between 1831 and 1837, Alta California had as many as fourteen governments. With each turnover, dissatisfied parties angled for something better. All of this compounded the difficulties of keeping a rancho.

What Californios wanted was something akin to fee simple ownership: the right of a landowner under Anglo-American law to exclude competing claims and dispose of real property at will. The history of the Domínguez family's Rancho San Pedro exhibited the mix of acquisition and frustration. When the original grantee, Juan José Domínguez, died in 1809 without any direct heirs, both his estate and debts passed to a nephew, Cristóbal Domínguez, who neglected the claim for eight years more. The lapse nearly forfeited the rancho to a tenant, Manuel Gutiérrez, who grazed his herd of one thousand steers there. Gutiérrez paid off Juan José's debts, then claimed the ranch for himself. He, in turn, bestowed rights on José Dolores Sepúlveda, a Domínguez relative who managed his own herd at the base of Mount San Pedro (today's Palos Verdes Hills).

Starting in 1825, Cristóbal's children moved to reassert his claim, and eldest living son Manuel Domínguez emerged as the family's great champion. The twenty-three-year-old petitioned the territorial governor to clarify his family's rights and to evict all others. Manuel would continue the effort for two decades, during which he served as Los Angeles mayor (*alcalde*) and council member; in the provincial legislature at Monterey; and in 1843 as Los Angeles prefect. Despite his political ties, Manuel's petition proved inconclusive. In 1834, Governor José Figueroa affirmed the Domínguez grant. But he also gave Gutiérrez the right of lifetime occupancy and awarded 31,629 acres, formerly Domínguez land, to the Sepúlveda family. Domínguez protested. He only got rid of Gutiérrez when the latter died in 1839. That same year, Governor Juan Bautista Alvarado ordered the Sepúlvedas evicted. But they resisted, and the politician backed down. In 1846, the two families finally agreed to a partition. This left forty-three thousand acres under Domínguez family claim. California's last Mexican governor, Pío Pico, issued new grant documents just months before the United States invasion.[6]

At the same time, Manuel sought to consolidate rights within

*Manuel Domínguez, circa
1854. The heir to Spanish
conquest would prove
remarkably astute about
US conquest.*

the Domínguez family. Each of the siblings held equal, undivided share in the rancho. One brother, Nasario, traded his for a herd of horses. But Manuel struggled to convince youngest brother, Pedro, and sisters, Victoria Domínguez de Estudillo and Elena Domínguez de Rocha. Of the three, the sisters seemed most amenable. Both lived out of the area—in Monterey and San Diego. Neither needed their husband's permission to sell, as under Mexican law a woman retained her property after marriage. But Victoria and Elena demanded cash payment. They understood that new, capitalist forms of wealth were preferable to land and livestock. Manuel learned, too. Like most Californios, he lacked liquid capital, so he could not buy them out. Meanwhile, brother Pedro earned a reputation for his gambling losses. He took out loans and used his land rights as collateral. In spite of Manuel's victories, it seemed his dream of property—like his siblings' shares—would slip out of reach.[7]

WHILE MANUEL DOMÍNGUEZ JOSTLED against the limits of his society, distant events brought him cause for optimism. Rumors arrived in early 1846 that the United States might annex Alta California, as it had Texas months before. Domínguez had grown more preoccupied by the colonial estate and his tenuous hold there. Having exited political office, he relocated full time to the Rancho San Pedro with wife Engracia Cota de Domínguez and children. They expanded the one-story adobe to a comfortable size. The home sat on hills overlooking the ocean and Catalina Island, with an estuary and grasslands in between.

From their hilltop vista, the family spotted the US warship *Congress*, which entered San Pedro Bay on August 6. News had reached Los Angeles the previous month of a military conflict on the Rio Grande. Word followed of the Bear Flag Revolt in Northern California. Then came the alert that US marines had taken Monterey and San Francisco (called Yerba Buena for its unbuilt pastures). The *Congress* at last brought the American invasion to Domínguez's world. On the morning of August 11, commander Robert Stockton and 350 soldiers began a march inland. Manuel perhaps watched the column form in the distance with its uniforms, colorful flags, and glint of shouldered muskets. It was better armed and larger than the militia he had commanded several years before.

What the rancher witnessed was the western extent of the US-Mexico War. Political disputes over the international border and the United States' annexation of Texas had erupted into skirmishes in April. Yet the conflict had its roots in the expansion of American slavery and cotton agriculture, the instability of the Mexican Republic and local resistance to it, and the military power of indigenous tribes on the Southern Plains. The US government, led by Tennessee Democrat and plantation owner James Polk, took this as pretext to seize the Pacific coast. Looking far beyond Texas, Congress declared a broad war against Mexico. Neither government had any idea how important San Pedro would become.[8]

Southern California's Mexican officials were undecided on how to respond to invasion. Some abandoned the pueblo to regroup in the countryside. Some fled south with Governor Pico. But not all.

As the US force stepped around the estuary, a group of locals led by vice-mayor Benjamin Wilson rode out to announce the region's surrender. One citizen had thought, kindly enough, that a commodore should not have to walk. The party presented Stockton with a fine riding horse befitting a conqueror. The animal was a gift sent by Manuel Domínguez from the Rancho San Pedro's herd.[9]

What are we to make of the horse? Was it a plea for mercy? Or a confident overture? The answer is clouded, as its mystery continued to grow. Angelenos soon took umbrage at the US occupation. In late September, they rebelled. The Yankees made haste to evacuate from San Pedro's shore. But at the last moment, the officer reconsidered. He took reinforcements from an arriving navy vessel, and the combined force turned to retake the pueblo. On October 9, after camping near the Domínguez home, US soldiers were surprised by cavalry attacking on two sides. The Californios, led by José Antonio Carrillo and Andrés Pico, were armed with lances—surely lethal at a horse's gallop. They also wielded an artillery piece once used for ceremonies at the pueblo. The undersized cannon became known as the "old woman's gun," in honor of a matriarch who had kept it hidden from US troops beneath the dirt of her garden.[10]

On that morning, buckshot and cavalry charges tore through the American ranks. Fourteen US soldiers died; others were wounded. The Battle of Domínguez Ranch became the rebellion's great triumph. Over the next several months, the Californio insurgency ejected US invaders from Los Angeles, and nearly from Southern California, too. Americans would not return until January 1847, when armies led by Stockton, Frémont, and Stephen W. Kearny overwhelmed the region. The Treaty of Cahuenga, signed at the mountain pass northwest of Los Angeles, brought resistance to an end.

But it did not answer what the peace might look like. Those details took another year, as war spread from the Rio Grande into mainland Mexico. One US army, led by Zachary Taylor and John E. Wool, captured the northern city of Monterrey. A second, led by Winfield Scott, fought from the coast at Veracruz to Mexico City.

After the US captured Mexico's capital in September 1847, negotiations began. Meanwhile, debate over the war's legitimacy and results—the potential growth of slavery, in particular—raged within the United States.

Manuel Domínguez's role in the Californio insurgency remains unclear. He took no active part in resistance, nor in the fabled battle on his family's estate. US soldiers thought it safe enough to encamp there. They procured—or rather took—beef, food, and a small wagon from the premises. Apparently, the Americans trusted Domínguez. Even so, it is hard to believe that a surprise attack could occur without his tacit approval.

The ambivalence of the gift horse and battle at Rancho San Pedro is the most meaningful detail. Domínguez, like other Californios, had reasons to accept conquest. Land had little value in and of itself. But alienable property promised to change this and enhance his prospects. So, too, did US talk of developing California into a hub for Pacific trade. Perhaps American expansion held a manifest destiny for local Mexican elites, too? Some made that leap, but not without conditions. None knew the future. Ranchers were proud heirs of a previous invasion, which lived on in their land grants and Catholic sacraments. National flags mattered less. As war closed, they waited to see what the government in Washington, DC, would do.

Chapter 3

~~~~~~~~~~~~~~~~

TO FIX A
CONTINENT ADRIFT

IN JULY 1850, THREE WEEKS AFTER ARRIVAL AT SAN Francisco, George Davidson and his party began reconnaissance of Point Conception. They landed with baggage and instruments through the waves at El Cojo (or Coxo), a beach and arroyo east of the point. For two months, they worked along the bluffs, covered by dry grass and stands of evergreen live oak. Mountains behind separated the sea from the inland valleys. On clear days, kelp forests and a trio of islands appeared offshore. Otherwise, they faded like ghosts into the fog.

Davidson's team had not come for the view. They scouted a potential lighthouse site and set up an observatory. They measured latitude, longitude, and magnetic variation. The latter is the difference between true north and magnetic north caused by the sphere of the Earth. Astronomical readings required Davidson and Rockwell to work through the night. By day, Lawson and Harrison mapped topography using a plane table. Nearly all tasks required strenuous concentration and use of the eye. Some involved physical strength. The surveyors had to move lumber, supplies, and equipment that weighed three hundred pounds. They had no pack animals, nor hired hands besides one cook, a Spanish-speaking man from Santa Barbara whom they befriended on the schooner that brought them. Davidson paid him $125 per month, which exceeded his own salary. The party quickly fell sick—no fault of the cook. Their bodies had yet to acclimate to local water and shellfish. Nonetheless, Lawson

Point Conception and Coxo, where Davidson began the
US geodetic survey of the Pacific coast

would recall Coxo fondly for its clean air, balmy weather, and morning "surf baths."[1]

Point Conception was the surveyors' first taste of the uncharted Pacific. At the same time, it introduced them to the tense society that Spanish and American conquests had created. Lawson reported that the area was uninhabited other than by "a couple [of] Indian spectators."[2] But Coxo remained an indigenous place. Its bluffs and coves once were home to a Chumash village. It became the Rancho Concepción, granted to the Carrillo family in 1837. Indians seen by Lawson were Chumash survivors and ranch hands. Anastacio José Carrillo y Lugo—whose brother, Antonio, had commanded at the Battle of Domínguez Ranch—claimed the area and 4,500 cattle. The United States did not think to ask his permission to enter. Incensed, the elderly rancher sent his son to accuse the trespassers of plotting to steal the land. Davidson, either by his broken Spanish or his cook's translations, resolved the matter. The Carrillo family later hosted the Americans at their adobe in Santa Barbara, then a seaside village of one thousand persons. The majority spoke no English, quite different than San Francisco. For Davidson and his colleagues, the visit indicated new challenges ahead. In Southern California, scientific unknowns were additionally fraught with property disputes and cultural conflict. Pacific and Southwest worlds overlapped to make the region a frontier like no other.

Davidson and his team spent the winter months engaged in mathematical computation, platting maps, and hand-duplicating documents so they would not be lost. They likely did this at the Survey's rented office near San Francisco's Rincon Point, which today lies buried beneath the Bay Bridge. After more fieldwork in early 1851, Davidson had three critical landmarks: Point Conception, Point Piños at Monterey Bay, and Point Loma at the entrance to San Diego Bay—his first steps in primary triangulation of the West Coast. The surveyors returned once again to San Francisco in time to witness the city's purge of criminals by public tribunal, hangings, and forced expulsions.[3]

THE VIOLENT TURN at San Francisco foreshadowed Davidson's next assignments. He and Lawson moved to the far Northwest to determine geodetic points and lighthouse sites. Most of this work took place in the Oregon Territory, a remote, windswept expanse of rocky shore and sea stacks that stretched to the border of what is now British Columbia. The absence of flat, open terrain made telescope and plane table sightings harder. Forests, mountains, and clouds obstructed views of the sky and horizon. The team often had to observe from water, while afloat in small canoes, which produced unreliable numbers. In the absence of strand beaches, they had to land with fragile instruments in tight coves of cobblestone or rock ledge (slippery and razor-sharp with barnacles). The chronometer—a friction-free, rotating-gear clock made practical as of the 1780s—became indispensable. Older pendulum clocks lost their accuracy on water, due to the rocking of a vessel. They also became inaccurate on land if the ground was uneven. Chronometers, by contrast, kept "true time" no matter the conditions. Navigators and surveyors used them to determine longitude by measuring time against a known "meridian" (the prime meridian, in Greenwich, England, for example) and by the position of the sun in the sky. Davidson prized his chronometers, scrutinized their quality, and demanded that Bache send better ones and more of them.[4]

Adding to the challenges, European-American settlement was

*The tools of geodesy: The chronometer (left) enabled
surveyors to determine longitude. The theodolite (right)
measured angles and azimuth.*

sparse and Indian violence common. Newspapers sensationalized
events like the 1847 Whitman Massacre, which occurred three hundred miles up the Columbia River, and the resulting Cayuse War.
Along California's northern coast, Pomo, Mattole, Whilkut, and
Wiyot peoples chased away settlers entirely. The absence of colonists heightened surveyors' physical insecurity. But their presence did as well. When Davidson's crew traveled to Port Orford in
the fall of 1851, the same ship carried troops sent to "punish" the
Coquille River people for killing a group of miners. Army officers
included George Stoneman, California's future governor, and Robert Stockton Williamson, a military engineer later prominent in the
Pacific railroad surveys. The soldiers massacred fifteen Coquille and
destroyed the tribe's salmon traps. Davidson and Lawson were not
present to see this, however. The paths of science and war diverged
for the time being.[5]

   Nature itself held the greatest peril. During the Coast Survey's first
two seasons in the Pacific Northwest, field parties naively attempted
to work through the mild, temperate winter. It was a severe mistake.
Teams exposed themselves to the full energy of the Aleutian Low.
This pattern of cold, low-pressure air generates winter cyclones and

giant ocean swells. Circulating clockwise, it hurls destruction at the shore from mid-Oregon down to central California. Encamped near Port Orford on the night of December 1, 1851, Davidson's crew suffered "almost total demolition," as hurricane-force winds and rain fell suddenly upon them.[6] Waves exploded over the beach and into the forest. The surveyors fought to keep tents and equipment upright and tethered to the ground. They took refuge in their makeshift observatory, which nearly collapsed in on them. Davidson saved the most valuable instruments, but he reported the aftermath to Bache: "Tents and flies torn and driven amongst the trees—blankets and bedding ditto—tarpaulins torn and tossed about like pieces of paper—provisions (which are precious now) spoiled—books wet and damaged." The storm lasted four days and left Davidson in shock. "Immense trees were torn up by the roots—and the beach strewn with dead sea fowl." He shuddered, "I have never seen such a storm on the Atlantic or Gulf coasts."

The terrifying experience distinguished two borderlands that Davidson and the Survey needed to move between and unify into national territory. The South Coast had a gentle climate, with open hills and plains. But its society was complex and turbulent, as it grappled with the Gold Rush and the transition from Mexican to US rule. The North Coast, by contrast, was home to hostile indigenous people and the maelstrom of nature. Connecting these spaces would require stabilizing coastal points. Davidson had yet to identify the most crucial of these. He soon did. Eager to avoid the north and its traumas, he proposed several projects that would allow the USCS to work elsewhere during peak storm season (roughly November to March). He offered to survey the Central American isthmus. Bache agreed, but postponed this project. Davidson suggested, too, a survey of California's South Coast and its unique feature: the Channel Islands.[7]

The Southern California archipelago consists of eight islands dispersed over ten thousand square miles. Those nearest the mainland resemble it, with their rolling hills, soft peaks, sage scrub, and oak.

Gabrieleño-Tongva and Chumash made their homes here until Mexican authorities removed them to the missions. Land grants and cattle took over. The islands gained importance as maritime commerce increased. They provided a sheltered sea and place marker to navigators. At the same time, they posed a danger in heavy fog or rough water. Knowledge of their exact position and harbors would be of great benefit.

But Davidson saw something more in the project. The Survey could use the islands as offshore geodetic observatories. That would be far superior to typical practice, in which one or two points of a triangle required ship-based observation. Measurements taken at sea were always less exact. With stable platforms, however, the Coast Survey could determine global coordinates as close to perfect as possible. Southern California had the only significant group of islands between the US-Mexico border and Puget Sound. That made it the ideal place to anchor the national map, the bureau's reason for being ever since its founding. Davidson's mood lifted with anticipation. "It will be delightful," he joked, "being seasick all day and observing all night!"[8]

Davidson's plan pointed him toward San Pedro Bay. Triangulation depends upon a reliable starting place or baseline from which surveyors build out their triangles. Mistakes here will throw off the entire sequence of geometry. An optimum baseline required level terrain and a minimum length of six miles. Its two endpoints needed direct sight toward an island, with angles of no less than thirty degrees. The hamlet of Santa Barbara sat directly across from Santa Cruz and Santa Rosa islands, largest and nearest to the mainland. But Davidson already knew that Santa Barbara was too hilly and compressed by mountains to lay out an acceptable base.[9]

Instead, Los Angeles provided the better location. Despite lying seventy-five miles southeast, and away from the channel's most treacherous waters, the Los Angeles plains were broad and flat with open country and few obstructions. The third-largest island, Santa Catalina, sat directly offshore (by twenty-six miles) and provided sufficient angles. The island also contained an important cove, Catalina (or "Cat") Harbor. However small, it was the only fully protected anchorage between San Diego and San Francisco. Knowledge of its location would be a saving grace to ships in need.

Davidson volunteered to locate the "San Pedro base." But he did not plan to survey it in its entirety. Nor did he plan to conduct the channel's triangulation. That task would fall to staff below his rank and specialized skills. He recommended his aides: John Rockwell, for geodesy, and James Lawson, for topography. Bache had other plans. He assigned the triangles to Sub-Assistant Joseph S. Ruth, another protégé from Philadelphia's Central High School, who had done this kind of work in Oregon. He kept Lawson and Rockwell with Davidson. They would go to the 49th parallel to locate the boundary with present-day British Columbia.[10]

ON BOTH the Southern California and North Coast work of 1852, Davidson's crew received ferriage aboard the USS *Active*. The 175-foot, masted side-wheel steamer had become the flagship of the Coast Survey's hydrographic division. They also came to know its commander, James Alden. Alden was a forty-two-year-old navy veteran with a litany of accomplishments. He had been an officer on the US Exploring Expedition (Ex. Ex.), which sailed eighty-seven thousand miles around the globe between 1838 and 1842. He led its survey of Tierra del Fuego and took credit for sighting Antarctica in January 1840. He was also the first prosecution witness in the court-martial of Ex. Ex. chief officer Charles Wilkes for misconduct and cruelty. Alden circumnavigated the Earth a second time with the USS *Constitution* and "Mad Jack" Percival. He then served in the US-Mexico War.[11]

Alden's hardened experience, salty demeanor, and sense of rank caused him to clash with Professor Bache, and even more so with young Davidson. Their conflicts came quick. Alden reached California during the summer of 1851. He waited for delivery of the *Jefferson*, a state-of-the-art surveying ship. But a powerful storm off East Patagonia destroyed the vessel on its maiden voyage. Alden hurriedly purchased a replacement from the San Francisco Custom House. Recycling a name from his Ex. Ex. days, he christened it *Active*. Bache was outraged over the price paid and that Alden had dared, without his approval, to close the deal and offer the *Ewing* for sale. But the superintendent could not stop the purchase without losing face with Treasury Secretary Thomas Corwin. Instead, Bache and Alden exchanged angry letters—angrier after it became known that the *Active*'s boilers and copper bottom required costly repairs. "If you expend money which we have not got," Bache steamed at Alden, "you will have to get it for yourself."[12] The navy man, who was accustomed to demanding satisfaction, did not balk. He called Bache's words an "unjust censure" and promptly mailed back his challenge. But worse than a pistol duel, Alden threatened to quit, which would have set back work twelve months or more. Bache relented.

The fight prefaced Alden's collaboration with Davidson. During April 1852, the *Active* sailed for San Pedro Bay. It was the Survey's— and Davidson's—first official visit to the Los Angeles area. Here, on a prairie dotted with cattle, he set up the San Pedro base. Alden's crew, meanwhile, took first soundings of the bay and coastline. None ventured inside the estuary, which remained uncharted and invisible to the US government. Next, Davidson and Alden examined the channel to plan Ruth's upcoming triangles. They chose six stations, or vertices—prominent points on the mainland, Catalina, and other islands. Davidson predicted that Ruth would need two years to triangulate between San Pedro and Point Conception.[13]

By May, Davidson had returned to San Francisco discontented. He complained to Bache about Alden's imperiousness. And he wrote that a new theodolite—a telescope specialized for measuring angles on both horizontal and vertical axes—had sustained damage when charged at and knocked over by an angry donkey. Good instruments were rare in California. They were handmade in artisans' shops, often in Europe. A lesser or damaged one could compromise a surveyor's career. A headstrong colleague might do the same.[14]

Considering his tasks at San Pedro finished, Davidson prepared for the 49th parallel study. At the end of July 1852, *Active* landed the party at Cape Flattery, inside the tip of the Olympic Peninsula. The 1846 Oregon Treaty between the US and Britain mandated the work. Davidson and Alden would locate an international border somewhere among the gray seawater, forested islands, and rock promontories that separated Vancouver Island from American claims. The project also tied to Davidson's primary triangulation. The area (later, Port Townsend specifically) allowed for numerous offshore stations. Together, the 49th parallel and Southern California's channel became the cornerstones for Davidson's geodetic map of the US Pacific.

Not that the work was easy. Cape Flattery provided the scientist with his first encounter of Indian violence. Initially, he took to it with bravado. Like many Americans, he was raised on frontier tales, whether in the popular press or the literature of James Fenimore Cooper. He assured Bache that he would "risk every hazard"

Cape Flattery and Neé-ah, home to the Makah people

to complete his work.[15] At "Neeah" Bay, those words must have lingered on Davidson's tongue. The *Active* put the surveyors on the beach and promptly sailed off. The shore party was all alone on a slim margin between rainforest and sea. Davidson broached friendly trade with the local Makah people. But the intruders also threatened to punish the tribe if it attacked or stole anything. The Coast Survey group of nine soon found their cotton signal flags missing and their camp outnumbered, as several hundred Makah and allies amassed in force. To defend themselves, Davidson's men built a breastwork of logs. They kept armed and awake, in fear for their lives, with sixty rounds ready to fire at all times. Upon the *Active*'s return, Alden— who during the Ex. Ex. voyage had carried out a massacre of eighty-seven Fijians at Malolo—summoned the Makah and vowed to raze their village if aggressions continued. No violence took place. But the encampment lasted to October, a duration that left Davidson sleep-deprived and shaken.

The brush with death gave the surveyors a sense of the bloodletting brought on by settlement. It also raised the question of how their scientific work provoked violence. They did not accept answers easily. Upon landing at Neah Bay, Lawson tried to convince the Makah that a mapping survey would bring the tribe commerce and prosperity rather than loss of lands or independence. He later recounted of his "peaceable mission": "They would share by having a trade with the whites in selling their fish, oil, &c, in return getting such articles, blankets, clothes, beads, &c, as they desired."[16] But the Makah's doubts proved more correct. Ships brought in smallpox, which killed three-quarters of the tribe by the next autumn. Settlement increased, and with it the mood of imminent war. Within a few years, the Makah had a reservation treaty forced upon them. Their neighbors fared worse.

Time would show that the Coast Survey was indeed part of US imperial expansion, with all its intended and unintended consequences. Benign measurements could bear destructive, even murderous effect in the hands of others. And that suggested a deeper problem. The scientists had come west to create order. But as the western continent grew more fixed by geodetic lines, its troubles intersected and accumulated. Davidson would experience this personally. A tragedy in Oregon soon forced him back to the San Pedro baseline and the rancho property through which it passed.

Chapter 4

PROSPECTING
IN PROPERTY

B Y THE TIME THE COAST SURVEY SHIP *ACTIVE* ARRIVED offshore, Manuel Domínguez grew anxious and impatient. Four years had passed since the end of war and the international treaty that obligated the United States to uphold Mexican-era land claims. The American government had yet to do so. In the meantime, rapid and unexpected changes tore through California. The whirlwind hit first in the distance. It undercut ranchers' position and their influence. It also brought a multitude of foreigners who sought land for themselves. Their presence upset old accords and expectations.

That was not how things had started. When the US completed its annexation of California, its diplomats assured landowners that their grants and social status would continue. The Treaty of Cahuenga, which ended the ranchers' rebellion in January 1847, guaranteed them "protection of life and property" as well as "equal rights and privileges."[1] The Guadalupe Hidalgo Treaty superseded this thirteen months later. Real estate remained its focus. The document transferred nearly 340 million acres—over half of Mexico's national territory—to the United States for the price of $15 million. Specific sections (Articles VIII and IX) reiterated the ongoing right to property. In Article X, the US agreed to perfect Mexican land titles, that is, to sanctify them under its law. Mexico's negotiators wanted the treaty to spell out such a process. In the end, the American republic pledged to do so unilaterally, through Congress, at a later date.

That proved difficult in the details. Washington politicians preferred to open up California's acreage to US settlers. Racial and cultural attitudes strengthened that imperative. But many conceded that the US needed to keep the old elite in place, at least for the moment. Land grant holders provided local leadership and taxes. Their cattle were the region's economy. Their agriculture and water projects employed a mass of landless laborers, dispossessed Indians especially. Their justice restrained the underclass, kept order, and punished outlaws. Ranchers guarded against raiding tribes and foreign powers like Great Britain. Hence, Mexican-era land rights remained a strategic necessity. Politicians, who otherwise disagreed over the extension of slavery, would come to terms on this. Shared interests bound together Californios and the United States.[2]

But mutuality came and went because of an unanticipated discovery. The US made its treaty promises in February 1848. At that moment California's non-indigenous population was ten thousand persons. Independent tribes vastly outnumbered settlers, including Hispanicized Indians, US troops, and Anglo civilians. No one expected this to change any time soon. Treaty signatories were unaware of the discovery made at Johann Sutter's mill on the Rio de los Americanos just ten days before. Those gold nuggets remained little known in California for weeks afterward, and unknown in the US for several months longer. Then, massive immigration began. By 1850, California had a non-indigenous population of 100,000 persons. Two years later, it approached 225,000.

The Gold Rush became the surprise epilogue to the Treaty of Guadalupe Hidalgo, and it entirely altered the calculus. With many thousands of US and global migrants populating California, Mexican-era ranchers and their claims became less essential to the social order. Prospectors overran them as they searched the Sierra

*Prospectors at the Sacramento River. This image shows
how the Gold Rush attracted diverse participants,
locally and from across the globe.*

foothills and downstream valleys. Once frustrated, they turned to
squatting, in a desperate attempt to survive. Sacramento and San
Francisco became inundated by property seekers, who formed into
mobs and menaced authorities. To meet the demand for food, trespassers stole cattle. Indians received the blame, but no doubt others
had incentive to steal. Cattle theft—actual or accused—provoked
shocking violence against indigenous people.[3]

Californios were now one small constituency among many, all
fighting for access to scarce resources. But the fights were complex and
made tenuous allies. Some miners and merchants became wealthy off
the Gold Rush. Looking to invest their winnings, they lined up to purchase Mexican-era claims. By necessity, these new proprietors disapproved of squatters, crime, and disorder. Or, rather, they disapproved
of the excesses. If they arranged to purchase from a grantee, would-be
buyers needed the original claim to be valid. If they arranged purchase
from anyone else, buyers needed the claim to be in doubt. They orga-

nized into factions based on whichever side they took. Civil courts did not yet exist to resolve the matter. That gave Mexican ranchers some room to maneuver. They aligned themselves as best they could.

How an individual grant holder might fare depended upon location, that most crucial variable of real estate. The Gold Rush economy took form around an economic center: San Francisco's peninsula and bay, inland along the delta to waystations like Stockton and Sacramento, and farther up the rivers to mining camps. Here, land grants crumbled rapidly. Once-powerful claimants fell into ruin or hastily sold what they could. But crowded riverbeds and boomtowns were a world away from Southern California and the far Northwest, both of which became supply peripheries to the Gold Rush. At Los Angeles, immigration remained small. The county's population scarcely reached six thousand persons. Ranchers and land grant families remained atop this world. In fact, their position increased because of the downfall of grant holders elsewhere. Southern Californios like Domínguez profited by shipping beef, wool, and leather northward to meet the insatiable demand.[4]

In this cruel, zero-sum economy, one rancher's loss became another's gain. This was true in politics as well. California's population growth put it over the threshold of fifty thousand citizens, meaning that it could organize as a state far sooner than expected. Statehood created laws, courts, and services. It conveyed rights: most important, representation by elected officials (rather than by military or federal appointees), a share of the public lands, and a full delegation to Congress. Some predicted statehood would bring a transcontinental railroad. Yet fee simple property and US land patents promised to be the greatest blessing. Thus, while it sowed conflict and discontent, the Gold Rush also drove consensus and commitment to the new order. A person might ride the turmoil to new heights and better days, even as others lost out—that is, if the person happened to be in the right place, with the right connections.

———

THE GOLD RUSH SUSTAINED Manuel Domínguez's optimism. He watched the changes from afar and seized upon their possibility. He began to act like a prospector, searching for greater fortune in California's new era. This would lead him, often correctly but sometimes mistakenly, down a path of collaboration. In 1849, Los Angeles elites sent Domínguez, among seven prominent citizens, to represent the region at the constitutional convention in Monterey. He joined neighboring ranchers José Antonio Carrillo, Hugo Reid, and Abel Stearns when proceedings opened in September. As a sign of demographic shifts, Domínguez was one of only seven delegates (of forty-eight total) born within California—thus, of Mexican or Spanish heritage. Together they drafted and signed the state's founding document.

Whereas Anglo delegates saw the Monterey convention as a departure from the past, Californios worked to bridge two eras. They tried unsuccessfully to preserve Hispanic institutions. They did block proposals that would have underrepresented Southern California, such as by discounting or disenfranchising its assimilated Indians. Californios receive praise today for their defense of male voting rights. Domínguez, with indigenous ancestry visible in his features, was among those defenders. Yet ranchers did so, in large part, as a way to preserve their clout. (The legislature would abrogate the convention accords in March 1850 and limit suffrage to *white* male citizens, with some Hispanics like Domínguez included.) Californios' most lasting contribution to interethnic democracy was the constitution's requirement to publish laws and public documents bilingually. Otherwise, the gold-settler regions were simply too populous. This led some ranchers to push for a division of the state into halves, an effort they maintained up until the Civil War.[5]

Domínguez took on the role of moderate and dealmaker, consistent with his view that conquest was a mutual project. He was attentive in general sessions and active in committee meetings. Some conventioneers were impressed enough to speak of him as a candidate for governor. He did not pursue the honor, although being considered no doubt pleased him. Instead, the highest offices went to American immigrants. Voters elected as governor Peter H. Burnett, Democrat and immigrant from Tennessee by way of Missouri and Oregon. For the US Senate, legislators chose John Frémont, the famed "Pathfinder" and instigator of conquest, along with William M. Gwin. Both Frémont and Gwin were Democrats. Both were deeply interested in the fate of Mexican grants. Both were expected to advance the cause of land patents with gusto. That placed Domínguez and his interests clearly on the side of power.[6]

He took a similar course back in Los Angeles. Domínguez partnered with a variety of Americans in pursuit of property: lawyers, surveyors, mortgage lenders, merchants, and prospective town developers. As much as they maneuvered to take advantage of him, he found them useful, too. And that cemented collaborative relationships. His most important was with Joseph Lancaster Brent. Brent arrived in Los Angeles in the first days of 1851, at the age of twenty-five. He had trained as a lawyer at Georgetown College in Washington, not far from his native Baltimore. Southern California desperately needed attorneys. Brent immediately found his niche. He was Catholic and learned Spanish. He also was daring in the courtroom. Weeks after arrival, Brent took on the case of Francisca Pérez de Silvas, who suffered violent domestic abuse. Her petition resulted in the county's first-ever highly publicized divorce case. Shortly thereafter he defended sons of the prominent Lugo family, charged with murdering two Americans.[7]

With his reputation as a friend to Mexican-era landholders, Brent built a successful practice along with the coalition that would dominate Southern California's politics for the next decade. Ranchers became his main clientele, as they sought to settle debts, evict counterclaims, and get paperwork in order. Domínguez hired Brent to represent him. Cash remained in short supply—there were no banks in Los Angeles. Instead, the attorney accepted land as payment. He amassed sizable holdings and dreamed of retiring to a rancher's life.

At least one Californio—left unnamed but possibly Domínguez—offered payment of a daughter's hand in marriage. An Anglo lawyer in the family would be advantageous, especially in turbulent times. Brent declined the father's offer. But with support from ranchers, he became leader and kingmaker—alongside Benjamin Wilson—of the Los Angeles Democratic Party. Domínguez joined them and became a prominent Jacksonian. Besides finding a son-in-law, that was exactly the kind of cross-cultural enterprise he hoped for.[8]

By his distance from the Gold Rush, and with US immigrants as allies, Domínguez increased his hold on the Rancho San Pedro. Profits from the cattle boom enabled him to acquire his siblings' shares in the estate. In June 1852, he acquired 2,900 acres from the children of sister Elena Domínguez de Rocha, who had died of tuberculosis. The parcel lay northeast of the San Pedro estuary. In the following month, Manuel acquired four thousand acres at the center of the rancho from older sister Victoria Domínguez de Estudillo. Unfortunately, he was too late to purchase brother Pedro's lands, which were lost to gambling debts. Even so, Manuel now claimed twenty-nine thousand acres, covering two-thirds of the grant as it existed

A rodeo, or roundup. While they were distant from the goldfields, Southern California's luckiest ranchers still gained by selling their cattle products.

in 1846, when US troops arrived. His strategy was paying off. It had costs, too, of course. Land acquisition required payment and legal fees. This, plus property taxes, forced Manuel to borrow money and, a few years later, to sell select parcels. Brent and Wilson, with several associates, would purchase the most significant of these.[9]

The US era had been good to Domínguez. Regardless of disruptions, he pieced together a world he wished to occupy. Yet one large piece was still missing. As of 1852, Mexican-era grants like the Rancho San Pedro remained without patent, their boundaries imprecise and unsettled. This left all real estate transactions in limbo. Meanwhile, the turmoil of the Gold Rush approached closer. A correspondent for the San Francisco newspaper the *Daily Alta* observed the sudden proliferation of squatters near Los Angeles:

> In some cases they have gone upon lands which have been cultivated for many years, have surveyed it into plots of 160 acres, and have actually given notice to the proprietors residing upon the soil that they must vacate their premises—warning a man to leave his own house.[10]

Stories like this set off panic. Property was a necessity, but one without legal protection. It was a source of stability and instability, security and avarice, rising and sinking fortunes, friendship and resentment. Such was the flaw of a society steeped in it. And so far, the federal government had done nothing to fix things.

Prospecting involved pursuit of the unknown. But ranchers like Domínguez were prospectors by circumstance, not by choice. They wanted certainty; they always had. Without it, victories rang hollow, even for the most adroit and well-connected negotiator of conquest. Domínguez waited on US officials to act true to their word. Soon, good news arrived. Or so it seemed.

Chapter 5

DR. GWIN BRINGS
GOOD NEWS

W HEN CALIFORNIA SENT OFF ITS APPLICATION FOR statehood, it set in motion the American republic's most tumultuous decade. On January 2, 1850, Senators-elect William Gwin and John Frémont departed aboard the steamer *Oregon* to deliver the petition to Congress. Both men knew the capital and were eager to return there. So it was a surprise when Frémont delayed his journey and remained in Panama. That put all eyes on Gwin. He relished his office and the consideration it earned from other passengers. Among them were soldiers from California's outgoing military government, including twenty-nine-year-old William Tecumseh Sherman, and a fellow officer named Edward Ord. Gwin regaled them with swashbuckling tales of profit and politics. Already, his expected arrival in Washington caused much of a stir. Gwin would hold sway as the Pacific coast's chief emissary, a promoter of himself and of causes or friends he championed. He would rise fast in the Senate and take an outsized role in how his state's Mexican-era land grants entered the union. That was good news. But for whom, exactly?

The answers seemed many at first. Gwin's opacity enabled him to be all things and, for a time, to evade the capital's most divisive issue: slavery and its status in the West. After decades of calm, slavery reemerged as a political divide during the 1844 campaign to admit Texas to the union. Two years later, the start of the US-Mexico War escalated the disagreement. Legislators who opposed an extension of slavery tried to pass amendments, known as the Wilmot Proviso, to bar slaves from all former Mexican territory. By 1848, they had formed a bipartisan coalition under the banner of "free soil." Their demands met with opposition from the formidable Senator John Calhoun. He argued for the common right of all citizens to populate the West and to do so with their property, including slaves. To deny this, Calhoun said, was to deny states and their people equal status under the Constitution. His argument—that slaveholders were protected by principles of equality and minority representation—seems nonsensical today. But it held great power at the time, increased by the senator's oratory and penetrating stare.[1]

Acrimony threatened to block all discussion of California. Adding to the impasse, Congress faced a task unprecedented in three decades: it considered a free state for admission without having a slave state available to admit concurrently. Since the Missouri Compromise of 1820, the federal government expanded the number of states on a one-to-one ratio to keep Senate parity. Most recently, it preserved fifteen slave states and fifteen free states by entering Florida and Texas along with Iowa and Wisconsin. But there was no territory to pair with California. New Mexico had sufficient population, many favorable to chattel slavery. However, because only eight hundred US settlers lived there (out of a population of sixty thousand, most of them Hispanos and assimilated Indians), even pro-slavery politicians hesitated to give it statehood just yet. President Zachary Taylor announced his opposition to admitting California unless it was paired with a slave state. Any bill would fail without the president's signature.

Debate polarized the nation and made casualties of Washington's old guard. Senator Henry Clay, who brokered compromise in Congress for nearly forty years, tried for another grand bargain. This would give the federal government greater responsibility for the capture of runaway slaves. It would organize two new territories, New Mexico and Utah, both open to slavery. However, Clay's colleagues voted down his omnibus bill on the last day of July. He was humiliated and suffering from tuberculosis. President Zachary Taylor had died weeks before. Calhoun had died in March.

After the failure of Clay's bill, younger leaders emerged who would dominate the antebellum decade. Senator Stephen Douglas of Illinois demonstrated his parliamentary skills and naked ambition by securing passage of Clay's compromise, this time split among separate bills. President Millard Fillmore promptly signed them. California gained admission on September 9. Douglas had outmaneuvered opponents on two sides. Foes included William H. Seward of New York, champion of the Whig Party's antislavery faction, and Jefferson Davis of Mississippi, who stepped forward to fill Calhoun's shoes. Davis swore he would block California's statehood. (He even pledged to tear the bill "to pieces" before the secretary of state and president could endorse it.) Among the Senate's rising

The Senate, led by Henry Clay, debates California's
admission to the union, 1850.

stars—once the 1850 compromise passed—was its newest member, William Gwin.[2]

WITHOUT MUCH of a voting record to make enemies, Gwin became a partner to the various sides in Congress. His fleet-footedness had a history, one that recurrently crossed paths with Jefferson Davis. Gwin's father, a Methodist preacher and frontier soldier (under Andrew Jackson), moved the family to Tennessee and, by the 1830s, to Mississippi's Yazoo delta. Like the Davises, the Gwin family set up plantations among the bayous. They owned tracts on the side of Warren County opposite from Davis's own Brierfield plantation. Gwin properties stretched into neighboring Issaquena County. William left the management of field slaves to others, but profits made by slavery allowed him to train as both lawyer and a physician. He opened a medical practice in Vicksburg. Soon, he aspired to public office. Dr. Gwin, as both friends and enemies called him, gained a commission as US marshal, and in 1841, at age thirty-six, won election to the House of Representatives.[3]

As it turned out, Gwin had a knack for scandal. And this shaped a knotted link with Davis. In 1843, he lost his congressional seat because of alleged financial misdeeds. If not dishonor enough, he tried to regain the office, only to see Davis win it instead. Failure hurt Gwin's pride. Then he poured salt on his wound. In November 1845, Gwin and wife, Mary Elizabeth Hampton Bell, attended a ball held in Vicksburg in honor of Davis, the congressman-elect. John Calhoun, who had known Davis for years and authored his recommendation to West Point, was on hand. So, too, was Davis's wife, Varina Howell—nineteen years old and newly wed—accompanied by her brother. In family correspondence, the Howells told how the Gwins tactlessly upstaged the evening's honorees. This began with Mrs. Gwin's "indecent" appearance in a low-cut, sleeveless gown of black velvet. (She had a penchant for eye-catching Parisian styles.)

Offenses extended to Mary's energetic dancing with bosoms half-exposed for all to see. "I could not look at her without blushing and feeling ashamed," the letter scorned.[4] "It would be better for any lady to have Mrs. Gwin for an enemy than for a friend." Jefferson Davis likely agreed. He continued to view Dr. Gwin as corrupt and irresponsible, so much so that he dismissed as a "joke" Gwin's claims to be anything but. With this talk circulating through Mississippi society, the Gwins needed a fresh start.

Oddly, it was Gwin's failure and exile that landed him in an equivalent office to Davis. After a stay in New Orleans supervising its US Custom House, William moved to California. There, he and Mary rebuilt their reputation with remarkable speed. He gained added wealth as a mine owner and real estate investor. In the coarse world of the Gold Rush, he commanded respect for his education and political experience. Six months after immigrating, he was a delegate to California's constitutional convention. It was Gwin who brought copies of the Iowa constitution, which became the delegates' template. State legislators rewarded him with the office of US senator. Fate, in its bizarre twist, scheduled the Gwins and Davises to share many Washington dinner parties and dances.

Gwin's reversal of misfortune shows how quickly life could change in nineteenth-century America. Most of all, it speaks to his calculated adjustments in politics, which alternately accorded with and antagonized Davis. Gwin's roots in Mississippi slaveholding, at first glance, put him out of step with California's free-soil constitution. But like all Monterey convention delegates—including fifteen who hailed from slave states—Gwin voted in favor of Article I, Section 18, which prohibited the practice. (The convention nearly barred free African Americans from residence, revealing the racism and moral wrinkle within free-soil ideas.) What distinguished Gwin most, however, was his response to the proposed separation of Southern California. Some delegates offered to create a boundary along the Missouri Compromise line (36° 30′), which reached the Pacific just south of Monterey. This likely would have allowed slavery below. Many ranchers approved, but other delegates refused. Southern politicians like Davis thought the split more palatable. Gwin found a novel reply to all sides. He argued that the Missouri line was irrelevant to California due to nature and aridity. Yet it was relevant east of the Sierra Nevada, even in the more arid Southwest.[5]

An improbable contortion of words, actions, and associations enabled Gwin to be antislavery for California, but not so for the nation. It allowed him to stand for free-soil union among California's diverse counties, yet stand on opposite principles for federal union. Gwin lived in a house divided. He called this moderation. But it presented opportunity above all. He knew the times might change, and he determined to change with them. Other politicians, such as Davis or John Frémont, argued from absolutes and traded insults. Gwin became a shape-shifter. This fed a mythology he later insisted upon: rather than pro- or anti-, he was "non-slavery."[6] It also placed an air of mistrust around the good doctor.

WHEN THE SENATE BEGAN session in January 1851, Gwin was Lazarus revived. He had reached the nation's preeminent chamber, its most exclusive club. He represented a state renowned for mineral wealth and access to the Pacific. Most impressive, he navigated the social fractures of the Gold Rush to become California's most potent figure. After his allies deposed Frémont from office, Gwin took control of the state's federal patronage—jobs like postmaster, customs collector, surveyor general, Indian agent, district judge, and US attorney. He consolidated influence by nominating those loyal to him. Meanwhile in Washington, Gwin became the authority on all matters California: gold, naval affairs, transportation, mails, and coastal mapping.

Yet the Senate's deference to Gwin extended most of all to the issue of land and property. His California Land Act to "ascertain and settle" Mexican-era claims was the result. Nearly 750 private grants issued by Mexico covered twelve million acres of the state.

Non-Hispanic immigrants took control of a good number in the mining region. But their holdings were contingent on US approval of the original concessions. Gwin's bill proposed do this. It created a three-person board of commissioners, appointed by the president, to evaluate applications. The US had used such a system for Spanish or French grants in Florida, Louisiana, Illinois, and Missouri. In those cases, findings were sent to Congress for review and confirmation. That proved a failure, resulting in corruption and a backlog of unconfirmed claims.[7]

Gwin's bill streamlined things. Instead of saddling a congressional committee with the task, his Land Act delegated confirmation to the Land Commission, the same body that gathered evidence. California's federal district courts would certify the confirmation by decree. The claimant then would hire a US surveyor to map the parcel and resolve any boundary disputes. This done, the surveyor general for California would sign off and forward the application to Washington, DC. Finally, the General Land Office would issue a US patent.

On March 3, 1851, Gwin's Land Act became federal law. He made a stunning Senate debut, which he extended with a series of successful bills. He also triumphed over one rather strong-willed opponent: Missouri senator Thomas Hart Benton. Benton, via his son-in-law John Frémont, had acquired substantial grant acreage in California. Together, they worked to replace Gwin's bill with looser terms. Benton's strategy was to catch Gwin among his competing constituents. He denounced the act as hostile to Hispanic ranchers and the mass of settlers who had acquired (or taken) land from them. But Gwin sidestepped the trap. The 1848 treaty, he reminded colleagues during floor debate, required the US to *protect* the rights of Mexico's former citizens. ("A more fearless or braver people does not exist," he effused.) His Land Act would do this. "Instead of disturbing, it will quiet their titles," he predicted: "They and their descendants will *ever hereafter* retain the possession of them in peace."[8] Gwin won the vote. And because the Missouri legislature had declined to renominate Benton, as of March 4, both he and Frémont were out of office and out of the way.

For the moment, Gwin was master of his state's fractious politics. But not entirely. His praise for land grant claimants contradicted earlier remarks at the Monterey convention that "it was not for the native Californians we were making this Constitution" but for the "great American population."[9] Gwin attempted to walk this back. But it proved truer with time, as the senator looked to trade the support of ranchers for that of the growing number of landless settlers. He introduced bills to allow preemption on Mexican-era claims. One of these proposed a "floating" claim, that is, the ability to exchange lands for equal acreage elsewhere in the public domain. Gwin's stated rationale was to compensate claimants for any losses to US squatters. But because Mexican-era claims involved the best lands, his proposal seems a gambit to push Californios toward lesser property. The senator was testing political winds.

That made the Land Act more flawed than it at first appeared. In Gold Rush country, where Californios urgently needed the leg-

William Gwin found success in the Senate and evaded its contentious issues—for a brief moment, at least.

islation, it came too late. Southern ranchers faced better odds. But that depended on a functional and timely process. Instead, problems arose. The law gave claimants a two-year window to file cases. The Land Commission had one additional year to hear and decide them. President Fillmore had to find appointees willing to serve at an undesirable and faraway post. Next, they had to survive a long, dangerous voyage to San Francisco. As of December 1851, the commission lacked a quorum to hold meetings. Once it gained quorum, the third member was absent for another month. Newspapers grew hysterical. According to San Francisco's *Alta*, appointee Harry I. Thornton had assaulted an election adversary in Alabama and lacerated his hand in the fight. The wound festered, so a surgeon amputated the hand. The incision grew infected, so the surgeon took off the arm. The missing commissioner then developed a fatal sepsis. The *Alta* exaggerated this last detail. Thornton arrived in January 1852, weeks after his predicted demise.[10]

The Land Commission convened with just two years remaining. In haste, it prioritized the most impacted areas of Northern California and refused all other applications. Claimants and investors from the southern counties were outraged. Commission staff, meanwhile, complained of overload. By May 1852, they had 250 cases in docket. Clerks struggled to translate Spanish-language documents. Finances ran short and the office had to be downsized. The commission's secretary publicly threatened to quit if his salary was cut, as the *Alta* reported it would be.[11]

Faced with a worsening crisis, Gwin hurried behind the scenes to rescue the Land Act. He passed a supplemental funding bill of $22,000, and later won a two-year extension of the commission's deadline. The process finally opened to Southern California applicants. Within weeks, claimants filed one hundred new cases. Among these, on October 19, attorney J. Lancaster Brent filed a petition in the name of Manuel Domínguez and eleven family members for the Rancho San Pedro. It became case number 398. Two days later,

depositions began. José Antonio Carrillo, hero of the 1846 Domínguez Ranch battle, provided the first testimony.[12]

Still, Gwin did not rest. As newspapers noted with curiosity, the
senator pushed Congress to fund a study of the San Pedro–Santa
Barbara Channel. Papers struggled to understand this action in light
of the Land Act. The *Los Angeles Star* conjectured that the appropriation aimed to extract the islands from rancho grants. That seemed
plausible. Mexico's provincial government had granted two islands,
Santa Rosa and Santa Catalina, to the Carrillo family.[13]

But something more elaborate had happened, and Senator Gwin
was at its center. The results would take several years to material-
ize. In the process, they would reshape Domínguez's land case and
determine how a collection of seemingly disconnected events—
surrounding the doctor-senator, Coast Survey, and San Pedro
ranchero—sorted out. Time would reveal the meaning of good news
and for whom it truly arrived.

Chapter 6

A PROFESSOR'S GAMBLE

G EORGE DAVIDSON AND HIS COAST SURVEY PARTY
spent the late summer of 1852 in a standoff with the Makah
people of Cape Flattery. Meanwhile, Superintendent Bache lobbied
Congress to fund further research on the Pacific frontier. He had
been remarkably successful so far. Survey operations in California
and Oregon enlarged to six field parties laboring on triangulation,
hydrography, astronomical-magnetic observation, and topography.
Tidal observation soon added another branch of study. By 1853,
Bache had seven assistants and four sub-assistants stationed on the
western sea. Military support grew as well. The navy loaned sixty-
one officers to the Survey, and the War Department sent fourteen
army officers.[1]

The republic's expansion had proved a boon to coastal science,
as Bache correctly wagered. One decade before, he had inherited a
moribund and beleaguered federal office. His genius was to revital-
ize the Coast Survey by enlisting it in the service of manifest destiny.
Equipped with funds and powerful friends, the superintendent sent
his handpicked staff into the distance. Their work brought discovery
and acclaim; the respect of European science, even. In Bache's mind,
the Survey could gain all in pure pursuit of knowledge while politicians respectfully followed. But he was mistaken. Politics remained
the preponderant force. Soon, it would send both the Survey and the
nation careening toward great crisis.[2]

The superintendent abetted the coming troubles unknowingly.

As of August 1852, he circled through Virginia and Washington, DC, visiting with patrons—most especially with Senator Gwin. To date, Gwin had secured $290,000 for USCS projects. He was among a select group of officials most responsible for the Survey's reinvention. Southern California's channel gave Bache reason to seek more money. He explained to Gwin how its triangulation would benefit science and improve navigation safety. The senator listened, particularly when it came to islands nearest the mainland.[3]

Bache thought it was his staff's research that dazzled elected officials. The proposed channel study certainly was marvelous, a grand project to connect the borderlands of the US Pacific and integrate them by ocean commerce to a solidifying nation. However, Gwin took interests of his own in the project. And he was not alone. Politicians had benefactors to please. Their votes set the terms for legislation. Gwin drafted a special appropriation bill that would allow the Coast Survey to proceed with channel triangulation straightaway, rather than waiting for normal budget sessions the following spring. But the bill greatly altered the project's timetable. It required work to be completed *within six months* rather than the two years Davidson proposed.[4]

On September 24, after Congress approved $16,000 in funds, Bache sent revised instructions to Pacific coast assistants Davidson, James Alden, and Richard Cutts to prioritize the channel. He asked Davidson to return to San Pedro to verify the baseline—that is, the first side of its first triangle. Preliminary measurements had to be correct before Joseph Ruth could take over. Bache alerted the schooner *Ewing* to transport the field parties as soon as possible.[5]

That was easier said than done. Davidson already had too many responsibilities, and he was falling behind. On the return trip from Cape Flattery, he planned to stop at Cape Mendocino to determine latitude and longitude, but high surf blocked his landing. The job had to be pushed back to the next year. Meanwhile, the channel study postponed his other tasks: testing new chronometers, and exploration above San Francisco Bay, between Point Reyes and

Tomales Point. Delays of this kind had become typical. They added
to the expectations placed on Davidson and made it impossible to
pause. The *Active* returned him to San Francisco in early October.
Instead of taking much-needed rest, he raced to start on the San
Pedro baseline.

A mounting workload pushed Davidson's mind and body to the
limit. He indicated to Bache that the life-threatening venture to
Cape Flattery had left him exhausted. "I was taken suddenly ill last
night," he told, "and cannot write in detail."[6] More details would
follow. But not before news reached him that Joseph Ruth had perished in the Columbia River on October 17, after his boat capsized
near Astoria. Davidson and fellow officers issued a statement published in the *San Francisco Alta*:

Resolved, . . . the Survey has lost one of its most able and
accomplished members. Resolved, That we most sincerely sympathize with the widowed mother of the deceased, who by this
sudden dispensation of Providence has been bereaved of an
affectionate and devoted son.

As former schoolmates in the same profession, Ruth and Davidson
often exchanged letters. Just weeks before, the two had a written
spat over fifty dollars Ruth had borrowed and failed to repay. Davidson kept their final letter as a reminder—of what he never said. What
did the sum matter now? He asked friends in Philadelphia to call on
Ruth's mother and tried to recover any remains to send home.

RUTH'S DEATH WAS A reminder of the dangers that surveyors
faced. The men were experienced with the outdoors and accustomed
to physical labor, but they did not exercise or condition their bodies
in the present-day sense. Nor did they know of the most basic life-
saving practices. For all his time spent on boats, *Davidson did not*

even know how to swim.[7] This was true generally of USCS staff
and mariners at the time. That knowledge had limited value anyhow. A skilled swimmer still lacked materials beyond wool and—if
lucky—a struggling campfire to keep warm after exposure to cold
Pacific waters. Sustained exposure of a few hours meant irreversible
hypothermia. Brief exposure might prove fatal. It is remarkable that
field parties did their work willingly and for little pay. They looked
to the rewards of adventure, science, and innocent lives saved. They
put faith in things loftier than themselves.

As a result, death stalked the Survey along the wild western coast.
James Lawson nearly was killed when his whale boat broke apart
on a sandbar off Tomales Point in late 1853. Earlier that year, he
survived a capsizing in frigid shore break near Drakes Bay. Davidson almost drowned in October 1854, when he was knocked off a
wharf into the swift, outgoing tide of San Francisco Bay. Lawson
thought quickly enough to save him. Davidson had drawn up a will,
but no one would sell him life insurance. The actuaries were correct:
coastal exploration was perilous in the extreme, at sea and on dry
ground. Archibald McRae, the navy lieutenant in command of the
Ewing during 1855, took his own life with a pistol after irrecoverable
depression. Several years later, Davidson lost an aide and dear friend
in a mountain accident.[8]

While he wrestled with timetables and his own mortality, Davidson experienced pronounced physical and emotional decline. He
returned to San Francisco with post-traumatic stress caused by the
storms and violence he had witnessed in Oregon. But he also damaged his body by his habit of working eighteen-hour days with precision instruments. In just two years on the Pacific, Davidson developed
rheumatism from exposure to cold and damp. This combined with
heart irregularity and recurring diarrhea. He manifested neuralgia
and overuse injuries, including to his "observing eye."[9] The latter
he blamed on a theodolite eyepiece, but he had no other, and the
pain became chronic. The added burden of the San Pedro triangu-

lation only deepened his struggles. "I have been sick for five weeks," he desponded to Bache, "and have been getting worse." Regardless, Davidson pressed on to the brink of collapse. By late 1853, the now-twenty-eight-year-old confided to his sister, Belle, that he was "tired to death." His letter brightened to describe the pleasures he took in elk hunting and fly-fishing in the Humboldt wilderness. But it promptly returned to the gloom and idleness imposed by thick fog, chilling rain, and endless redwood-cedar forest.

He was not alone. The North Coast had ways to break even the most steadfast of spirits. Five months after Davidson wrote to his sister, Captain Ulysses S. Grant arrived at Fort Humboldt to take command of an infantry company. The same cruel conditions drove the thirty-one-year-old officer—later renowned for his calm in the midst of bloody battle—into depression and alcoholism. By mid-April, Grant resigned from the army, presumably never to return. Davidson remained sober and committed. But the price was suffering. He informed Bache in November 1853 that rheumatism left him "unfit for even office duty."[10] "Am now under the doctor's advice,"

Yurok village at Trinidad, near Fort Humboldt—
where Davidson, Ulysses Grant, and others suffered
on the Pacific frontier

he told, "taking the most powerful medicines" (perhaps the opium tincture laudanum).

American expansion could exact a severe toll on individuals in lowly government service. In that regard, the Coast Survey and its professor were as much the problem as the punishments of frontier life. Davidson's health worsened because of his chief's ambitions. Bache was a mentor, counselor, and taskmaster. He and his wife, Nancy Clarke Fowler, had no children, so he dealt with staff in fatherly ways. That meant rare praise (as was the era's convention) and constant correction. In 1850, he promised Davidson and the others leave after one year's service. Yet the superintendent thereafter declined all requests for time off because of the staff's indispensability and the sheer volume of results demanded by Congress. Bache expressed regret and encouraged his underlings to persist, but his position and theirs became irreconcilable, creating a downward cycle. When denied, they worked longer and faster, yet the list of tasks only lengthened. In the spring of 1854, Davidson wrote Bache, "I am now suffering so much that every movement becomes painful in the extreme."[11] The professor gave in, alarmed by his pupil's deteriorating condition. Davidson would take his first furlough starting November 1. Just barely. He had almost drowned off the wharf in San Francisco two weeks before.

Bache had gambled on American expansion. But he gambled *with* the lives of others. That caused dissension among his Pacific coast staff. Davidson complained that his salary and budget had not kept pace with those of other assistants. The superintendent increased his pay but remonstrated him also for his "constant grumbling" and "spirit of discontent."[12] At the same time, Davidson developed a poisonous rivalry with James Alden that inhibited their cooperation. As a naval officer, Alden was accustomed to being the sole authority—even an autocrat—on ship. He dealt with Davidson accordingly, despite the fact that both held equal rank within the Survey. Alden pronounced that, regardless of titles, the younger civilian was a subordinate while aboard *Active*. That caused them to squabble over credit for charts and discoveries made while at sea. Davidson resented his dependence on Alden for transportation. "This being at the mercy of the other parties," he protested to Bache, "is prejudicial to the work and humiliating to me." The Coast Survey later purchased a vessel for Davidson's use. He would rename the eighty-three-foot brig *R. H. Fauntleroy* to honor his deceased mentor and the family he later married into. The partnership with Alden continued from a cold distance.

WHEN DAVIDSON EMERGED from bed rest in December 1852, he traveled to San Pedro to begin the Channel Islands project. The baseline would run from a point ten miles south of Los Angeles pueblo to two miles west of the Domínguez family's adobe. However, Davidson found the task impossible to accomplish. Winter storms covered the Los Angeles plain with thousands of acres of additional wetlands and mud. Once the rains began, roads were "in dreadful condition."[13] He could not ride out to take coordinates. On one attempt, his horse sank into quicksand nearly up to the saddle. Animal and rider escaped narrowly with their lives. Alexander Harrison reported only four good days when he could survey topography. After returning to San Francisco on December 30, Davidson counseled with Alden. The two agreed to suspend the South Coast work until the end of winter.

Davidson returned in the spring of 1853, this time to clear skies and good results. To lay out the baseline, his crew meticulously turned four-meter iron bars end over end. It took 2,500 bar rotations and ten days to run the full length, 6.3 miles. Davidson had to calculate the temperature expansion of iron throughout the day and adjust his numbers for accuracy. He set the base in true position with a confirmed geodetic reference point, "San Pedro observatory."[14]

Some conditions remained trying. This time, Davidson complained of the heat. It created refraction waves and mirages that caused survey flags to disappear during sighting. On numerous occasions he walked toward a marker, thinking it had fallen down, only to see it reappear. Davidson also faulted Manuel Domínguez's "immense herds of cattle" for obstructing flags or trampling them under.[15] Then there was the problem of floodwater and pests. Wading through cold ponds caused Davidson's rheumatism to flare up. Lawson, who worked on elevation profiles (horizontal views of the baseline in relation to sea level, a necessity for geodetic readings), particularly disliked the beetles and scorpions that invaded his bedroll.

From his hilltop vista, Domínguez watched the federal surveyors. Government maps would validate his claim and increase its value. On the opposite side, Davidson could see the Spanish-colonial home and the past it symbolized. On May 14, he notified Bache that San Pedro's initial triangle was complete. More shapes would grow forth beyond the coast and onward to his scientific achievement. Domínguez and Davidson glimpsed different futures. Yet their lives had met.

TRIANGULATION OF Southern California's channel now could begin. But who would lead it? The United States in 1853 entered a moment of ferment brought on by its sudden expansion across the continent. Vast opportunity and rewards tempted science to great ambition, to push at known limits, almost recklessly so. The same things tempted politicians to reach for mastery over the West. They would do this through competing efforts to establish a Pacific railroad. Their gambits caught Bache among his various ties in Washington, DC, and complicated the channel study in ways he had not counted on.

First, events in the Northwest wrenched the Coast Survey deeper into that volatile region. In March 1853, free-soil proponents in Congress established the Washington Territory. They installed as governor Isaac Ingalls Stevens, a dashing military officer with no political experience. Stevens was, however, a veteran of the top-flight Army Corps of Engineers. Since 1849, he had also been a member of the Coast Survey and managed its central office in the District of Columbia. Aside from Washington's governorship, Stevens would oversee a War Department railroad survey, command militia, and consult on the USCS's study of potential harbors. He would serve as the territory's ex officio superintendent of Indian affairs, which empowered him to seize Native lands by treaty or by force. Bache held Stevens in high regard as a soldier-scientist. He had been the superintendent's number two, helping to lobby politicians and overhaul the Survey's administration. Yankee expansionists had played a formidable hand.

Stevens knew too much about the Coast Survey, enough to jeopardize its nonpartisanship. His appointment transformed the Northwest from a maritime space into a political region, and that brought the contest over slavery into the superintendent's domain. In mid-May 1853, Bache issued confidential orders to the Pacific assistants. Davidson and Alden were to proceed to the Washington Territory to cooperate with Governor Stevens's plans. These soon included a number of belligerent campaigns against indigenous people. The goal of science and safe navigation had borne bloody consequence. The Survey no longer could avoid the pull of empire and war that its own work had invited.[16]

Yet views on slavery and Western railroads were multiple, cer-

The USS Active *(right), with indigenous fishing boats in Bellingham Bay. Despite their feuds, Davidson and James Alden together witnessed the violent birth of the Washington Territory.*

tainly more so in 1853 than before. Proponents of alternate routes countered the Washington Territory's creation and Stevens's moves with their own. Two of Bache's congressional allies were prominent in the effort. Senator Gwin peddled his Pacific railroad and telegraph bills, as well as a line of transpacific steamers. Meanwhile, Jefferson Davis had become secretary of war. In contrast to Stevens and his backers, Gwin and Davis favored a railroad into California. As legislators, they had won the Coast Survey large sums of money. If Bache deigned to owe anybody a favor, Gwin and Davis topped the list.

Competing railroad schemes trapped the superintendent, his prized bureau, and his staff, too, on a transcontinental battleground. He had made a number of deals to expand the Coast Survey. Entanglement in the sectional conflict now proved to be the bargain's true cost. That jeopardized the agency's mission. Rails and slavery threatened to pull apart the western borderlands, north and south, before Davidson's triangles could unite them. Still, the professor insisted he could maneuver through the hazards. He watched as partisans aligned themselves for and against various routes. They articulated incompatible visions. Yet each and all required a Pacific harbor and precise maps. *They* depended on the Coast Survey and its expert science, did they not?

But Bache would find that his influence extended only so far as friends in Washington, DC, allowed. They looked to advance their political causes by way of his accomplishments. And they were too sharp to miss how recent events had left the professor vulnerable. With Joseph Ruth dead and Davidson sent to the Washington Territory, Bache struggled to find a replacement to lead the San Pedro–Santa Barbara Channel study. Congress required that he do so, or else forfeit the appropriation. For months, even as late as June 1853, he held the position for William Greenwell, a seven-year veteran of the Survey and its triangulation along the Gulf of Mexico. Unfortunately, Greenwell fell severely ill. Doctors warned that he would not survive a voyage to California.[17]

Science could wait; politics could not. In July 1853, Bache abruptly wrote Davidson that he would send out an army officer and new-comer. Edward Ord had joined the Coast Survey on the War Department's recommendation only months before, at the end of December 1852. He was West Point–trained but had only brief experience with geodetic mapping. By the time Davidson received word, Ord nearly had arrived through the Golden Gate. It was a landmark he had seen before. For Ord was the fellow traveler who had sailed from San Francisco aboard the *Oregon* at the start of 1850, with William Gwin and California's statehood petition.[18]

Ord was less known to Bache, certainly compared to the protégés he had dispatched previously. But Ord was quite well known by Senator Gwin and War Secretary Davis. The army officer also knew Manuel Domínguez and the lucrative opportunities that awaited where continent and ocean met. As Bache and the Coast Survey gazed off through trusted instruments—and as Domínguez looked for the blessings of US conquest—new imperatives gathered around.

Migrations

GEORGE DAVIDSON TRAVELED TO CALIFORNIA DRAWN
by the logic of numbers. His triangles, stretched over the
Earth's surface, aimed to give the western coast meaning within
national territory. But San Pedro Bay already had meaning as well
as great purpose. Its 3,400-acre estuary was a hub of ecological
activity, providing sanctuary for the production and reproduction of
life. The bay beckoned things to it from land and sea, in some cases
across tens of thousands of miles. Biology formed its own network
and shapes governed by the logic of survival.

Nature organizes by scale. Endemic species might meet all of
their needs locally. At the estuary, they lived and died amid mud,
marsh, sand, and water. This included topsmelt fish and slough
anchovy, along with specialist birds such as the savannah spar-
row and hen-like Ridgway's rail. Their short-distance lives enabled
the peripatetics of others. Perennial visitors entered the bay, often
blurring the distinction between land and sea. Seals and sea lions
basked on dry beaches under the sun. Curlews needled at tide lines
with long, downcurved beaks. Brown pelicans glided from shore
to scoop fish from the bay. Squawking seagulls followed. From the
uplands, animals came but just as easily left again. Coyote, wolf,
puma, and bears (black and grizzly) stalked or scavenged. Hawks
and owls did as well. From pelagic waters, fish and marine mammals moved in with the rising tide. Croaker, round stingrays, and

angel sharks probed the bottom. Sea otters did, too, when not for-
aging the kelp beds. Then, they rafted leisurely on the bay's surface.

Long-distance migrants arrived with the changing season. Osprey
pairs and flocks of American white pelicans made their winter flight
from as far as the Rocky Mountains. Other sojourners came by the
Pacific flyway, a north–south thoroughfare of watercourses and wetlands. Every autumn, they fled the declining food supplies at higher
elevations or latitudes. In spring, they retraced their paths. As the
largest estuary for a four-hundred-mile stretch, San Pedro Bay provided an important stop for coastal ducks, egrets and herons, and
various shorebirds. Perhaps most remarkable was the smallest: the
sparrow-sized sanderling, which flies a circumpolar route and one of
the longest known migrations.

Offshore of the estuary, great marine herds navigated the channel.
Baleen whales, most notably the gray, passed by on their fourteen-
thousand-mile annual journey between feeding grounds in the
Bering Sea and the warm nursery lagoons of Baja California. The
humpback, known for its intricate songs, followed the same course.
Schools of tuna—albacore, bonito, bluefin, and yellowfin, which
later made San Pedro a center for the canning industry—migrated
during summer. Some tuna circuited across the Pacific to Asia. These
carnivores fed upon smaller resident fish, which themselves moved
in and out of the estuary. The tuna's prey included steelhead trout
and Pacific lamprey. Such anadromous species entered the estuary
from the ocean (steelhead from hundreds of miles away) to access
spawning gravels upstream. They, most of all, joined continent and
ocean into one.

For eons, non-human migrants had connected Southern California into a separate sovereignty. The map they made was whole, sub-
lime even. So much so that scientists like Davidson had to ignore the
humbling shapes in order to supplant them. Yet there were further
possibilities. Others soon arrived who hoped to connect San Pedro
to far-off plantations and towns of the Mississippi River. Their plans
would unsettle nature and nation alike.

San Pedro Bay, with estuary at center, painted by an artist and officer of the Coast Survey ship Active *in June of 1859.*

Part Two
RAILROADS

Beware of Swindlers

I N September 1854, Manuel Domínguez received a letter from Edward Ord of the Coast Survey. The two men had grown acquainted in the year since Ord arrived to triangulate the offshore channel. Their friendship symbolized the possibilities of post-conquest California. The Mexican rancher introduced Ord to the multiethnic world of Los Angeles. Ord connected him to the US government and to the unfamiliar mindset of Americans now populating the state. Each served as cultural intermediary for the other.

Ord certainly thought so. "Many congratulations on the confirmation of your rancho," he wrote in imperfect Spanish.[1] He had read about the Land Commission's verdict in the San Francisco newspapers. It, he explained, "guarantees all of the land within the original boundaries." Yet he warned Domínguez to beware. Among the new immigrants to California were hordes of squatters and swindlers. To guard against them, he advised that the rancher sell a small parcel to a trustworthy "group of Americans." This would win powerful, English-speaking allies who would defend Domínguez's claim and support his application for a US patent. "Regards to your wife and daughters," he closed. "Con que, quedo su amigo, Eduardo Ord."

The officer's words reveal the mixed character of Los Angeles in the 1850s. Here, former Mexican nationals and Anglo immigrants needed each other in order to prosper. That required trust and time spent in shared company. But friendship resulted from tensions as well. A greater common foe existed: others who could not be trusted. Domínguez and Ord's bond was forged in this three-sided combination.

Bonds were not quite as they seemed, however. The rancher maneuvered to validate his property, and Eduardo, his friend, professed to help. But Ord remained an agent of the United States. He represented imperial expansion and its sharpening divide between North and South. Ord brought these entanglements to San Pedro Bay. Soon, they would undercut Domínguez's land claim along with the two men's friendship. Their shared world did not last.

Despite Ord's caveats to beware of others, his actions would remove thousands of acres, including most of the San Pedro estuary, from the rancher's estate. The following chapters examine why. They tell how a transcontinental railroad—the ambitions to locate and build one—provided the thread connecting people, places, and conflict to come.

Chapter 7

PERIL IN THE SIERRAS, INTRIGUE IN WASHINGTON

L IKE MANY OTHERS, EDWARD ORD CAME TO CALIFOR-
nia by ship. But an unseen network of railroad schemes brought
him as well. By 1850, contending rail plans had emerged, backed
by rival political interests. These focused increasingly on the southern portion of California, which promised year-round access to the
Pacific as well as the possibility of a compromise in Washington.
Debates and disagreements remained, however. Who would claim
credit for the rail line, determine its course, and take hold of its
rewards? The questions initially were about the legacy of one man,
John Frémont, and his discoveries as the United States' famed "Pathfinder." They rapidly escalated into much larger stakes, the future of
North and South most of all. At a number of stages, the contentions
pulled Ord into the railroad story. That would carry him also into
the story of San Pedro Bay.

Steam railways had become the basis for Americans' designs on
the continent by the early 1840s, when promoters first talked up a
line between the Great Lakes and Oregon's fertile valleys. To support such a plan, Congress sent Frémont to map the Oregon Trail
and passages through the Rocky Mountains. Best-selling memoirs compiled by his wife, Jessie Benton, popularized the exploits.
On a second expedition, that of 1843–44, Frémont's party crossed
illegally into Mexican California. Here, guide Joseph Reddeford

Walker showed Frémont a mountain portal that indigenous people used to cross the lower Sierra Nevada, moving between the Great Basin and California's Central Valley. Frémont renamed it Walker Pass. He thought it suitable for a railroad, and considered the lands lying west of it more desirable than any in Oregon.

Frémont's musings helped veer US expansion southward, inadvertently sending the railroad issue, and Frémont himself, into political trouble. He again set out for the Sierra and Walker Pass in late 1845, just as the US prepared to annex the disputed territory of Texas. Frémont's journey ended with him in command of settlers in California during the Bear Flag Revolt against Mexico. By the late summer of 1846, navy ships arrived to claim the region for the United States. Congress began to consider railways to access San Francisco's harbor instead of the Columbia River. That required a gap through the mountains. Walker Pass, Frémont argued, remained the best choice.[1]

As railroad plans shifted their focus to California, they ceased to be a point of political consensus. A railway through the Southwest would speed the creation of new slave states and favor their access to gold and Pacific commerce. Slaveholders understood and embraced that goal. For the same reason, antislavery politicians opposed it. Resistance to a Southwest railway became the next best means (in the absence of a legislative ban) to keep slavery out of the Mexican cession, or at least to arrest its growth. Iron rails were no longer a means of national connection. They became the crux of disagreement.

The sudden divisiveness of the rail issue was not what Frémont had wanted. But he had added to the controversy by his conduct in the US-Mexico War. The officer and popular hero believed he was destined for ever-greater things. Despite his mid-level rank, he took to calling himself "Commander of United States forces in California," and next, "Governor." General Stephen W. Kearny objected to this once he arrived to suppress the Southern Californio rebellion in early 1847. When Frémont refused to step aside, Kearny marched him east to stand trial for mutiny. A court-martial conviction hardly ended the episode. Frémont's father-in-law, Senator Thomas Hart Benton, procured him a presidential pardon. Frémont then resigned from the army on his own terms. He had enough fame and connections to do as he pleased. Plus, he acquired a massive Mexican land grant, Rancho las Mariposas, in the foothills below Yosemite Valley. The property held significant gold deposits soon to be discovered. His star continued to rise. And he planned to get even with those who had wronged him.[2]

Frémont's mind settled on Walker Pass. He set out to prove it viable and to grab it for a prospective railroad company he would lead. To underscore his heroics and command newspaper publicity, he thought up an audacious scheme to chart the pass on a route from St. Louis *during the dead of winter*. The conceit became his fourth expedition, that of 1848–49, funded by investors. The party entered the southern Rockies in late autumn. Soon they became trapped by snowfall. Multiple rescues arrived from Taos sent by friends and former guides, Edward F. Beale, Christopher "Kit" Carson, and Alexis Godey. But not before eleven members perished. Two more were killed by Ute Indians when they returned to reclaim a cache

John Frémont. His conflicts with the army led to a new search for railway passes into California.

of equipment, money, and papers. Survivors mourned. Some came closer to death than they could forgive. Whispers of cannibalism haunted them. Several, including topographer Charles Preuss, never worked for Frémont again.

THE ARMY HAD every reason to deny Frémont further glory. It would launch its own railroad explorations, allowing an unknown figure to enter the saga. Lieutenant Edward Ord had arrived in Monterey at the start of 1847 with his company of the 3rd Artillery. The soldiers arrived too late to take part in California's conquest—a misfortune, because it greatly reduced an officer's chance of promotion. They had little to do but drill, wait for news of the war in Mexico, and serve as local police.

Yet Ord and his compatriots were in the right place for the Gold Rush and the immense land speculation it set off. Military education included proficiency in surveying and engineering. Such skills were in high demand, as mining, investment, and immigration sparked a boom in property claims and construction. During furloughs, Ord became a prominent surveyor for hire. He used the work to acquire land near Santa Cruz and a "ranchito" in Monterey.[3] In spring 1849, fellow officer William T. Sherman was at Sacramento, platting the fast-emerging town for Johann Sutter. He invited Ord to join him in a survey to the Cosumnes River for land baron William E. P. Hartnell. In exchange, the soldiers received speculative parcels, which they sold for $3,000 each—far above one year's army pay. Ord regretted that he could not pan for gold. Nonetheless, he learned to prospect for land among California's old elite. In July, he took his services south and made the first survey of the Los Angeles pueblo and environs.

Ord's skills involved him in the search for rail routes, as the army looked to take charge of the issue from Frémont. One group of surveyors entered the northern Sierra to map passes rumored to exist

above the Feather River. Ord did not join them. But while near Sacramento during spring 1849, he and Sherman relayed messages to
the party commanded by William H. Warner and R. S. Williamson. The survey made progress, and it rescued numerous overland
migrants lost or in distress. But their work came to a sudden halt in
September, when unknown Indians, perhaps Pit River Achomawi or
Atsugewi, killed Warner.[4]

By this time, the army had sent Ord to make a corresponding study
of Southern California. In October 1849, he rode with a guide into
the backcountry. Their mission was to find crossings that avoided
high elevations and snows, like the one that had fatally trapped the
Donner-Reed Party near Lake Tahoe three years before. Military
officials also hoped for an alternative to the Gila River route. With
its scorching desert and Apache and Yuma raiders, the Gila Trail
visited equal horror on emigrants.[5]

Ord examined several sites in the mountains of Southern California. First, he visited "El Cajon de los Mejicanos" or Cajon Pass. This
lies at an elevation of roughly 3,800 feet, in a break between the forested San Gabriel and San Bernardino ranges, east of Los Angeles.
The army believed gangs of Sonoran horse thieves and "Pa-Utah"
Indians used it to raid from the Mojave Valley. Interstate 15 and historic Route 66 cut through the gap today. Ord described it as "stony
and barren," "without grass," and filled with chamisa bushes.[6] He
wrote, too, of "Corvilla" Indians—perhaps Serrano people—living
in dirt-and-bulrush huts, and "Jurupas" living on nearby ranche-
rias. The pass featured steep canyons, boulders, bends, and shadowy ledges. It was a prime spot to be ambushed or robbed. Ord
ventured three miles in before wisely turning the horses back to
exit. He then gathered information on two more passes. These were
San Gorgonio, east of Los Angeles, and Tejon, to the north. Of the
pair, Ord thought "Gorgona" less viable. Its trail lacked water and
tended toward the Gila route. Tejon Pass, by contrast, led to the San
Joaquin-Tulare Valleys. Canyons offered passage at an elevation of

five thousand feet. Old Tejon Pass sits about fifteen miles east of today's Interstate 5. Ord surmised that any road here would need a significant troop presence to guard against bandits and hostile tribes.

Most important, the army had Ord research Walker Pass. He could not see this firsthand, nor even by telescope. Its entrance lay beyond the Tejon, in a remote section of the San Joaquin Valley. From there, trails passed out of sight behind tule fog and the Sierra's sharp angles. Deep within the mountains, the pass crests at five thousand feet elevation amid a sparse and rocky landscape—more desert than forest. Then, the grade lowers gently east to the Mojave. Based on interviews and Spanish-era travelogues, Ord learned that the western approach ascended by trails above the Kern River gorge. Today, State Highway 178 weaves from Bakersfield, up the gorge to Isabella Dam and reservoir, then through the pass.

Ord concluded that either Cajon or Walker passes might serve for a rail route. In late December 1849, he submitted findings to the Tenth Military Department, headquartered at Monterey. The passes, he wrote, allowed access to the Colorado River and Santa Fe on the upper Rio Grande. High-elevation desert in between had sufficient water to support a road. Native American "hostiles" like the Mojave and Navajo presented the "only difficulty."[7] Ord suggested a system of forts to safeguard travelers and "civilised Indians." Yet he predicted that settlers would be the greatest source of violence and trouble, even as he accepted their inevitable migration. It was a circular logic shared by many army officers.

Ord's report provided conclusions independent of Frémont's. But on its own, it did nothing to advance the transcontinental rail project. The West was awash in military and emigrant maps, all claiming important discovery. Remarkably, however, Ord's report would find powerful champions just as California's statehood application landed in Washington, DC. After submitting his findings, the soldier packed up and left California. In January 1850, he boarded the Pacific Mail side-wheeler *Oregon* with a friend, William T. Sher-

Mojave Indians—an artist's sketch from
the 1853–54 railroad surveys

man, who was set to be married in the East. Ord carried a copy of his
railroad report in his baggage. Coincidentally, John Frémont was on
board, along with his fellow senator-elect William Gwin.

A SEA JOURNEY from California to New York meant four weeks
in close quarters, interrupted by bumpy transit across Panama. The
experience is so alien to our sense of travel today that it merits description. The *Oregon* carried three hundred persons segregated by ticket
class. Within a tossing and creaking hull, passengers had continuous
encounters. Opportunities abounded—on deck, in drawing rooms,
and at meals—to make business deals or swap stories. En route, vessels transported large, eastbound shipments of gold, newspapers,
and mail. That brought out crowds, celebration, and sutlers at every
port. But time at sea was unpleasant, too: seasickness, discomfort,
spoiled food, the stench of bodies and poor sanitation, boredom,
disease, and even death. Then there were the eroding relations on
ship. Courtesies grew tired. Gossip, whispers, intrusive eyes and ears
were all around. Yet the experience dragged on.[8]

Who talked to whom mattered a great deal. Passengers had items to share. But they also had secrets to keep. How each traveler managed this could change their lives. So it was for John Frémont and Edward Ord. One cannot say whether the two traded thoughts on Walker Pass. But it is unlikely. Frémont had informants enough to know of Ord's report. And the two had met at least once before. While stationed near Washington, DC, in early 1845, Ord heard that Frémont was hiring for a mission to Mexican California. He called on the Pathfinder in hopes of joining the party. Frémont turned him down. But Ord was soon thankful. By 1847, he disapproved angrily of Frémont's quarrel with General Kearny. Some army staff wanted the defiant explorer put before a firing squad. Ord suggested to family that he agreed.[9]

William T. Sherman gave Ord additional reason to steer clear of Frémont. Sherman was a courier for the military's papers, which included charges and evidence against the Pathfinder. He was to deliver the cache to army general-in-chief Winfield Scott. Sherman also had links to President Zachary Taylor, no friend of Frémont's, via his foster father (and soon-to-be father-in-law) Thomas Ewing Sr., the nation's secretary of the interior. But the soldiers were wary for one more reason: Frémont had proclaimed himself a free-soil, antislavery Democrat alongside his father-in-law, Senator Benton. Ord was a Democrat, Sherman a Whig. Neither was critical of slavery, although they refused to partake of it. ("Don't own a slave and don't want to," Ord wrote at the time.)[10] They certainly were not abolitionists. Such positions were thought too radical—career-killing, in fact—within the army.

Frémont, the popular character of books and serials, had become much less so in person. The *Oregon* voyage was likely unfriendly and uncomfortable. No wonder he and wife Jessie reportedly fell ill. When the vessel reached Panama, they opted for a one-month layover (until the next ship departed). It was a preview of difficult times ahead. California Democrats would sack Frémont in 1851, after he had worked just twenty-one days in the Senate. Missouri

Democrats would expel Benton from his seat, ending (almost) his three-decade career in Washington. The family would scramble to gain back power and influence.

By contrast, William Gwin proved popular aboard the *Oregon* and gave the soldiers, Ord especially, an audience. The soldier and senator-elect had numerous affinities. Ord was from a Democratic family and raised in Maryland, a border slave state that prided itself on political mediation. He also admired Gwin for his lucrative gold mine and knowledge of Mexican land grants. Most of all, Ord could benefit from a patron in high office. Military officers often cultivated such arrangements to advance in rank. Gwin could also provide patronage jobs.

By the time the travelers disembarked, Gwin and Ord had found common cause in the officer's railroad report. Both made their way to Washington City, where their connection entered a new phase. Ord returned to the embrace of family; Gwin returned to a place of friends and enemies, where his prior humiliations were well known. He needed to reestablish himself with authority. California's potential statehood already was the talk of the capital, and he now had valuable information to broker on a most contentious matter. Gwin used Ord's report to enter Washington's inner circles and to gain desired allies. Foremost among them was Mississippi senator Jefferson Davis, who demanded a southern transcontinental railroad in exchange for California's admission.

Gwin's ambition thereby became Ord's as well. Davis found out about the report and, as chair of the Senate's Military Affairs Committee, obtained a copy from the War Department. During July, amid fierce debate and fisticuffs over the Compromise of 1850, Davis submitted Ord's report to the congressional record. This transformed what had been an internal study into a public document and conversation topic. Ord began to correspond with Davis as well as Secretary of War Charles M. Conrad, another proponent of a southern route. In reply, the officer praised Walker Pass for its proximity to the Colorado River delta and Sonora, two territories of Mexico that were being eyed by Southern expansionists. Davis thought highly enough of Ord by February 1851 to recommend him for an instructorship at West Point. The application was not successful, but Davis would keep the officer in mind.[11]

Gwin had some role in these introductions. He knew Davis from the 1845 congressional election, and farther back from Mississippi's planter society. They were two years apart in age, never friends, but rivals and occasional partners. Both attended Transylvania University in Lexington, Kentucky. From there, their paths continued to cross until they met in the Senate. Davis was a military hero of the US-Mexico War. Observers spoke of him as Calhoun's political heir, destined to lead the chamber's Southern delegation. Davis also had links to President Taylor, his former army commander and father of his deceased first wife, Sarah. Based on their past, Gwin had few things that Davis wanted. But he now possessed one Senate vote and an outsized say in the matter of a Pacific railroad. Over the next decade, Gwin and Davis would share several mutual interests—including their client, Edward Ord.

Unfortunately for Davis, California's statehood did not include stipulations for a railroad. But that was good news for Gwin. The issue continued. And as Senate leaders tried to conquer it, they would need his help. Gwin introduced his first Pacific railroad bill in December 1852. When it advanced to the Senate floor, he read Ord's 1849 report into the minutes. Gwin's plan was to construct a transcontinental line via Walker Pass, through northern New Mexico, and across a central latitude. The line would split into northern and southern branches as it neared the Mississippi River. He expected this to please all sides. To his chagrin, Southerners like Davis withheld support. They had no objections to Walker Pass, but they insisted on a route approaching it from a lower latitude. Meanwhile, Northerners noted that Gwin's so-called central route passed mostly through areas open to slavery. The bill failed.[12]

*The fearsome Sierra Nevada, as seen
from California's San Joaquin Valley*

By March, Gwin had gathered enough votes for a compromise. He led colleagues in amending the army appropriation bill to fund several concurrent expeditions that would become the famous Pacific railroad surveys. The Senate intended these to be conclusive; they were anything but. Topographical engineers set out in the spring of 1853, under the management of an ad hoc office within the War Department. The office, in turn, answered to the recently appointed secretary of war, Jefferson Davis.

As the engineers departed, the rail question seemed to move on without Ord. He was beyond the pool of officers eligible for the Topographical Corps. At peak size in the 1850s, it consisted of only thirty-six members, an elite circle of soldier-scientists and the top graduates of West Point. Frémont had been the most unlikely member, having never attended the military academy. He won entry regardless because of patrons like Senator Benton. Ord graduated in the middle of West Point's 1839 class. (Isaac Stevens was the year's top cadet.) Nor did he have advocates inside the government. Or did he? In 1850, shortly after arriving in Washington from California,

Ord wrote to a brother that his transfer to the Topographical Corps was imminent. It never happened, but the letter shows that someone gave him expectations, perhaps even a promise.[13]

Hence, Ord may have anticipated a return to the West after all. He had just started with the Coast Survey at Savannah, Georgia, on a transfer dated December 30, 1852. Superintendent Bache kept a sharp eye on his new hire. Yet neither Jefferson Davis nor William Gwin were far removed.[14]

Chapter 8

ORD FINDS AN AUSPICIOUS APPOINTMENT

SEVENTY MILES INLAND OF SAN PEDRO, US DEPUTY Surveyor Henry Washington ascended to a ridge atop San Bernardino Peak. He recorded the date, November 8, 1852. Over three days, Washington and his party had trekked to an elevation of ten thousand feet. Cold winds foretold snow. From this height, the fields of Yucaipa Adobe, a Mexican-era rancho, appeared as small geometric shapes. Villages of Tongva, Cahuilla, Serrano, and Luiseño peoples caught his eye from across the valley. The Mormon settlement at San Bernardino sat twenty miles away. East of the peak, beyond San Gorgonio Pass, lay the vast monochrome of the Mojave Desert. Different historical moments stood in miniature before him.[1]

Washington was a herald of change. Using solar compass and theodolite, he took coordinates while his crew erected a twenty-five-foot log monument. They fastened the pole with tin reflectors so it could be seen from the valley floor. The marker provided a cartographic "initial point." From there, Washington set forth the San Bernardino Meridian, a reference line running north to south. He laid out its corresponding baseline, a perpendicular bearing west to east between the ocean and the Colorado River. This done, Washington registered his measures with the US surveyor general for California, an appointee who oversaw federal lands. Of the initial points established in the contiguous United States, Washington's remains the highest and most difficult to reach—and a destination for hikers today.

The San Bernardino Meridian provided a milestone in US expansion. It brought to Southern California the Public Land Survey System (PLSS). The General Land Office used the PLSS to locate private and public lands west of the Appalachian Mountains. Unlike geodesy (which maps on a sphere), the PLSS orients its surveys on a Cartesian plane, a theoretical space created by two perpendicular number lines. The principal meridian gave each locality a y-axis. The baseline served as the x-axis. Subordinate perpendiculars—tract lines, township lines, and section lines—created a grid of varied quadrangles, each made up of regular units. Tracts measured twenty-four miles square, while townships were six miles square (23,040 acres). Sections were one mile square (640 acres). A surveyor might divide a section down to one-sixteenth of a mile square (two and a half acres). Right-angled lines allowed a person to find or describe any point within the plane using precise, consistent numbers.[2]

Grids were not the land system's only effect, however. Meridians and baselines caused real estate markets to erupt. Property became Washington's legacy, even as his log pole disintegrated to ruin. His initial point allowed US surveyors to swarm Southern California and transform landscape into commodity. Yet, because Mexican land grants covered the area, the Land Commission was also required to settle claims. As stipulated by the 1851 Land Act, US surveyors needed to locate each rancho on the PLSS grid. Only then could the applicant's petition proceed toward a US patent. With Washington's initial point in place, commissioners began in early 1853 to work through the backlog of Southern California cases.[3]

Manuel Domínguez saw the arrival of surveyors as a good omen. In the spring of 1853, George Davidson determined geodetic coordinates along with a baseline for the Channel Islands study. His work gave the Rancho San Pedro a place on federal maps. Then, in October, US Deputy Surveyor Henry Hancock arrived to plat the Domínguez grant in relation to PLSS township lines. One year later,

Hancock dispatched his assistant, George Hansen, to locate the rancho on PLSS section lines and map its topography. The presence
of surveyors—dirt-stained, sunburnt men bearing notebooks, colorful flags, and delicate instruments—inched Domínguez closer to
the secure property he had wanted for so long. He eagerly followed
his Land Commission case. Hancock and Hansen would testify on
the rancher's behalf and remain important to the estate through the
next decades.[4]

From Domínguez's point of view, the surveys appeared to be halves
of a single story. But they were not. They represented an incoherence
within US conquest. PLSS work took place by conventional methods rather than the Coast Survey's geodesy. The latter existed for
the sake of exactitude and scientific advance. The PLSS did not. By
employing a local grid for each region, conventional surveys worked
only well enough to create real estate. The PLSS borrowed the trap-
pings of science. But it did so for the purpose of parcels and markets.
US conquest took shape in the aggregate of actors and actions. It
brought together discordant elements pulling in varied directions at
once. From the outside, Domínguez saw only the entirety. But inside,
parts and pieces were in conflict. That made his prospects less pre-
dictable than he imagined.[5]

THE RANCHER WAS not alone. Edward Ord and Superintendent
Bache, too, found themselves caught in between: one by choice, the
other by surprise, and both by prior ties. Triangles and grids, science and politics, rails and ships—all mixed, but not so well. As
of November 1853, Ord was at San Pedro Bay, organizing his outfit. With supplementary funds from Congress, Bache promoted Ord
to the rank of USCS assistant. This gave him a budget of $7,700
per six months, thirty-eight percent above what Davidson deemed
necessary. Ord would hire a crew of eight men and a pair of mules
and horses. The party functioned as two because they often worked
opposite sides of a triangle.[6]

Ord's task was to map the offshore spaces. His attention wandered, however, toward Southern California's burgeoning real estate market and its potential rewards. Before he had even reached San Pedro, Ord wrote Bache to raise the issue of land grants on the Channel Islands. He wondered what he might do to break up the grants. "The islands are all claimed," he noted. "But I think the claimants in nearly every instance have but a slip of paper signed by some revolutionary Mexican who gave what he had no right to, for a little money in hand."[7] Ord did not hesitate to offer his premature findings: "The US government ought to give or sell them to actual settlers—such will be my report." This must have perplexed Bache. The phrase "actual settlers" came out of preemption policy—that is, laws that empowered settlers to seize lands in the West in exchange for improving or cultivating them. It had no relevance to coastal navigation or geodetic science. Bache said as much in reply. He thanked Ord, but added that he had referred the matter to the US Land Office.

Ord never again stated intentions so clearly. Instead, he tried new tactics. Subsequent letters detail his attempt to buy the schooner *Frolic* from the San Francisco Custom House. He planned to use this vessel to explore the Channel Islands at his own pace. The *Frolic* came at a price of $8,000, already above his semi-annual budget. Ord arranged to spend a further $1,500 on naval stores, provisions, and equipment. He also recruited a ship's captain at $150 per month and a crew of five "kanakas"—Hawaiian sailors and watermen. Because Ord did not have the funds, he arranged a stopgap. He would draw credit from the San Francisco bank of Lucas, Turner and Company. This was the bank near Jackson Square managed by William T. Sherman, newly retired from the army.[8]

Ord's scheme alarmed the superintendent. Ship-based observations would negate the plan to use the islands as stable geodetic platforms. Perhaps the officer knew that would be a problem? Ord tried to enlist Assistant Richard Cutts to advocate on his behalf and lend money from *his* budget. But Cutts knew better. He warned the

*Catalina Island, seen from San Pedro Bay's mainland. It
provided the crucial vertex for the Coast Survey's geodetic study.*

superintendent, who scolded Ord for his extravagance and the "very
extensive scale" of his proposals.[9] To block the purchase, Bache
agreed to allow $6,000 for a vessel, an unrealistically low amount. "I
trust," he advised, "I shall not be obliged to disapprove of your purchase." Ord gave up on the *Frolic*. Bache put him under closer watch.

Initial fieldwork further displeased the superintendent. Ord
rushed through San Pedro Bay's measurements as if he had other,
more pressing concerns. Over fifteen observation days ending in
December 1853, he sighted the offshore triangle extending from
Davidson's baseline to Catalina Island and back again to station
"Cerito." He predicted he would wrap up after three weeks and
recommended that the USCS declare Catalina "unfit" for settlement.[10] Such determination would not bar ranchers or farmers,
but it would allow the General Land Office to dispense with the
island in very large parcels, if subdivided at all. (Today, Catalina
is the *only* Channel Island with salable real estate and a significant human population.) Most important to Ord, the declaration
would allow him to skip out on time-consuming observations
and tracing of the shoreline.

The army officer seemed particularly intent on extricating himself from "secondary triangulation," a geodetic task indispensable to
Davidson's long-distance work. Because of the length of secondary triangles, surveyors had to sight from elevations of 1,000 to 1,500 feet
to clear ground obstructions, fog banks, and the Earth's curvature.
Sightings relied on the heliotrope, an instrument made of mirrors and
magnifying lenses, designed to reflect a beam of sunlight to a distant
observer. Ord had instructions to sight between Catalina and land-

marks near San Diego and "Point Duma" (Point Dume), sixty to eighty miles away. He complained, however, about the island's steep inclines. Hardship, he said, was not worth the low price the isle would fetch for the US government.[11]

As Ord extricated himself from the San Pedro–Catalina triangle, he was all too eager to chase land elsewhere. He opined to Bache that cattle ranching and farming, not navigation, should guide USCS activity. Ord pronounced the northwestern islands of Santa Cruz, Santa Rosa, and San Miguel to be most valuable in that regard. Newspapers recently had reported the arrival of a "wild" Indian woman from San Nicolas Island, the "last of her race."[12] There, she had survived two decades alone after the forced removal of her people. Baptized now as "Juana María," the channel's last indigenous holdout would die after just weeks on the mainland. To acquisitive minds, that emphasized the islands' availability.

Ord requested a vessel once again. If he sailed quickly, he explained, the Coast Survey might bring the northern islands into the public lands system. The government could then settle (or reject) their Mexican-era claims and sell them into private hands. For the same reasons, Ord declared that he would skip the smallest islands of Santa Barbara and "Anna Cappa," which he dismissed as worthless. Bache paused on the words. Several months before, Pacific Mail's *Winfield Scott* had smashed against the cliffs of Anacapa in heavy fog. The islets' size was the exact reason why they were so dangerous.[13]

As he fielded these dispatches, Bache confronted a navigation corridor that was more destructive than ever. The number of wrecks, near-accidents, property losses, and persons dead or bereaved increased in proportion to ship traffic. The *Tennessee*, which had carried Davidson to San Francisco for the first time, beached at the bay's entrance in March 1853. One month later, the *Lewis* mistakenly sailed past the Golden Gate in darkness and hit a reef outside Bolinas Bay. *Lewis* passenger William T. Sherman describes in his memoirs the terrifying experience (and the rescue he organized). In

June, the *Carrier Pigeon* ran aground on a headland thereafter named Pigeon Point. No lives were lost. But this was by Providence alone. One year earlier, the *General Warren* had broken up at the Columbia River's entrance, causing forty-two deaths. In February 1853, a San Francisco-bound vessel, *Independence*, wrecked along Baja California, killing 130 people. Further catastrophe awaited. In October 1854, the Vanderbilt steamer *Yankee Blade* collided with rocks off Point Arguello, near the San Pedro–Santa Barbara Channel's north end. Forty souls perished and $150,000 in gold dust sank. Between 1850 and 1855, at least twenty vessels wrecked along the West Coast. With each accident, pressure mounted on the superintendent to demonstrate his agency's progress.[14]

It was as if Ord and Bache were involved in different projects. But the officer's artless requests hinted at deeper problems. The bureau's methods, so attuned to science and minute detail, could not relent to property markets without compromising themselves. And without geodetic methods, the agency had little reason to exist. Con-

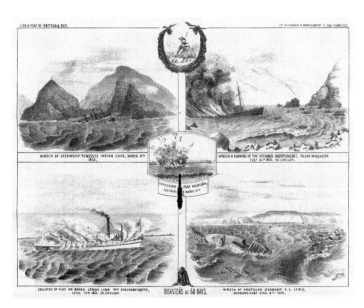

*The graveyard of the Pacific. Frequent shipwrecks
put pressure on the Coast Survey to show results.*

gress assumed that mass land hunger and elite geodetic science could marry. But despite a shared propensity for straight lines and right angles, they were fundamentally mismatched. The superintendent tolerated, even cultivated, this assumption in exchange for political support. Yet the nation's democratic, expansionist ethos—its land hunger especially—threatened to overpower him. Most distressing, the threat appeared in the person of Edward Ord, the officer he had entrusted to lead such an urgent and foundational study.

BACHE HAD CHARGED headlong into a conundrum. Yet he faced others as well. War Secretary Jefferson Davis had signed off on Ord's detail to the Coast Survey. The superintendent's personal ties to Davis would complicate his efforts to rein in the officer. It is unclear when Bache and Davis first became acquainted. Davis's papers record a meeting in May 1846, months into his term in the House of Representatives. On promotion to the Senate, Davis became a vocal defender of the Coast Survey—years before Gwin arrived to do the same. It was Davis who advocated for the agency's entry to the Pacific coast in 1849. That meant taking on the navy as well as Senator Benton, who tried to set up a rival bureau under John Frémont's leadership. Davis forwarded copies of his Senate speeches to Bache and checked to make sure they arrived.[15]

Political alliance strengthened into devoted friendship by the time Davis became secretary of war and began to live in Washington year-round. The Mississippi slaveholder and Philadelphia professor found they had much in common. Both were West Pointers—they overlapped for two years and recalled shared experiences and classmates. (Davis may have encountered Bache as an instructor, for Bache was so brilliant a cadet that West Point enlisted him to teach underclassmen.) Their bond extended farther still. Bache was a scholar of natural philosophy. Davis had a college-level education before he entered West Point. He was fluent in science, languages, and classics. A rare mix of military and scholastic training set Davis and Bache apart from most federal officeholders.

They soon regarded each other as worthy company. Davis began to dine at the Survey's headquarters, a barrack just off Capitol Hill, and at the superintendent's home. In August 1853, Bache and Davis traveled for three weeks through New England. The two stopped in Boston, camped in New Hampshire's White Mountains, and hiked to the USCS observatory on Maine's Mount Blue. It was the arch-Southerner's first visit to a region he regarded as a foreign land. And its strangeness crescendoed because of protests against the fugitive slave law. Later, during the summer of 1858, Bache invited the Davises—Jefferson, Varina, and children—to join him at the Coast Survey station on Lead Mountain, near Beddington, Maine. Varina wrote of pleasant clambakes, of New England's lack of biting insects, and of the two grown men jesting like schoolboys. Possessed of a Philadelphia education, she relished Bache's discussions of science as much as her husband did. She recalled the superintendent as "one of the greatest savants the country had produced" and

Patrons of Yankee science: Jefferson Davis as war secretary,
and wife Varina Howell Davis

wife Nancy Clarke Fowler as a woman of "epigrammatic wit" who deserved equal credit for the bureau's achievements.[16] Respect was mutual. In 1854, Bache commissioned a new Survey schooner, USS *Varina*. More so than her husband, Mrs. Davis enjoyed socializing with erudite Northerners. One of her favorites was Massachusetts senator Charles Sumner, who, she noted, was so polite as to never voice his antislavery views in the presence of Southern ladies. Sectional politics still had limits in refined society and among friends.

Trust and loyalty became the fiber connecting Bache and Davis, as the two sought to professionalize science and raise its influence in government. While war secretary, Davis served alongside Bache as a regent of the Smithsonian Institution. He joined Bache's "Saturday Club," which gathered for discussions and to mentor politicians and military officers. One was army engineer Montgomery C. Meigs, whom the war secretary assigned to construct the Capitol legislative chambers. (Bache and Davis advised Meigs on design.) Davis also became involved in turf wars that Bache pursued with enthusiasm and occasional spite. The war secretary became an ally of the "Lazzaroni," Bache's secret society of intellectuals and scientists. The group included Benjamin Peirce, Louis Agassiz, James Dwight Dana, and others based around Harvard and Yale colleges, as well as the Smithsonian's Joseph Henry. They maneuvered to deny scholarly credit or funding to opponents. Matthew Fontaine Maury, head of the US Naval Observatory, emerged as a major foe. Maury provides a strange case in which Davis sided with a group of Yankees against a fellow Southerner and proponent of slavery. It offers a measure of his regard for the superintendent.[17]

Edward Ord fit somewhere between two Washington titans. That limited how Bache and Davis responded, while enabling the officer's waywardness all the more. The result was counterintuitive. Their hands were tied; his were not. Ord would juggle his loyalties and test just how far his patrons might trust one another.

Chapter 9

A FAMILY ON THE MOVE AND ON THE MAKE

OUTHERN CALIFORNIA'S CHANNEL STUDY EPITOMIZED
the messiness of democracy. The project emerged in the last days
of August 1852, just before Congress adjourned from session. It was
a last-minute amendment—one among many—inserted into a bill
to fund navigation improvements. In such arrangements, the people's representatives picked like vultures at leftover money. Chambers bustled as politicians traded votes in search of support. Or they
banked a colleague's favor, intent to recoup this in the future.

The most skilled legislators might do all of this and advance
several pet projects at once. None more so in this case than Senator Gwin, who authored the channel study appropriation. When it
came to federal spending, Gwin kept numerous irons in the fire. He
worked to open the public lands and was himself one of California's
most successful speculators. He led efforts to improve Pacific commerce. And he presented himself as the capital's expert on Mexican-
era land claims. With the bill's passage, he achieved a victory on
all accounts.[1]

For the same reason, the senator was the cause of growing disputes between Ord and Bache over the purpose of the channel work.
Throughout the spring of 1854, the officer pushed for greater liberties. He still hoped to leave San Pedro and Catalina Island behind.
Yet Ord's new strategy was to argue that his plans were "more con-

formable to the will of Congress—[and] disposition of the money."[2]
In a letter of April 29, he dared to explain to Bache that the "ostensible object" of Congress was not geodesy, but rather a "survey of
shorelines" in order to "sectionalize and sell" the islands. Bache, he
warned, would be the one responsible if legislators were displeased.
This forced the superintendent to take a peremptory step, rare in
his dealings with USCS assistants. He commanded Ord to resume
observations at Catalina post haste. "This," he underscored, "you
will regard as the definite instruction desired." Given Bache's stature
in Washington, DC, Ord's pronouncements certainly were insolent.
They were insubordinate, too—surprisingly so for a soldier accustomed to orders. But perhaps he had insider knowledge beyond that
of a typical assistant or army captain? His stubborn claim to know
the appropriation's intent makes sense only if he had an understand-
ing with its author, Senator Gwin.

But the dispute had longer roots, too. For decades, the Ord family
had pursued political clientage to obtain land in the American West.
Father James Ord (b. 1786) believed he was the son of the Prince of
Wales, later King George IV, from a secret, illegitimate marriage to
the Catholic widow Maria Anne Fitzherbert. Historians accept the
claim. Put up for adoption, James became American and middle-
class, his life punctuated by bouts of insecurity. He attended Georgetown College and considered joining the priesthood. He then worked
as a shipyard artisan—the craft of his adoptive father—before joining the navy. James became an officer in the War of 1812, for which
he received a land bounty. Soon, he owned modest farms in Allegh-
eny County, Pennsylvania (where Edward was born in 1818), and
later in Washington, DC, and Maryland.[3]

Noble lineage but middling existence gave the Ord family an irrepressible longing. Like many Americans who desired wealth and
social mobility, James leapt into land speculation and expansionist
politics. By the early 1830s, he had found a patron in Lewis Cass,
Democrat and secretary of war. Cass procured Edward's admission
to West Point and gave James a federal job. For a decade beginning

in 1837, the son of a king served as a US Indian Agent at Sault Sainte Marie, on Michigan's Upper Peninsula. James used the position to acquire choice property. Ord Street remains near the town's waterfront today, below Lake Superior. Then, in the mid-1840s, James's luck turned. He failed to move up in the state's Democratic Party and lost election to Congress. He plied Cass (now a senator) for jobs, but without result. There were too many seekers, too few spoils. Struggle and precarity returned.[4]

As the father aged, his sons took up his habits. They pursued real estate and political connections, and soon they did so on the Pacific. Edward arrived in California in January 1847. Younger brother James Lycurgus ("Cursy") arrived, too, as an army physician. Both received military land bounties. Older brother Pacificus ("Siffy") arrived in San Francisco. Along with William Gwin and Manuel Domínguez, he took part in the state constitutional convention. The legislature appointed Pacificus to a term on the state supreme court. Edward later brought brothers John Stephen ("Jack") and William Marcellus ("Marcy") to California for

Edward Ord (left), with siblings, photographed in Washington, DC, 1851, between assignments in California

his Coast Survey crew. He offered James Placidus ("Placy") the same work.

Meanwhile, Edward's ties to Gwin opened access to high-level appointments. In 1854, President Pierce named Pacificus as US Attorney for the Southern District of California. Several years later, Edward advised his still-struggling brother Placy that the "influence" of "Doct. Gwin" might find him an office in one of the federal territories.[5] He explained that Gwin, like other purveyors in patronage, typically "pledged" appointments to friends, then maneuvered Congress to authorize the jobs. In exchange, the politicians expected "service." He also intimated in an 1853 letter that Gwin could broker a Coast Survey commission for an army officer in need. Ord knew from experience. It marked one chapter in a long family history.

THE ORD FAMILY wasted no time parlaying their appointments into California real estate. Edward's training as an army surveyor brought him into contact with buyers and sellers. As a US attorney, Pacificus scrutinized Mexican-era estates. Expansive holdings, shaky titles, and loose record-keeping made it easy. As a rule, Siffy regarded rancho claims as fraudulent and saw his work as being in the nation's interest. It also served the family's interest. Parents James and Rebecca made plans to migrate to California with daughter Georgiana ("Sissy"). In September 1854, prior to sailing for the Pacific coast, James wrote Edward to discuss opportunities at Sacramento, Monterey, and San Diego. He inquired about similar possibilities if the US, as rumored in Washington City, annexed the Hawaiian Islands. Upon arrival in 1855, James and Rebecca's appetites were rewarded. Their sons presented them with part of the Soquel Rancho in the forested hills of Santa Cruz. There were further acquisitions. James Lycurgus, the physician, married into the landed de la Guerra-Carrillo family in 1856. Two Ord brothers, Robert Brent ("Bob") and Placidus, set up farms near Santa Cruz and Sacramento. Today, the place-names Ord Ranch and Ord Bend persist in Glenn County, near the Sacramento River.[6]

Edward gained one more advantage for the family. In October 1854, just before his thirty-sixth birthday, he wed Mary "Molly" Mercer Thompson. She was the twenty-year-old daughter of Robert Augustus Thompson, a former Democratic congressman from Culpepper, Virginia. Thompson arrived in California as a Pierce appointee to the US Land Commission. This, of course, was the body that reviewed each and every Mexican-era land claim. In fact, the Domínguez–Rancho San Pedro claim passed across the commissioner's desk shortly before the Thompson-Ord wedding.

By Edward's own admission, he struggled for years to get married, to the point of "crisis." His dark hair grayed prematurely, giving him the semblance of an older man. His father insisted that he find a Catholic mate, preferably a Jesuit. Not that Edward complied. He nearly married into the Van Zandt family of Manhattan's Washington Square district. He wrote of twenty-five-year-old Miss Serena as "my future wife" and marveled that the Van Zandts were "ten times richer than want."[7] Something frustrated his plans, however. Lonely bachelorhood continued. It even caused him to transfer between army posts. "I am disgusted with the Bostonians," he complained. "They won't admit officers to their firesides—at least they don't like to have them too much with their daughters." Ord gave the army a different rationale: rheumatism and his need for a drier climate. That was the summer of 1852, just before he joined the Coast Survey. In the end, he married into a family of Episcopalians, Southerners, and slaveholders.

Edward's marriage may have disappointed his father's sacred wishes, but it fulfilled the family's secular faith. The Ords now had ties to a US senator, a US attorney, and a land commissioner—all involved in the fate of Mexican-era property. The potency of this network is remarkable. So, too, that Ord never disclosed it to Bache. He wrote nothing of his brother's motions before the district court nor of his link to Commissioner R. A. Thompson. On September 1, Ord wrote to Bache that finding a "companion" *previously* had inclined him to return East. "But," he added, "I hope soon to do so without the necessity of going so far."[8] No details followed. Would it have made a difference? The notion of conflict of interest did not yet exist. Some, like Bache and Davidson, seemed to have an innate distaste for it. Bache forbade staff from side work that might reward speculations or discredit his agency. In 1850, Davidson declined a hefty offer of $3,000 to plat the town of Santa Barbara. That was his honest loss. All around, public office and private gain cohabited. Others availed themselves happily.

THE ORDS' OPPORTUNISM provides valuable context, both then and now. It explains why Edward insisted on using his Coast Survey work to test Mexican-era land titles. The dream of buying low and selling high gripped him, narrowing his thoughts. As seen from today—a time when real estate continues to dominate personal fortunes and the cult of success—the same dream points to unpleasant facts of history. US expansion resulted from microeconomic practices as much as from government decisions. Racial and cultural attitudes were part of this. Yet material insecurity mobilized ideas into action. In an era of brutal financial cycles, land was a secure investment, but also stubbornly illiquid. This meant that even comfortably middle-class people like the Ords were cash-poor and only several steps from being unable to make purchases or payments. No government safety net existed, so land served as insurance. A speculative property might be the last resort to stave off economic ruin or provide throughout old age. Harsh realities helped make the US a republic of land pirates who found the rationales they needed. Security in an insecure world—for the Ords and for countless more—depended upon the misfortune and downfall of others. That world was not their choice. But the Ords determined to make the better of it.

Government work served similarly. Army commissions and

patronage offices, both fed by Western settlement, enabled middling Americans like the Ords to find steady salaries while avoiding farm and factory work. By the 1850s, a small white-collar sector had taken shape. It paid a premium in the West. Edward explained to Placidus that a frontier appointment paid *twice* as much as a minor clerkship in Washington, DC. He acknowledged the price: painful separation from family, wives, "little ones," and friends.[9] But it opened an inside track to land speculation. Duty and real estate: Ord brought this duality to his Coast Survey assignment. It served as compass and lodestar. Bache grew wise to this. But because of his privileged background and devotion to science, he did not understand Ord's singlemindedness.

That was the essence of their difference. Other staff bristled at the Survey's exacting schedule and spartan budgets. But Ord's irritations grew beyond bounds because of his lack of commitment to the agency. He began to write that the Channel Islands study was impossible, given the meager resources. And in March 1854, *six months* after he started on the project, Ord asked for return transfer to the army. Bache said no. But the officer made it clear he was not seeking permission. He had put the request by "private letter" to War Secretary Davis. This broke the USCS's chain of command. It rebuked the superintendent's civilian authority as well. Citing the rigors of work, "exposure," and lack of "society," Ord explained, "if I am to remain here I should prefer to be with my [army] company."[10]

With his ties to the Survey now irreparably damaged, Ord slowed progress on the channel study. He prepared to spend the summer in San Francisco, where he would attend to family and his courtship of Molly Thompson. He spent time off duty around San Pedro and Los Angeles. Here, Ord met merchants, landowners, and speculators. Many were backers of Senator Gwin. Ord drummed up side work, despite Bache's proscriptions. He wrote of Southern California, "The vineyards and olive groves are few but very pretty and productive, and I am on the trail of one for the Ord family."[11] By coincidence, his brother the US attorney began to review local rancho cases. So did the Land Commission. Opportunities abounded. But cash remained scarce. "I regret," he lamented, "that I did not make money too in the early days of the gold mining." Fortunately, Ord stumbled into some paying work. Manuel Domínguez prepared to sell a sizable parcel. He needed a surveyor.

Bache realized his agency's exemplary project was in the wrong hands. At the end of May 1854, he messaged Davis. The professor did not mention names, but he announced that he would discuss two army officers on loan to the Coast Survey. He had just received Ord's insubordinate letter of April 29 on the "object" of Congress. Perhaps Davis had Ord in mind, too? His transfer request had arrived. There was another reason, however. More so than real estate, this cause— made of iron rails and slavery—was of national importance. One friend would reassure the other, for now.[12]

Chapter 10

SECRETARY DAVIS'S MASTERSTROKE

D URING THE SUMMER OF 1853, FOUR EXPEDITIONS OF Army Topographical Engineers set out to find routes to the Pacific. A pair of these aimed toward Southern California. One made its way along the Sierra Nevada, after arriving at San Francisco by ship. The second departed Fort Smith, Arkansas, and pushed west along the 35th parallel. Both railroad expeditions would cross paths with Edward Ord. This is surprising. On its face, their work seemed unrelated to the Coast Survey's research at San Pedro Bay. The region's unprotected coastline made it a doubtful terminus for a transcontinental railroad. Surely, no topographical engineer would say otherwise.

But, far away in the East, politics and ambition stirred. Commercial centers of the Mississippi River Valley—New Orleans, St. Louis, Memphis, Chicago, and others—each pressed for their favored route. Disagreements over slavery drove an additional contest between North and South over the rail line's location and which section would gain most by it. Legislators conflated the railroad with the highest of stakes. War Secretary Davis did, too. He believed the expeditions would answer profound questions about the American union and its future. In his mind, California's statehood had imbalanced the Constitution. Iron rails might restore the balance and ensure the republic's survival—that is, if the railroad was located optimally. He determined to see it done.[1]

To the politician who could wield them, the railroad expeditions

*Washington, DC, facing west—the Capitol under
construction and the Washington Monument in the distance.
With the start of the Pacific rail surveys, politicians looked
across the continent as never before.*

promised great rewards, especially because of the authority the work
conveyed. Survey parties would cover four hundred thousand square
miles of country previously little known. They would gather astronomical data, biological specimens, and ethnographies of Native
Americans. They would catalog flora and fauna as well as geology
and water resources. Artists would sketch and paint landscapes.
Findings were hard won, the product of long and dangerous labor.
By the end of the decade, they would generate nine thousand pages
of reports and illustrations, across twelve printed volumes. It was a
staggering project.

The Pacific railroad surveys marked the first time the US government attempted to use science to decide a contentious issue. The
idea was for empirical data to inform legislation while Congress and
the War Department remained impartial. Today, this logic seems
familiar, as do its faults. No matter how carefully gathered, data
soon enters the less scrupulous world of officeholders, parliamentary maneuvers, and negotiations. Findings must withstand popu-

lar passions. Officials interpret research through preexisting beliefs or commitments. When cloistered within institutions, their views may become stubborn and less empirical. Science and politics might operate with autonomy, but they are too rarely independent.

ORIGINALLY, the proposal to send out multiple expeditions was Senator Gwin's attempt to bypass worsening sectionalism. But the proposal was a symptom that could not cure the ill. The former physician tried anyway. Frustrated by the defeat of his central line bill, Gwin in 1853 proposed funding only the terminal portion of track within California. This failed. And again, he pursued an alternative. By the new year, Gwin had secured creation of a Select Committee of the Pacific Railroad, which he would chair. The ad hoc group would serve as the clearinghouse for information arriving from army field crews. It would hash out agreement before a bill reached the Senate floor. In the light of day, Gwin maneuvered himself into the role of Senate mediator. As always, however, his actions had shadowy depths.[2]

He was not the only one. From all around, forces mobilized to preempt the surveyors' results. In particular, politicians hurried to organize Western territories. This would create areas of direct federal jurisdiction, allowing Congress to issue land grants and tax settlers. A new territory might sway the placement of a transcontinental line; the Washington Territory took form early in 1853 for this reason. Not to be outdone, Illinois senator Stephen Douglas introduced a bill to organize a Kansas-Nebraska territory when Congress reconvened in December. He had the backing of interests in Chicago and St. Louis. And he was a towering figure, particularly among the Democratic Party's Northern delegation. To gather votes, Douglas allowed an amendment to repeal the Missouri Compromise ban on slavery above 36° 30′ latitude. This set off explosive debates over whether his prospective territory would allow slavery or not. Douglas seized the moment to offer the compromise of popular sovereignty: local voters would decide the question. Congress, more acrimonious than ever, passed Douglas's Kansas-Nebraska Act in May 1854. Yet trouble worsened. Popular sovereignty resulted in a disputed vote. Soon, it sparked civil unrest and murderous violence. The bill fractured the Democratic Party along sectional lines, while it destroyed the Whigs all but in name. A new antislavery and Northern-based party, the Republicans, would coalesce from the pieces.[3]

The Kansas-Nebraska Act thoroughly absorbed Congress and the nation. By doing so, it sidelined Gwin's select committee and kept the Senate from taking up the ad hoc group's railway bill. As is true today, Congress of the mid-nineteenth century worked on "legislative days" and not by a conventional work week. Within a two-year cycle, lawmaking took place on a complex schedule of sessions and breaks, some as long as nine months. Every session lost caused considerable delay. Gwin was furious with his friend Douglas for sidetracking the chamber. However, Gwin's own gambits had contributed.

WAR SECRETARY DAVIS faced fewer restrictions. His authority carried on year-round, whether legislators were present in Washington or not. While Congress descended into the Kansas-Nebraska mess, Davis attempted to clear a path for the transcontinental project and to achieve fast results. He had instructed officers to find the shortest, most direct, most practicable, and most economical courses. Concerned about timing, he asked them to submit preliminary reports by February 1854. He hoped to recommend a single, outstanding route to Congress by the close of that year.[4]

Yet Davis's drive and determination stemmed from foregone conclusions. His instructions favored the southernmost, 32nd parallel course most of all. That route avoided mountain terrain and heavy snows, which were endemic to the central and northern routes. Steep grades extended track mileage by forcing engineers to use switchbacks. They also required expensive bridges, trestles, and tunnels.

Level desert topography had opposite effect: It shortened track distance and lowered costs considerably. The only drawback was the lack of timbers for construction or fuel, but wood could be imported. Davis's preference had merit as the simplest route to engineer. Still, the secretary privileged a path through the Southwest, and one connected to slaveholding states and their would-be settlers. He prided himself as a man of science. But for all this conviction, he, more than anyone else, evinced the dilemma of science in American politics.

For Davis, the field data was necessary to show others he was correct. This led the war secretary to a set of stratagems. He asked the Pacific Railroad Office to compile a highly detailed composite map. It was an unprecedented endeavor. The office contracted to reproduce the document using the new technology of the large-format chromolithograph press. The map would show the pathways and elevation grades of all routes on a single scale document. Fine hachures and detailing would enliven the flat, linear image so that it leapt off the page. The map would visualize data and enable lawmakers—many of whom Davis thought too dim to grasp the technicalities of an engineering report—to compare findings. In his view, the map would place his conclusions beyond dispute. It would remind naysayers that, despite sectional implications, Davis had led an empirically-based inquiry. The map would also outshine his intentions, so that they were not seen. It was to be his masterstroke.[5]

Nonetheless, Davis needed to convince his foes. And they discerned motive in his every action and word. Suspicions heightened as the War Department joined the chase for new territories. In the latter half of 1853, the United States acquired the Gadsden Purchase—thirty thousand square miles of Mexico located between the Gila River, Guadalupe Mountains, and Mesilla Valley on the Rio Grande. The War Department gave several reasons for the purchase, but the railroad was prominent among them. The purchase contained areas that the Topographical Engineers and US-Mexico Boundary Survey already had deemed necessary to a route along the 32nd parallel. Davis did not participate in negotiations, but he secured appoint-

ments for a number of allies, including James Gadsden, to the diplomatic legation. Then, he supplied them with army reconnaissance
about the most desirable terrain. The Gadsden Purchase was Davis's
answer to the creation of the Washington and Kansas Territories.
It epitomized his skill at pursuing sectional and national interests
at once. In skeptics' eyes, he slyly advanced the former in guise of
the latter.[6]

In organizing the rail surveys, Davis demonstrated just how
careful an illusionist he could be. From his years in the army and
leadership of the Senate's Military Affairs Committee, he knew the
Topographical Corps rather well. Some were classmates and close
friends. Davis also knew that opponents would scrutinize his selection of commanders. Hence, he appointed Northerners to lead each
main survey. He ignored John Frémont, who was out of the corps
but nonetheless synonymous with transcontinental exploration.
Frémont was in Paris with Jessie and their children when he read
news of the rail survey appropriation. He cut short the family's stay
and rushed to Washington, DC, feeling assured of a commission.
(He even purchased scientific instruments, expecting to be reim-
bursed.) That commission never came. Davis made no effort to hide
the snub.[7]

The war secretary introduced one more gambit. He shaped the
expeditions' designs to favor his outcome of choice. He understood that, regardless of courses found farther east, the selection
ultimately hinged on the mountains of California. But rather than
extending each latitudinal survey to the Pacific Ocean, Davis created
a distinct study to evaluate passes within the Golden State. He set
this up specifically to connect to the 35th parallel at the Colorado
River, or 32nd parallel via the Gila River. The survey would focus on
Southern California. Davis assigned the work to twenty-eight-year-
old Robert Stockton Williamson.

Williamson is virtually forgotten today. Outdoor enthusiasts and
historians may know him for Mount Williamson, at 14,379 feet, the
second-highest peak of the Sierra Nevada and the backdrop to the

Manzanar Japanese internment camp during World War II. In his own time, Williamson held a reputation as one of the Topographical Corps's most promising young officers. Named for the navy commodore (also family friend and US senator) who in 1846 led the invasion of Mexican Los Angeles, Williamson grew up in New Jersey. He graduated from West Point in 1848 and entered the army's most elite body. In photographs, he appears dark-haired, gentlemanly, and thin, a sign of worsening tuberculosis.

For Davis, Williamson had one attribute that made him a top candidate. In 1849, he conducted an army study of passes and roadways through the *northern* Sierra. During the project, Edward Ord and William T. Sherman brought dispatches to Williamson and his commander, William H. Warner, who would die in a Native American attack. Much of their work remained unpublished as of 1853. As Davis looked to disqualify the central and northern transcontinental routes, he knew he needed an officer whose expertise on the upper Sierra Nevada was beyond doubt. Williamson was the choice. Others objected. John James Abert, chief of the Topographical Corps,

Robert Stockton Williamson. Davis assigned him the most politically fraught of the railroad surveys.

wrote in April 1853 to protest Williamson's appointment. Abert thought him too junior to lead such a high-profile expedition. Davis was not swayed. On May 6, he issued Williamson instructions. The two met in Washington City prior to the officer's departure by ship.[8]

Williamson arrived at San Francisco in June 1853. Two young engineers accompanied him: John Grubb Parke, his second-in-command, and George McClellan, who journeyed on to Fort Vancouver (north of the Columbia River, opposite Portland) to assist on Isaac Stevens's rail surveys. The same vessel carried a member of the US Land Commission, Robert Augustus Thompson, and his daughter Molly, who would marry Edward Ord fifteen months later. The world of post-conquest California was vast, yet also improbably small.[9]

Chapter 11

"SAN PEDRO HAS
NO HARBOR"

H E HAD RECEIVED THE ASSIGNMENT OF A LIFETIME.
Yet, with each day, controversy surrounding the Pacific railroad engulfed R. S. Williamson. Other topographical expeditions
followed numerical lines of latitude (such as the 49th, 38th, and 35th).
Williamson's probed several, and it might well determine which one
the War Department and Congress would select. That was sure to
displease the interests behind each route, even some of his fellow
commanders. His work, more than anyone else's, might disappoint
Jefferson Davis. And the war secretary was known to hold grudges
in his management of the army.

Such burdens weighed on the engineer as his party rode out from
Benicia Barracks toward the Sierra Nevada. Four hundred miles
long and seventy-five miles wide, the range separates California,
like a massive granite wall, from the continent beyond. It includes
the highest peaks in the contiguous United States. Even in late July,
they were capped with snow. The expedition did not consider the
northern portion, where trail elevations reached nine thousand feet.
That stretch was passable only between summer and early fall. By
contrast, they regarded the central Sierra as utterly impossible. Even
today, a two-hundred-mile stretch contains no through highways.
Instead, the survey crew set out for the South Sierra, and further to
the mountains behind Los Angeles. The region was new to Williamson, but it had been the subject of Edward Ord's research in 1849.
Walker Pass, near the Kern River, was the first site of interest.

To get there, Williamson's column traversed the San Joaquin and Tulare valleys—at the time, one of California's most remote sections. Here, only a small post, Fort Miller, and minor islands of European settlement existed. Williamson reported that the "town" of Woodville consisted of a single house. Otherwise, the valley was a humid tangle of marsh, lakes, willow, and oak with dry plains and desert between. Summer temperatures exceeded one hundred degrees Fahrenheit. Tule elk moved through the mist. Grizzly bears stalked fish. Feral horses roamed in the distance. Clouds of "venomous mosquitoes" buzzed and attacked. Fur trappers had introduced malaria, which devastated local Native Americans. Nevertheless, the First Peoples held on to their independence and, as of 1850, numbered four thousand. Williamson recorded their names: "Ton-Tache," "He-ame-e-tah," "Cowee," and "Cho-e-minee." Under the late-summer sun, they harvested crops for the coming winter.[1]

At the margins of Indian country, culturally mixed bands of trappers, rustlers, and thieves operated freely. A well-equipped surveyors' party made an attractive target. Just as Williamson's team reached the area, state bounty hunters killed Gold Rush desperado Joaquin Murrieta in a gun fight at Arroyo Cantúa on the valley's west side. For protection, the army gave Williamson an escort of thirty mounted troopers led by George Stoneman. Williamson and Stoneman had worked together in Oregon. There, in fall 1851, they crossed paths with George Davidson of the Coast Survey. That episode ended in a massacre of Coquille River villagers. This one would be different—at first, that is.[2]

Williamson's party met no violence in California's Great Valley. But less obvious dangers found them. On the road to Walker Pass (near modern Visalia), they found Senator William Gwin waiting with a column of soldiers. The wealthy politician did not often venture into the wilderness. He explained that he had toured prospective railroad sites and wanted to share thoughts, in particular about Walker Pass. Gwin's commitment to his Sacramento- and San Francisco-based supporters had turned him away from the South

Sierra. He now thought a railroad might best access California via Noble's Pass, a significant detour located near Mount Lassen and the Oregon border. The senator asked if Williamson might include Noble's in his official study. He would put the same request in writing to the War Department.[3]

Gwin hoped to prepare legislation and receive credit for its success. But War Secretary Davis continued to obstruct him. Gwin's latest bill advocated for a central line, like the one he had proposed unsuccessfully the year before. He had not forgiven Davis for leading its defeat. (He never did. Late in life, Gwin claimed the bill's passage would have averted the Civil War.) The two men again grew at odds. After moving to the War Department, Davis kept the California senator in the dark about the army's railroad work. Gwin's response was to intercept Williamson in the field. He intended to pry useful information from the young engineer.[4]

Williamson seemed prepared for the interference. He hastily discussed railroad matters with Gwin. Then, claiming to be "too unwell," he cut short the interview. Williamson promised to share his thoughts by letter. When the officer did, weeks later, he informed Gwin that the South Sierra proved more difficult than he expected. That was what the senator wanted to hear. However, the engineer divulged little more. Even so, he had written too much. In December, Gwin read Williamson's letter aloud in the Senate, a step that caused problems for them both.[5]

Gwin was not alone in the attempt to outflank Davis. Farther down the road, near Woodville, Williamson encountered mountaineer and trapper Alexis (or Alexander) Godey. Godey's presence seemed fortuitous, as many considered him an expert on the area. ("Better acquainted with that country than any man living," one San Francisco journalist wrote.)[6] Godey explained that he was looking for the Indian superintendent, Edward F. Beale. At the start of the year, Beale had sailed to Washington, DC, to protest the Senate's rejection of treaties negotiated with California's Native Americans. He rushed to save the agreements and a modicum of protection for the tribes. That would preserve, for suppliers and ranchers, lucrative annuity contracts. He was expected to arrive any day with word of the government's decision.

Godey, too, signaled intrigue. Despite his preference for life in the rough, he had extensive political connections. Both he and Beale were part of networks tied to John Frémont. Godey was a veteran of the Pathfinder's transcontinental explorations, including those through the Sierra in 1843–44 and 1845–46. He had defended Frémont during his court-martial and rescued him from the San Juan Mountains in 1849. More recently, Godey managed Frémont's mining and cattle operations at Rancho Mariposa. That Godey was searching for Beale during the summer of 1853 also tied back to Frémont. Earlier that year, Thomas Hart Benton had tried to place Beale with Frémont on the War Department's 38th parallel expedition. Denied this, he bankrolled Beale's current journey from St. Louis to the Pacific as an unofficial rail survey. The Indian superintendent brought important news, and about much more than the California tribes.[7]

Now Godey just happened to appear in Williamson's path, and only several steps behind Gwin. Was the officer prepared to fend off the political interests behind him? Perhaps. Williamson did not hesitate to hire the mountain man as a guide. That positioned Godey to direct the government party on behalf of his patrons—or, just as easily, misdirect it. He also could monitor its staff, including Charles Preuss, Frémont's former cartographer who now worked for the War Department. Yet Williamson could do much the same. All depended on what one knew of the other and what they would find.

GODEY SPENT ONE WEEK guiding the expedition through the Greenhorn Mountains in and around Walker Pass. Based on what Williamson saw, he deemed this route "impractical" for a railroad. Key to his determination were interviews conducted with the Tübatulabal people near the forks of the upper Kern River. The indigenous community was kind and helpful. Williamson wrote of their large, seasonal village at a place called "Chay-o-poo-ya-pah," where Indians harvested sugar from bulrushes dried in the sun. The unnamed Native voices told of swift cataracts down to the valley and large boulders in the canyons. The descent was far too tight for switchbacks. Williamson later wrote of Walker Pass, "I consider it one of the *worst* of all the known passes in the Sierra Nevada."[8]

Given Williamson's expert company, his verdict is astounding. Godey not only knew the South Sierra-Kern River backcountry better than anyone, he had spent much of the summer promoting it as a transportation corridor. In July, he gathered information from Native Americans and traced a road through to the Colorado River. Word of the mountaineer's progress appeared in San Francisco newspapers, which favored the route, to great fanfare. "[Walker] Pass is free from obstructions of all kinds," the *Alta* announced. "A buggy can be driven through it without difficulty."[9] It would seem Godey led Williamson to an unimpressive, nonviable site rather than the best he knew: there were several nearby paths across the South Sierra, all informally known by the name Walker Pass. Both Godey and Preuss

The Greenhorn Mountains, circa 1890—Williamson's approach through the South Sierra toward Walker Pass

informed Williamson that the particular summit they visited was not the same one inspected by Frémont in the 1840s. That portal, they said, lay slightly southward.

That left other possibilities. Godey might have known that Frémont and Benton were preparing to claim the best of them. Son and father-in-law had organized their own railroad survey to prove the St. Louis–Walker Pass route superior to all others. Their greater purpose was to discredit the War Department and set up Frémont to emerge as the hero once more. In late September, the Pathfinder assembled at Westport (today's Kansas City) with a talented expedition team. Information about Williamson's doings—where he had been or not been—would be of great value if it could reach them in time. Godey's name does not appear in firsthand accounts of the survey (kept secretly because Frémont prohibited staff journals). But Jessie Benton Frémont's unpublished manuscript of her husband's life places Godey there. This has created some dispute among biographers. A likely explanation is that Godey's intelligence found its way to Frémont, whether he did in person or not. Results support this explanation as well. By April 1854, Frémont had arrived in California and announced the discovery of a new Walker Pass at 36° latitude—*not south*, but curiously north of the site to which Godey had led Williamson. Frémont had his moment, however short-lived.[10]

At the same time, Benton launched a campaign to keep Walker Pass under consideration in Washington. The old politician remained a force, having reentered Congress by popular vote to the House of Representatives. In November 1853, he began a barrage against Davis after word arrived that eight members of the 38th parallel expedition, including commander John W. Gunnison, had been killed near Sevier Lake in Utah (near today's US Route 50). Benton charged that the attack was a hoax, its purpose to scare Congress from supporting a central-latitude railroad. He demanded that Davis release preliminary findings, including Williamson's dispatches on the Sierra. When Davis replied that he would release only the final, edited reports, Benton accused him of unlawful secrecy. But theirs was a game of secrets. The Missouri politician clearly knew more about the Williamson survey than he could say. Davis called his bluff and refused to comply.[11]

Furthermore, Benton alleged that Williamson had leaked information to unauthorized persons, an unusual step brought on by the engineer's communications with Gwin. The California senator defended his actions. He and Benton began a public feud over the merits of Walker Pass that would last several years. Regardless, they found themselves in a similar position. Both desperately wanted to review Williamson's work; both were refused by Secretary Davis. Gwin tried repeatedly, even more so when his railroad committee convened in 1854. On one occasion, he walked to the War Department, a two-story brick-and-pillar building at 17th Street and Pennsylvania Avenue, and politely asked to see the reports. The office clerk disappeared, returned, and answered no. If Davis was on hand, he did not show himself. Gwin could do nothing but send a complaint.[12]

As DISPUTES RUFFLED national leaders in Washington, Williamson moved his exploration away from the Sierra Nevada, that cause for contention. During the fall of 1853, his team visited passes at the south rim of the San Joaquin Valley: "Tah-ee-chay-Pah" (Tehachapi, elevation 3,832 feet) and Cañada de Las Uvas (also known as Grapevine Canyon, elevation 4,160 feet). With further study, he charted two additional paths: "New Pass" and "San Fernando Pass" (elevations 3,164 and 1,940 feet, respectively). The combination of Tehachapi, San Fernando, and "New" Passes made for a route connecting the Tulare Valley east toward the Great Basin and then south to Los Angeles through the "Santa Clara" (today's Santa Clarita) Valley. The Southern Pacific Railroad would build by much the same path in the 1870s. But in 1854, Williamson only laid out a wagon road. The army would use this to establish Fort Tejon within the Grapevine in August.[13]

The surveyor's efforts opened the Tulare-San Joaquin Valley to

transportation and settlement. By the same token, it caused escalating violence against the valley's indigenous people. Beale, the
Indian superintendent, rushed to calm settlers and convince chiefs
to trade their ancestral homes for life on small (and temporary) military reservations. Land-hungry interests resented even that. Beale
soon found himself in command of a militia with orders to carry out
punitive strikes and forced relocation. This would include an 1856
attack on the Choinumni-Yokut villages, which Williamson's party
had observed peacefully by the Tule River. The Tübatulabal, who
had assisted the rail surveyor near the summit of Walker Pass, would
suffer massacre and removal as well.[14]

With trouble emerging in his wake, Williamson left the valley
to scout the deserts east. Half of his column followed the Mojave
River to determine whether it reached a confluence with the Colorado. They found it did not. The other half, led by John Parke,
searched out Cajon Pass. The Old Spanish Trail through Cajon had
become much-used in recent years as a pathway between San Ber-

The Mojave Desert, east of the mountains between the
Tehachapi and Cajon Passes, with the distinctive Joshua tree
(Yucca brevifolia). Terrain and scarce water led Williamson to
favor San Gorgonio Pass.

nardino and Utah. Parke pronounced it practical for rails, but noted it would need lengthy, expensive tunnels to render a safe grade. This fact, along with the waterless gap beyond, effectively disqualified the Cajon and 35th parallel route.

But Parke made a key discovery. While charting the road from Cajon Pass toward Fort Yuma (on the Colorado River), he visited San Gorgonio Pass. This is a low-elevation (2,800 feet) desert break between the San Bernardino and San Jacinto Mountains. Ord had discounted it in 1849, but Williamson and Parke thought different. Gorgonio lay one hundred miles southeast of Walker Pass and bypassed the Sierra Nevada altogether. It also provided direct access into Los Angeles from due east. Today, Interstate 10 cuts through the pass toward Phoenix, Arizona.[15]

Early in 1854, ill health forced Williamson to relinquish command. He filed a "skeleton report," preliminary and incomplete, which Congress received that March. Senator Gwin had secured a resolution demanding that the War Department release data to his committee. Davis complied, in letter though not in spirit. He sent Williamson's document, which outlined the study of seven mountain passes—Walker being the "most difficult" and San Gorgonio the best. But the report seen by Congress contained no cartography. Many suspected the war secretary of holding back information.[16]

Critics noticed as well how Williamson's work tipped the railroad issue in Davis's favor. San Gorgonio Pass made the southernmost, 32nd parallel route the most practical way to reach the Pacific. But its final selection hinged upon a new question: Was there an anchorage to provide a coastal terminus? A harbor in Southern California would enhance Williamson's findings considerably. (Everyone knew the line would continue toward San Francisco, but California's legislature, or a private contractor, might build and pay for this.) San Diego and San Pedro were the only possibilities. Williamson found the approach to San Diego too mountainous, but he hesitated to evaluate the harbors. That was the job of the Coast Survey. His final report, dated December 31, 1854, left the matter unresolved. "If the

question is simply how to continue that road to the Pacific," he wrote
of the 32nd parallel route, "the answer is at once apparent."[17] The
line would enter at San Gorgonio, move through San Bernardino,
and proceed "from there to San Pedro, or some other point" on the
coast. He cautioned, however, that "San Pedro has no harbor," and
the "only really good harbors . . . are those of San Francisco and
San Diego."

In writing this, the sickly army engineer ignored rumors in Los
Angeles about a prospective waterfront development. These intruded
on the railroad debate by other means: a second topographical expedition arrived shortly. Amiel Weeks Whipple led the important and
extremely dangerous task of charting a route along the 35th parallel, up the Canadian River, and across northern New Mexico. His
party ventured through homelands of the Comanche, Cheyenne,
Ute, Apache, Navajo, and Mojave—all fearsome indigenous powers. Aside from Williamson, Whipple was the only officer entrusted
by Davis to survey into Southern California.

Whipple believed his labors would have great consequence for
the railroad question. Instead, he reached California in March 1854,
eight months after Williamson did. The latter already had stolen the
attention of Congress and relegated the 35th parallel route. Further
complicating things, Whipple had overrun his budget. The party
raced to auction off wagons, mules, and sundries after reaching Los
Angeles. Days later, they learned that a steamer was set to arrive at
San Pedro on short notice. If they missed it, they would face several
more weeks of expense. By dusk and moonlight, a stagecoach rushed
them over a "fine prairie" and "well-cultivated valley" flooded with
rain and mud.[18] The next day, Whipple and his officers boarded a
ship to San Francisco, where they would sail for the East.

Unlike Williamson and Parke—maybe because the two had left
him little new to say—Whipple included information about San
Pedro Bay in his report. He conceded that the site was outside the
War Department's jurisdiction and was instead "within the limits"
of the Coast Survey. He also noted, consistent with USCS informa-

tion, that the bay gave too little protection to vessels. Nonetheless, he wrote of ships crossing a sandbar at high tide to enter the adjacent estuary, "where they are entirely sheltered." As Whipple explained, "A city has there been laid out, and several houses built, in the expectation of rivaling the pueblo of Los Angeles."

Given Whipple's hurried exit, he must have received information about the estuary and its city-to-be as hearsay. He certainly did not witness ships crossing the sandbar. That would require staying through two high tides, roughly twenty-four hours apart. Most important, no waterfront town existed, either in fact or on paper, except as a conceit of speculators who hoped to buy the land. Their plans had appeared in newspapers, but there were no houses for the engineer to see.

The origin of Whipple's statement, therefore, is a mystery. He did name one source, however: a fellow officer he knew from West Point. "*Captain Ord* . . . informs me that a breakwater about a mile in length would make a safe and commodious harbor even for a fleet" and provide San Pedro with "every advantage of a good port." Given the intense contest over rails and slavery—and the even more intense contest over information and secrets—it would be Whipple, the latecomer with a moot assignment, who divulged one of the most important details. He left California. Ord remained. And by his work, plans for a transcontinental railroad became clearer still.

Chapter 12

GUESTS OF THE DOMÍNGUEZ FAMILY

I N THE FIRST WEEKS OF 1855, JEFFERSON DAVIS announced that his report on the Pacific railroad explorations would be delayed. So far, the war secretary had given Congress some preliminary data, but no conclusions. He threw himself into the work, particularly after the death of his two-year-old son Sam. But the field studies made slow progress. Now, his term in office approached the halfway mark. Opponents and critics grew stronger. And Washington politics had become more dysfunctional. Legislators, regardless of which route they favored, were spoiling for a vote. They demanded that Davis provide his recommendation along with supporting evidence. Of course, he already knew his route of choice. He just needed the information to justify it. That information had not yet arrived.[1]

Then, something shifted. At the end of February, Davis notified Congress that field studies "conclusively" showed the 32nd parallel "is, of those surveyed, 'the most practicable and economical route.' "[2] The same route provided the shortest distance to the Pacific, and an estimated cost one-third less than any other. He ruled out the central route entirely, as the Sierra Nevada blocked access to San Francisco. The Pacific Railroad Office and the Topographical Corps endorsed this. Still, Davis offered one more conclusion: "It is, therefore, assumed," he stated, "that the terminus of this route should be at San Pedro." The war secretary assured legislators that, despite the bay's exposure to storms and swell, "construction of a breakwater" would "constitute it a safe harbor." This was no small claim.

Davis's swift turn from waiting to action, from caution to certainty, was the result of disparate pieces of knowledge flowing into the War Department. R. S. Williamson forwarded his official report. Isaac Stevens submitted on the northernmost route, and likewise Edward G. Beckwith on the 38th parallel. A. W. Whipple's was the lone Pacific railroad report yet missing. It would arrive by summer. But among the paperwork on Davis's desk was intelligence more privileged and off the record. Unofficial sources included Edward Ord's recent letters, which suggested the possibility of a harbor inside San Pedro's estuary.

Just as he connected the San Pedro–Santa Barbara Channel to speculation in public lands, Ord became the link between Southern California and the contest over rail and section in Washington, DC. During the summer of 1854, if not earlier, Ord assumed an escalating role in railroad politics. The issue took hold of him. As he extricated himself from Coast Survey tasks, he gained new freedom to maneuver on behalf of his own interests and for others more powerful and wealthy than he. In addition to contacts within the nation's capital, Ord built alliances with landowners and investors around Los Angeles. For a brief yet critical moment lasting into 1855, he balanced these entanglements before their tensions ruptured. In time, the outcome proved significant. Ord brought the Pacific railroad debate to bear upon San Pedro Bay. And through his actions, the Domínguez family would no longer control the site, or its future. It was Ord's scattered ties that drew coastal mapping, railway politics, and local land interests together in a narrow moment of synergy.

This began six months before Davis reported to Congress. Ord passed the summer of 1854 in San Francisco, where he compiled notes and looked forward to his wedding. With his brothers, he prepared to move their parents out to California. The officer had informed Bache that he would resign from duty and return to the army. The superintendent refused this, despite his displeasure with Ord, but by August, Bache had relented. He agreed to send out William Greenwell, his original choice for the job. However, Greenwell's illness kept him from traveling until the next spring.[3]

Ord would need to work one more winter. He determined to do this on his own terms. Most important, he stepped up his prospecting of land and politics. As Ord looked to return to San Pedro, he became especially interested in the affairs of Manuel Domínguez. The rancher's claims case, number 398, continued before the US Land Commission. On September 6, the *Daily Alta* printed a notice that the panel had approved Domínguez's petition in full. Ord likely had foreknowledge of this decision from Commissioner R. A. Thompson, soon to be his father-in-law. Nevertheless, he waited until newspapers broke the story. Then, he wrote to Domínguez. The letter, which opens part two of this book, conveyed his congratulations and told of Ord's upcoming arrival. He would proceed immediately to the Domínguez Ranch to reclaim the tents and equipment stored there. The friends—a rancher and a government officer—would see each other again. Ord's enthusiasm marked a change. Previously, he had pressed for permission to leave San Pedro—a place he dismissed as "desolate and adobe like." But the enterprising soldier had shifted designs. He found opportunity where it lay.[4]

NEWLYWEDS Edward and Molly Ord sailed to San Pedro in October and spent several days as guests of the Domínguez family. During the stay, the officer reoriented his government work. George Davidson's geodetic triangle, laid out two seasons before, ran toward Catalina Island via San Pedro Hill (today's Palos Verdes). Ord supplemented this with a number of associated triangles running from the hilltop to the island and back to the mainland at "Pt. Duma" on the Malibu coast; "Cerito," a cactus-filled highland overlooking the San Gabriel River (today's Long Beach); and "Las Bolsas," a mesa above marshes (today's Huntington Beach) some twelve miles southeast of Cerito. Yet, in the fall of 1854, Ord added his own vertex, which split an existing triangle. He placed this close to the Domínguez adobe and named it "Coast Survey Station, Domínguez Hill." He explained to Bache that the change was necessary to correct a sighting angle smaller than thirty degrees, which jeopardized the project's trigonometry. Ord implied that the mistake was Davidson's, but he hesitated to say so. Instead, he pledged to measure by both Davidson's method and his own, and then compare the results. Bache could not question this, at least not right away.[5]

Ord's post on Domínguez Hill enabled him to pursue his side interests. Under Davidson's design, triangulation passed over the rancho. All vertices were distant from the landowner's home and avoided close interaction or meddling. Ord's scheme reversed this. From the new observatory, he could spy the flicker of heliotrope signals set on Catalina. He also could spend time in the company of Manuel Domínguez and family. Edward and Molly's stay included much social activity, as well as a Californio-style celebration in honor of the guests' recent nuptials. Loud dancing, music, and festivities kept them up until three o'clock in the morning. Ord recorded the experience in his diary as "very unpleasant." "Won't be caught at it again," he scribbled.[6] What seems like ingratitude may have stemmed from his concern for Molly's health. She had contracted typhoid fever months earlier and suffered recurring symptoms. But the officer also hinted at cultural views, which he would articulate in time.

The experience did nothing to stop Ord from calling on the Domínguez family or taking meals or siestas at their adobe. He continued to do so through his final winter on the Coast Survey. Edward wrote to Molly of these visits with "Don Manuel," Doña Engracia, and their daughters. He told of Engracia's fondness for them, which grew after Molly addressed her as "Mother," and of affectionate kisses he placed upon the girls' foreheads. María Reyes, eight years old and the youngest, became his favorite. He chaperoned the older, unmarried daughters—possibly Ana and Victoria—on a visit to Los

(Left) Edward Ord's marriage to Molly Thompson
was advantageous, but caused him grief a decade later.
(Right) Engracia Cota de Domínguez (with daughter Reyes)
brought land rights from an adjacent rancho to
her marriage to Manuel.

Angeles pueblo. As the officer dreamed of reuniting his family and starting his own with Molly, he enjoyed feeling adopted by another.[7]

Manuel Domínguez had several real estate sales in the works, which had awaited the outcome of his land case. Ord knew of the transactions already. In his letter congratulating Domínguez on the claim's approval, he warned about possible difficulties ahead. He encouraged the Californio to defend his ownership rights by partitioning and selling some parcels to Anglos who, he explained, "would do all in their power to secure your lands, as the owner that sold it to them."[8] The idea was not Ord's alone. Domínguez watched other grant holders use subdivision to raise money and gather allies among the arriving foreigners. Subdivision became more common because of the US claims process, as applicants needed to offset fees and taxes. Still, Ord's advice was noteworthy for how it invoked the rancher's confidence. He told Domínguez that some Anglos were

enemies, but others, like himself, were friends. Ord asked Domínguez to trust him.

Two partition sales of Rancho San Pedro property followed
during the last weeks of 1854, while Ord conducted Coast Survey
tasks nearby. In early December, Domínguez sold 214 acres of *sali-
nas*, or salt ponds, at "west beach" (today's Torrance and Redondo
Beach). Los Angeles's assistant surveyor, George Hansen, mapped
the parcel. On December 23, the Domínguez family sold a second,
larger property. This consisted of 2,400 acres at the San Pedro estuary, along its northwest shore. Roughly half belonged to Manuel,
the rest to his sister's children. The sale netted $20,000. A syndicate
planned to develop a waterfront settlement, which they called "New
Town." The *Los Angeles Star* reported the prospective sale back
in February 1854. It was the supposed city referenced in Whipple's
Pacific railroad report.[9]

Domínguez and the New Town buyers hired Ord to survey the
estuary parcel and lay out its streets. Over the next months, he
became increasingly involved with both sides. The syndicate included
attorney J. Lancaster Brent, land grant claimant and orchard entrepreneur Benjamin Wilson, Los Angeles pharmacist (and future California governor) John G. Downey, and the merchant and freight
operator William T. B. Sanford. The group informally included
Sanford's protégé and brother-in-law, twenty-four-year-old Phineas Banning. Banning in 1854 could not contribute capital to the
purchase. But by the close of the decade, he would emerge as New
Town's major force.[10]

Ord's connection to the syndicate had begun during the prior season. To meet his government duties, the officer contracted with a
number of local businessmen. His papers show purchases of books,
soup plates, produce, coffee, firewood, soap, socks, sugar, beef,
bread, fish, candles, mules, and sheep. The wholesale and forward-
ing firm of Banning & Alexander extended Ord store credit, provided blacksmithing, and handled the officer's mail. Ord's spending

brought US dollars to businesses still deprived of cash and used to trading in scrip or promissory notes. Merchants and landowners looked to him to learn about federal policy. And they solicited land surveys. The officer arrived with a reputation that preceded him. Locals recalled the work he had done for Los Angeles city and county in 1849.[11]

Politics further linked the officer to New Town's investors. Brent and Wilson were Democratic Party leaders and part of Senator Gwin's political machine. Wilson had served as Los Angeles mayor and county supervisor. Brent and Sanford were on the city council. In 1854, Brent became city attorney. Through Gwin's patronage, Wilson was appointed as a US Indian agent in September 1852, and Sanford as postmaster. During Ord's time in Southern California, the group grew increasingly valuable to the senator. Gwin faced opposition from David Broderick's faction of Democrats, who hoped to seize control of federal jobs. Ord boasted to Molly of their efforts—and his own—to defend Gwin.

The New Town circle brought together businessmen and like-minded partisans. Nearly all had Southern ties. Several had connections to Davis's War Department. Sanford and Banning won a US contract in early 1854 to supply freight to Fort Tejon. This was the new army post and Indian reservation, which Wilson also had a hand in, located north of Los Angeles, in Grapevine Canyon. Officials in the military and in Washington had long regarded Wilson as a strategic asset. He arrived in Southern California among the first American settlers after years spent as a Santa Fe Trail trapper and Indian trader. He married into the Yorba family and acquired several land grant estates. But "Benito" Wilson remained an American, by citizenship and loyalty. He served as an emissary between US forces and local Californios through the war of 1846–47. It was Wilson who rode out to meet Commodore Stockton with tidings of peace and the gift horse sent by Manuel Domínguez. When Ord visited Southern California in the fall of 1849 to research mountain passes, he made sure to interview and name "Mr. Wilson" ahead of others.[12]

New Town's buyers were functionaries of American expansion on the Southern California frontier. At the same time, they inhabited a world deeply contoured by the Mexican era and cohabited by a Hispanic elite. To be successful, the two groups had to integrate as a ruling class. They came together as creoles, with shared politics and common foes, bound by an intercultural and imperial partnership. Domínguez became prolific at forging such bonds. Democratic Party leaders Brent and Wilson welcomed the Californio into their fold. Brent, of course, was Domínguez's trusted lawyer and steward of his land claims case. The rancher knew the other buyers through commerce and politics, and he later purchased a one-sixth share in their development. New Town symbolized mutuality. Grantees like Domínguez brought their abundant acreage and cattle to the partnership. Anglo investors contributed capital, knowledge of legal and business institutions, and the privilege as members of the US regime to choose Spanish-speaking allies. But creole relationships were unstable, a product more of necessity than goodwill or commitment. Anglo land investors, like the settlers and squatters around them, understood their actions in terms of conquest. Once they obtained land, what more did they need from grant holders? Despite some reciprocity, the partnership was rife with contradiction. It entailed the taking of advantage and the one-way transfer of resources and power. It was a fragile union that could not and would not last.[13]

Ord landed in this space between nations and cultures. He and his Coast Survey crew completed the parcel's partition line as side work during December 1854 and January 1855. The job caught Ord between his various employers, particularly given his friendship with Domínguez. Yet he represented US conquest in nearly every way, from his family's involvement with Western politicians and speculators, to his ties to the Land Commission and claims process, to his service in the army and Coast Survey. As Ord drew himself closer to Domínguez, the officer took even larger steps to integrate with the New Town syndicate and Senator Gwin. Throughout the winter of 1855, the Wilsons, Downeys, Sanfords, and Bannings appear in

Ord's correspondence to Molly. Letters tell of their offers for survey work. Ord enjoyed discussing land markets with Brent and sought his advice when the Ord family looked to purchase property near Los Angeles and at the "Monty" (El Monte), up the San Gabriel River. In 1854, Ord's brother, the US attorney, relocated to Los Angeles to investigate Mexican-era claims. Wilson became his neighbor in the pueblo. The families socialized and talked party politics.[14]

Ord was in the middle of the Rancho San Pedro–New Town deal, yet he began to break in a clear direction. He knew which way power and opportunity would tend—had to tend—as the region's borderlands and creole bonds fractured. He would side with the newcomers, with the Americans, with the conquerors.

IT WOULD SEEM this sequence of events came about quickly, within a span of a few months ending in the spring of 1855. But in fact, Ord foresaw this all four years earlier. The officer's desire to have Domínguez sell an estuary parcel was as old as their acquaintance. It is unclear exactly when Ord and Domínguez first met, but the rancher appears in Ord's diary at the start of 1851. In the entry for January 27, Ord wrote of arriving at San Pedro and riding by stage to the Domínguez home, where he ate supper and slept. Terse remarks suggest the two already knew each other. Perhaps they met during the summer of 1849, when Ord mapped the Los Angeles pueblo and countryside? Ord certainly took note of San Pedro Bay in his army railroad report. "I think the first class of men will buy up these lands," he remarked, "when they shall discover that, attached to these plains, at San Pedro is a good harbor for coasters and steamers."[15]

By 1851, Ord was friendly enough with Domínguez to lodge overnight at his home. Then, the two men traveled together around Los Angeles and San Diego. Ord's Catholic faith, then-marriageable status, and ability to speak basic Spanish ingratiated him with the Californio. They shared a number of interests as well: in land and the search for wealth and security. On January 29, Ord recorded their visit to San Fernando, near the foothills northwest of Los Angeles. The local landowners, Ord wrote, "want slaves here much."[16] Domínguez was no stranger to Indian peonage and convict labor systems prevalent in the region. Most important, the rancher and officer discussed a subdivision at the Rancho San Pedro. "Domínguez [is a] clever fellow," Ord observed. "[He] has a pretty and extensive ranch. Plenty of water on it. . . . Extensive holding and pretty place on shore of *Estero* for town." But, the army officer groused, Domínguez "won't have a map made or city laid out. And I am too dyspeptic to humbug with him about it. I shall wait and see what can be done." Ord had a clear idea for *what* should happen at the estuary. It only remained to answer *how* and *when*. As of late 1854, his wait ended. Now was the time to act.

Then there was the large matter of the transcontinental railroad and the War Department surveys searching for a path to the sea. Ord knew what results he wanted. In that regard, he shared a striking similarity to Secretary Davis. But the parallels ran farther and converged. In late January 1855, Edward wrote to Molly at San Francisco. He asked that she send him, *"by the first steamer,"* two "small maps of San Pedro harbour" that he had rolled up and cached in *her* bureau drawer.[17] This followed the officer's mention weeks before that he was busy writing "a report to Washington." The request breached several protocols. Coast Survey practice dictated that draft maps be stored in San Francisco bank vaults until hand-duplicated and forwarded. As singular copies in a town regularly charred by fire, they were as valued and irreplaceable to the government as gold. Storage in a staff member's home, in their absence, and hidden among a woman's personal effects was unthinkable. Moreover, January was not the season for reports to Superintendent Bache. USCS correspondence shows no report sent in by Ord until eight months later. But in early 1855, Ord's timing corresponded precisely with that of Davis and others who, by misdirection, reached to grasp their long-held schemes and distant objects.

Some would succeed more than others. By the end of winter, a map of the estuary reached Washington, DC. It would emerge bearing a peculiar title: *Map to Navigable Water of Inner Harbor, January 1855, by E. O. C. Ord.*[18]

Chapter 13

TRIANGLES OF A DIFFERENT SORT

A T THE VERY MOMENT THAT JEFFERSON DAVIS endorsed a transcontinental railroad to Southern California, Edward Ord sent word of a possible terminus inside the San Pedro estuary. By no coincidence, this "inner harbor" sat adjacent to both the Coast Survey's channel study and the Domínguez rancho, with its impending subdivision. However briefly, Ord linked these stories. His departure from Los Angeles would cause the connections to vanish. But the effect—his so-called inner harbor—would remain. Although the Coast Survey sent him to measure geodetic triangles, Ord proved far more adept at triangulating among people. He flourished in the role of middleman, informant, and negotiator. It was a remarkable achievement for a soldier who did not aspire to high army command, nor, as of the 1850s, hope to ever obtain it. Power and wealth were not quite his. Yet they were close enough to think otherwise. Even so, he had things he wanted. Would the carefully constructed shapes provide for him, too? That would depend on how long he could maintain confidences.

There was some good news. Ord's most important relationships, to Davis and Bache, seemed to stabilize in mid-1854. The war secretary and superintendent had to mind their ties as each pursued ends at San Pedro Bay. Perhaps they discussed the officer and came to terms? Bache's antagonism with Ord seemed to lessen. Likewise, Ord began to complete tasks without complaint. He hiked up Catalina Island's steep ridges to set signals. On the mainland, he camped with his crew, a dog named Spot, and a mule called Jim.[1] They slept in small, cold tents. Or, rather, they were kept sleepless by the yip-howl of coyotes, the call of burrowing owls (a stuttered *ha-ha*), and the honking of geese. Ord endured meals of dried beef, bad coffee, potatoes, and stale bread. He laughed as Spot rampaged in pursuit of *ardillas* (California's Beechey ground squirrel). On occasion, Ord encountered an "old Indian" still living by tradition. Meanwhile, his survey crew, multilingual and racially diverse, showed the opposite drift of time. It included a domestic he named simply as "Catharine," without surname or the obligatory *miss* or *missus*. Catharine (or Catarina) was likely from the region's lower class, perhaps a Gabrieleño-Tongva whose forebears had lived near San Pedro Bay.[2]

Just as suddenly as Ord complied with his duties, he changed course once again. By the early months of 1855, he had lost interest in the channel. After a long bachelorhood, his mind fixed on homemaking and family. He wrote to Molly of being "desperately in love" and of embraces and kisses that awaited upon his return to San Francisco.[3] "I find it very natural to want to stay with you," he confessed, "and not half as easy to live by myself as it used to be."

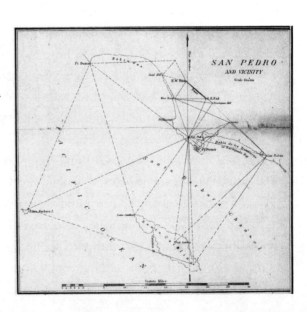

Coast Survey triangles, circa 1854. Ord made triangles of another sort, as land markets distracted him from duty.

Private passions ran so strong that they tested his loyalty to the army. As a low-level officer, he faced a career of middling pay and long separations from home and hearth. He pondered leaving the soldier's life to search out riches, as many of his West Point peers had. He insisted that Molly stop addressing her letters to "Captain." He preferred "Eddy" instead.

Desires found a familiar outlet. Ord looked for a quick profit in real estate to provide for himself and his wife. The officer maneuvered to acquire a piece of the New Town venture, or some other Domínguez land on the estuary. He boasted to Molly that he could buy the land "for less than almost anyone else" and predicted a tenfold appreciation within a few years.[4] That depended on the success of Davis's railroad project. Yet profit hinged most of all on friendship with Manuel and Engracia Domínguez. Both initially agreed to sell. Engracia offered to do so even if her husband did not, because of her fondness for the Ords. Edward encouraged Molly to keep up ties to secure the deal. In an age before the separation of home and work, households practiced business together. The model of companionate marriage, emerging in the mid-nineteenth century, made husbands and wives partners in strategy. Edward and Molly's professions of love overlapped in their letters with material aspirations. One can imagine the same for Manuel and Engracia. All hoped for a life of reward and comfort. "Prospects look bright for it," Edward promised.[5]

Ord's intimate dreams drove a final wedge between him and the Coast Survey's superintendent. Unlike their earlier conflicts, Ord kept his frustrations private this time. When his anger boiled, he wrote to Molly instead. In an undated letter, he complained that, within the Survey, "money is so scarce." His pay was overdue, forcing him to deny her allowance. She would need to sell a wagon to cover urgent costs, or call on banker William T. Sherman for a loan. Both steps were disgracing. More so because the situation, Ord explained, was "only a ruse" concocted by Bache "to hold me in check."[6] He assured his young wife that he would suffer such treatment no longer. He signed the letter "U.S. Coast Survey . . . but still a Captain."

Just as the soldier lost patience, he heard welcome news. In early February 1855, Ord received written permission to "report" and "give information, &c." directly to Secretary Davis. That was less than two weeks after he asked Molly to express him the maps of San Pedro's harbor. He promptly disbanded his field crew. Yet the officer did not rush to rejoin the army. Instead, he withheld news of his return from commanders in San Francisco and Benicia and asked Molly to do the same. Silence among the officers' wives would prevent parlor gossip and keep him from being redeployed too soon. In fact, it would be several months before Ord rejoined his artillery company. The couple's secrecy allowed him to preserve a direct line to Davis. As his government triangle collapsed, Ord would prioritize one relationship above all others.[7]

WHILE HE DELAYED his return to the army, Ord continued to pursue his real estate windfall. Much to his dismay, however, Manuel Domínguez pulled back. With the sale of the New Town parcel complete, Domínguez no longer had any urgency to sell. Since the 1820s, he had battled for property against neighbors, relatives, and squatters (perhaps, too, Gabrieleño-Tongva coming home to Suanga.) He knew he could speculate in a railway harbor, just like the Anglo newcomers. But was he willing to give his old friend "Eduardo" an easy land boon? In the absence of written sources, the rancher's thoughts are a mystery. Nonetheless, Edward wrote to Molly in disappointment. Manuel, he explained, needed to consult too many "hens."[8] It is unclear whether that was the rancher's deflection—an alibi that employed Anglo views of Hispanic marriage and womanhood—or an index of actual discussions with his wife, sister, and adult daughters. The term may have been Ord's alone.

At the same time, the officer started to voice hostility toward Mexican landholders. Such attitudes were on the rise, as settlers squared their land hunger with their inability to satisfy it. The *Daily Alta* observed of grants near Los Angeles, "Fast, fast is the property here . . . passing into the hands of Americans."[9] Former claimants "have so little thrift and foresight," the writer adjudged. "Want of any occupation, and the attendant vices of racing, play and drinking, &c, are enough to ruin the prosperity of any people." California's newer residents found comfort in such ideas, whether they succeeded at gaining land or not. The logic's appeal was that it explained either result. Ord, driven by his "temptation to speculate," proved no different. During Coast Survey work at Las Bolsas, he solicited Juan José Murillo (or Morillo) for some grant lands. Murillo demurred that debts and boundary disputes put the claim in jeopardy. He could not sell even if he wanted to. Edward returned empty-handed. Yet, in giving the bad news to Molly, he blamed his own failure on Murillo's turpitude. "I go into this detail," he wrote, "to show you what sort of people the Rancheros are."

Ord involved himself with all sorts of people. But it was land that he wanted most. That was what his triangles were about. Persons and institutions provided access. If not, he suddenly held them in less regard. So it would be with Domínguez. Between February and April of 1855, when Ord departed the Coast Survey, the officer's letters make reference to work at the rancho and New Town. But social visits ended. In choosing among his government connections, Ord lost what he most wanted for himself.

Was there an event that dissuaded Manuel and Engracia from selling to Ord? By March, the rancho–New Town partition neared completion. Buyers and sellers filed a US surveyor's map in federal court. They executed a deed in mid-April. (The court would finalize partition in December.) Documents showed two steps taken by Ord, which Domínguez may have noted unhappily. First, Ord included thirty-seven acres of the estuary, areas submerged perpetually or for parts of the day, within the boundaries of New Town. This gave Ord's Anglo clients, the syndicate, property to develop into harbor frontage. The rancher may or may not have agreed to this.[10]

Either way, there was a problem. Ord's survey declared *all other* parts of the estuary below the line of ordinary high tide, even those

located far outside New Town, to be henceforth removed from the
Rancho San Pedro. His map claimed this large portion (later determined as 1,100 acres) to be "navigable"—that is, containing water
deep enough for a small boat to move freely. The total set forth in
Ord's 1855 map became known as the "inner bay exception," as it
was excepted from the rancho and yet not part of New Town's limits.

Who, then, owned the estuary? Under the Constitution, lands
beneath navigable waters are within the public domain of the United
States. And lands *adjacent to* navigable waters that are exposed by
the daily tides belong to the state—in this case, California. By the
stroke of Ord's pen, both types of land—submerged and tidal—
ceased to be Domínguez's. That conflicted, however, with the
Land Commission's decree of confirmation, which put the estuary
squarely within the estate.

Something strange had happened. And it altered the land deal to
be far less favorable to Domínguez. Subdivision doubled the amount
of acreage removed from the estate. The family relinquished this
additional property without ever putting it up for sale. In fact, they
had paid for the privilege. The future value lost was sizable, given

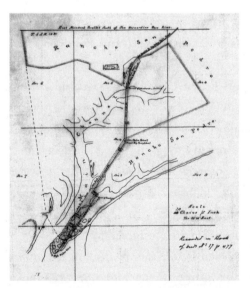

The inner bay
exception, from
a late nineteenth-
century map, showing
the boundary Ord
established in 1855
between rancho
and estuary.

talk of a transcontinental railroad. Ord brought the sting of conquest into Domínguez's private world. Perhaps that was enough to
change his mind about Edward and Molly.

There is another possibility, one that involves the officer's other
clients, the New Town purchasers. Ord liked to talk real estate and
politics with them. But he may have overreached. His September 1854
letter to Domínguez suggested that J. Lancaster Brent had made a
"deal" (or treaty) with a "squatter."[11] Did the rancher raise this matter with Brent? Did the attorney, in turn, expose Ord's schemes? The
question is without answer. But it reveals a key facet of the story.
Ord kept powerful company and fancied himself smart enough to
outwit them. However, he searched out this company because *they*
held greater advantage. And those advantages could turn on him,
especially when he was no longer needed. Such were the pitfalls of a
life of triangulation.

THE INNER HARBOR EXCEPTION was a triumph, betrayal, or
paradox for each respective side. For the Coast Survey, however,
it was simply inscrutable. Only it had the prerogative to call ocean
spaces navigable or worthy anchorage. Bache did not send Ord to
San Pedro for such a purpose. Instead, he paid hydrography crews to
do so. As of early 1855, they had not examined the estuary; yet Ord
assessed the waters anyway. And he based his claim of navigability
not on soundings, but on tide readings taken in March 1854—weeks
before the arrival of Whipple's Pacific railroad expedition, to whom
Ord talked up the future seaport. Ord's methods contained greater
flaws that would surface decades later.[12]

Like the Coast Survey, the War Department took no steps to
claim the estuary for federal use. Two military committees existed to
recommend harbor-related exceptions from the public lands or private property. One was the Board of Engineers for the Pacific Coast,
established in 1849, which reserved sites for shoreline forts, artillery batteries, and other facilities. The other was the US Lighthouse

Board, established in 1852. Secretary Davis was mindful of the committees' shared authority, which brought together the Coast Survey, Navy Department, and War Department. He refused lobbying to the contrary. In February 1855—in fact, on the same day Davis issued his recommendation of the 32nd parallel route—Senator Gwin asked him to endorse a bill for breakwater construction at San Pedro, the road's "certain terminus."[13] Ord submitted a petition from locals calling for the same. However, Davis replied that the matter fell "within the sphere of the Coast Survey." He forwarded the message to Bache to show he would not intrude on the superintendent's turf, even at the urging of others. Of course, through Ord's actions, Davis already had.

That was the essence and audacity of the inner bay exception. The federal government had no direct, official role in creating it. Instead, it took form by networks that Ord had assembled around himself. Documents do not give a full picture of how these networks might have functioned, yet they reveal an intersection of events, persons, and motives. It is important to remember that the authors of letters kept in archives today did not write for historians or their readers. Instead, they wrote to people with whom they had extensive communication and shared experience. All knew enough, but also enough to not say. Their letters contain shorthand and subtext, the meaning of which materializes only when placed in a mosaic.

Hence, in a January 1855 letter to Senator Gwin, Pacificus Ord raised the name of Manuel Domínguez. The US attorney described how Mormon settlers in San Bernardino had filed a bogus claim to gain double the acreage originally granted by Mexican officials. This and other crimes might succeed, he warned, because of "unscrupulous Mexicans who are banded in schemes to circumvent 'Los Yankees.'"[14] The attempted swindle, he seethed, was the fault of "an ignorant Alcalde" who now cost the federal government $50,000 to defend its interests in court. That man, the US attorney told Gwin, was "our old friend Don Manuel Domínguez." The letter gives a glimpse into how the Ords thought about Domínguez, and what

they believed the Californio owed to the United States and its emissaries. This was the background to the rancher's pleasant, inquiring
visits from Edward and Molly.

Networks of people—faint, linear shapes through society—made
all the difference. But what mattered most were those with power to
say what was legal or not. In June 1855, Domínguez received notice
that federal authorities would contest his Rancho San Pedro confirmation decree. The US attorney general initiated the appeal months
before, on recommendation of his local US attorney. Pacificus Ord's
signature appears prominently on filed papers alleging that Domínguez's claim belonged to the federal government "by virtue of conquest."[15] The phrase did not mean that the US intended to take all of
Domínguez's land. But it reminded the rancher that he was the sup-
plicant. By this time, Edward Ord had sailed away to rejoin the army
at Benicia Barracks. Surely, the officer came to mind when Domínguez considered how events had turned.

The attorney general's protest delayed Domínguez's land patent
by several years, enabling one last back-channel event. Under the
California Land Act of 1851, a grant confirmed by the Land Commission was passed along for review to the General Land Office and
its representative, the US surveyor general for California. In 1855,
that was Colonel John Coffee Hays. Like most federal appointees in
the state, Hays found his job through the patronage of Senator Gwin.
He was also an army officer and acquaintance of Edward Ord. The
pair were friends enough to contemplate starting a survey business
together. Their plan presented Hays with conflicts of interest that,
even then, could not be ignored. Perhaps this is why Ord wrote that
the venture was tentatively (and humorously, in reference to early
morning labor) called "Coffee, Ord & Company"—omitting Hays's
surname and effectively cloaking him as a silent partner.[16]

The Hays-Ord enterprise had murkier depths still. Ord's same letter spoke of an impending assignment Hays would "give" to him "to
survey," despite the fact that he was not a US surveyor. Two days later,
on February 9, the final map of the Rancho San Pedro–New Town

subdivision was complete and submitted to the district court. It bore the name of Henry Hancock, US surveyor, acting under the instructions of Surveyor General J. C. Hays. But the work belonged to Edward Ord. True credit was withheld in court, yet known to the parties involved. Perhaps some far away, in the nation's capital, knew also?[17]

When Domínguez's attorney, J. Lancaster Brent, returned from Washington, DC, in February 1859 bearing the rancher's US land patent, the document contained the inner harbor exception created by Ord. Domínguez was fortunate to receive only the second patent in the state for a Mexican land grant (John Frémont received his already), and the first in Southern California. His buyers were equally fortunate. Domínguez never disputed the exception or the patent. But in other ways, his family would have the last word.

The Coast Survey, too, gave final comment. The question of a harbor inside the San Pedro estuary was not closed. Ord had some loose ends to tie off, even after he transferred from the agency in spring 1855. Bache's clerk concluded that Edward had overpaid his brother Marcellus for service on the field crew. He demanded that the Ords return the excess pay. Edward assured the superintendent that all had transpired in good faith. "Please Professor," he wrote, "accept my thanks for the confidence and favor you have extended me and consider me at all times ready to forward the interest and promote the success of the Survey."[18] He knew to never burn bridges. After all, the nation's coasts were endless with opportunity.

But watchful observers had grown wise to him. When Ord submitted his report on the winter season, he asked Bache to reprint this in the annual report to Congress, an honor reserved for outstanding work. "I think such publication," Ord insisted, "would do no harm to the survey."[19] But Bache's annual report left out Ord's document. Instead, it included remarks by George Davidson about San Pedro. "The attempt to make an inner harbor within the lagoon," Davidson wrote, "will never be made by anyone who has experienced a southeaster in that vicinity"—that is, by anyone concerned with safe navigation.

Bache went a step farther. His report to Congress singled out army officers for special merit in service to the Coast Survey and the American republic. This was his token of gratitude to the soldiers, but also to the War Department and his war secretary friend. Davis, in his quest to micromanage the army, was sure to take notice. Edward Ord's name was conspicuously absent.

Chapter 14

DAMAGES DONE

I N   A P R I L   1855,   S A N   F R A N C I S C O ' S   *D A I L Y   A L T A*   C O N-
demned Jefferson Davis and his proposed Pacific railroad. "It
does not astonish us," editors wrote, "that the surveying parties sent
out by him should report that the Southern route for a railroad was
preferable to all others. . . . They were sent for that very purpose."[1]
Six weeks had passed since the war secretary issued his recommendations. With the lag in shipping, copies of his report had not yet
reached the Golden Gate city. But word of his decision set off an
uproar just as Edward Ord arrived from San Pedro. The *Alta* called
Davis "one of the last men in the country" deserving of national office
and public trust. "It is no source of astonishment," they inveighed,
"to find him using his position to advance his peculiar notions."

If Davis expected that science would overpower his political
opponents, that plan was wrecked sensationally. His words, even
when supported by the Topographical Engineers, failed to surmount
doubts. Instead, Congress grew impatient as slavery polarized the
nation. Critics decried Davis's cost estimates. They demanded that
the Pacific Railroad Office release data and maps, so that Americans might compare the routes for themselves. The *Alta* called for
its preferred line, the Frémont-Benton route from St. Louis. Others revived the idea of a central line to give Northern and Southern
interests equal access to the West. It was as if the railroad surveys
had never happened.

The setback connected unlikely associates. Despite their disagreements, Davis and the editors of the *Alta* found themselves deprived

of a mutual ally in the Senate. William Gwin publicly favored a central railroad, but privately offered to Davis to support his 32nd parallel line. Sadly, for all who took Gwin at his word, his Senate term expired in March 1855. Prior to the twentieth century, senators were elected by state legislatures, not by voters. Gwin's delicate balance—among mining companies, Californios, land speculators, squatters, slavery moderates, and proslavery ideologues—held no longer. His enemies, led by David Broderick, gained enough votes to stonewall his reappointment. Loyalists, like the Brent-Wilson circle of Los Angeles, failed to save him. Even so, the legislature could not agree on Gwin's replacement. California would have only one senator for the next Congress. The transcontinental railroad and Coast Survey languished without him. He would return to the Senate two years later, alongside Broderick. News soon emerged that Gwin had bargained with his rival—groveled, some said—to regain office.[2]

Faced with the unfavorable developments of 1855, Davis searched for all the potential allies he could find. Beyond Southerners in Congress, he looked to the realm of dispassionate science. He pressed the Topographical Engineers to complete the railroad reports. And he asked that they correspond with newspapers like the *Daily Alta* to rebut the allegations against him. Davis kept up ties to the nation's leading scientists, especially Bache and members of the Smithsonian.[3]

Most of all, the war secretary pushed the Pacific Railroad Office to speed production of its showpiece, the large-format composite map. He still believed the document would win him support. Unfortunately, the map met with a series of mishaps. One involved R. S. Williamson's 32nd parallel survey and its cartographer, Charles Preuss. Preuss was renowned for his prior work on Frémont's first, second, and fourth expeditions. Despite the association (or perhaps because of it), Davis hired Preuss for Williamson's crew. Preuss might have gotten in over his head. He wound up having to corroborate Davis's choice while showing Frémont and Benton's to be impossible. One side, or both, would be angry. In September 1854, after returning to his wife and children in Washington, DC, Preuss hung himself from a tree. His death is ascribed to sunstroke, homesickness, mental illness, or post-traumatic stress disorder. Few have considered the political context. Whatever his intentions in joining Williamson's party, Preuss caught himself amid opposing figures, any of whom could destroy his livelihood and reputation. In that sense, he contrasts with Edward Ord. Both men sought opportunity in between powerful forces. But the consequences could be severe. Preuss's suicide cast more doubt on Williamson's work.[4]

Next, the Pacific Railroad Office ran into trouble with its maps and elevation profiles. Proofs were painstakingly etched in reverse image onto limestone blocks that weighed 1,500 pounds. They were so heavy that only two lithograph presses, both in New York City, could handle the job. Negligence by the first printer delayed the completion of Williamson's map. The printer then claimed to have destroyed the stone proofs—but not before the Adams Express Company misplaced a shipment of profiles on their way to the House of Representatives. The military officer assigned to oversee the project wrote to superiors in frustration, "Is there no way of getting satisfaction? Can the secretary of war do nothing?" Davis likely asked the same.[5]

The composite map—a landmark in American exploration and cartography—also ran into misfortune. Some initial copies, half-sized and disclaimed as a "hurried compilation," were shipped in May 1855. Months later, the lithography shop's owners dissolved their business. Their court battle ensnared the map's revised editions. A new owner acquired the limestone proofs and demanded that the government pay a ransom to reclaim them. The War Department struggled to respond. Finally, an armed US marshal seized the tablets. Large-format railroad maps, measuring 41.7 by 45.7 inches, appeared in Washington late that year. Critics pointed out that these had significant blank areas. Not so for Southern California, where engravers added stunning detail based on George Davidson's longitude measures and Henry Washington's San Bernardino Meridian. Given the unknowns elsewhere, legislators refused to decide.[6]

The sequence of delays cost Davis valuable time. Then, in 1856, the war secretary's window of opportunity shut completely. The Bleeding Kansas civil war, a result of Stephen Douglas's railroad gambit, began. Free-soil and antislavery ideas gained strength in Northern politics. Davis's (and Douglas's) Democratic Party lost its majority in the House of Representatives to a coalition of Republicans, Know-Nothings, and Northern Whigs. It became impossible to fund a Southern transcontinental line, no matter how expertly surveyed or mapped. When Democrats regained the House majority, in March 1857, Davis's tenure as war secretary had expired. He returned to the Senate, along with Gwin. The two would resume their fraught connection as they joined the Pacific Railroad committee. Gwin's sectional politics remained elastic, while Davis's grew rigid. Yet the two continued to share some things in common. When Davis vacated his War Department office, he took with him a powerful memory for those who had served him. On his last full day, Davis recommended Edward Ord for a brevet promotion in recognition of outstanding military service.[7]

JUST BEFORE DAVIS LEFT his cabinet post, the US attorney general dropped the government's suit against the Rancho San Pedro claim. That coincided with the two-year window Manuel Domínguez had to protest. Southern California's US attorney, Pacificus Ord, continued in office for another eighteen months, until July 1858. By that time, his purview had diminished precisely for the amount of litigation he chased up. To deal with the backlog, President James Buchanan appointed a special counsel for California land cases. Lawyer Edwin M. Stanton (a future attorney general and secretary of war) took the lead in investigating Mexican-era claims, especially those at San Francisco that were later known as the Limantour frauds. Stanton also looked to end land speculation by US attorneys. With greater troubles to contend with, the Buchanan administration agreed to the Rancho San Pedro. In December 1858, the president signed a US land patent issued to Manuel Domínguez and family. While serving as secretary of state a decade before, Buchanan had presented the Treaty of Guadalupe Hidalgo to the Senate. Thus, his signature twice fulfilled Domínguez's goal of private property and colonial birthright.[8]

Edward Ord did not write to congratulate Domínguez this time. He walked away from Southern California in the spring of 1855. The officer passed up an offer from then Secretary Davis to observe the Russian Army in the Crimean War. Nineteenth-century militaries hosted foreign observers, as they sought to modernize and make known their ability. An observer mission was quite the honor, although it involved travel, arduous reports, and upfront expenses. Ord declined. Instead, his 3rd Artillery sailed for Oregon to suppress Native American insurgents. He lamented the carnage, but the assignment allowed him to search out property and patronage jobs.[9]

For three weeks during the summer of 1856, Ord had a respite from the bloody Pacific Northwest. The army sent him to Los Angeles to examine San Gorgonio Pass for a military fort and Indian reservation. He had first viewed the pass during his 1849 army survey but thought it impractical. Since then, Jefferson Davis had recommended it for a transcontinental line. A garrison at San Gorgonio would further justify the railroad. This time, Ord praised the site for being "famed among Pacific Railroad explorers" and "the only valley which cuts from side to side the great Cordillera or Sierra Nevada."[10] Local Indians, like those at the village of "Haynyo," were amenable. The army's harder task, Ord thought, would be to procure property title from the US Land Office. Speculative markets, he knew, burned hot. They promised the War Department a mess of lawsuits.

Ord's 1856 study allowed Molly and him to make a tour of Southern California, to catch up on acquaintances, and to gauge the region's progress. He published a travelogue, *City of the Angels &*

the City of the Saints. The book described San Pedro Bay, the Mormons of San Bernardino, injustices suffered by mission Indians, and the insurrectionary threat posed by Los Angeles's "Mexicans" and "Sonorians." Ord visited friends, including the politician and New Town investor Benjamin Wilson. Yet, if Edward called upon the Rancho San Pedro, he did not say. Instead, he gave a generic, unflattering description of Californio ranchers as "careless" and in "happy ignorance of taxes, lawyers, squatters, and moneylenders."[11] It was a narrative of conquest and appropriation that gathered force. "And no wonder they hate us," he added. Ord made no mention of seeing Manuel and Engracia Domínguez, nor of their dark-haired daughters, whom he once had kissed affectionately upon the head.

The promise of partnership in Southern California's borderlands had crashed against real estate, a railroad, and its harbor-to-be. And the breaking of creole union at Los Angeles became tied to distant fractures in the American union. Ord vanished from this world along with the fleeting networks he had built. Meanwhile, others

Los Angeles in 1854—a sparse village of single-story adobes.
The plaza and Catholic church of Our Lady Queen of the
Angels is visible at lower left.

were left with the consequences. Unlike them, Ord's path through history remained open. He would prove the most protean and full of possibility when the damages of disunion finally found him.

What did this mean for the friendship of Manuel and Edward? Did their bond break, or simply fade to be forgotten? Neither left a clear answer. But if Ord avoided returning to the Rancho San Pedro, Domínguez likewise moved on. The rancher, as he had before, took lessons and adjusted strategy. In October 1855 and July 1857, Domínguez acquired two Anglo, Catholic sons-in-law who served as advisors in the role once presumed by "Eduardo." James A. "Santiago" Watson and George C. Carson married daughters Dolores and Victoria, respectively. Carson, a successful merchant, member of the Los Angeles Council, and junior member of the Brent-Wilson faction of the Democratic Party, became especially important. He would guide the family's interests for the next four decades. As part of this, the Domínguez daughters would challenge the US government's removal of the estuary from their father's estate. Their case would hinge upon purported errors in Ord's 1855 survey.

This, more than anything, reveals that Manuel Domínguez never forgot the army captain and the times they shared. It hints, as well, at tales of friendship, power, and betrayal passed down among family. Domínguez kept very few personal papers. But the congratulatory letter he received in 1854 from Edward Ord, written in the officer's broken Spanish, survives.

All Waters Flow
to the Pacific

NEITHER RAILROAD EXPEDITIONS NOR CONGRESS
could find an easy path to the Pacific Ocean. But that was
no problem for the West's great watershed, which cut chasms down
mountainsides and broad valleys below. The first US explorers to
view the region had noted this. In August 1805, at Lemhi Pass near
Idaho's Salmon River, Meriwether Lewis inferred by the direction of
creek flows that his party had crossed the continental divide. They
now entered the basin of the fabled Columbia River. Longitude told
him that its descent to the sea, as compared to the ambling Missouri
River, would be fast and precipitous. The continent tipped westward. Past the point of balance, people—like rivulets of water—
would go with it. That favored the republic's expansion and the
arrival of settlers.

Southern California tilted more so. Its coastal plain emptied a
watershed of four thousand square miles. With numerous peaks
between eight thousand and eleven thousand feet, the encircling
mountains were higher than Lemhi Pass and much closer to the
shore. Their steep angle forced clouds to release rain and snow, often
in torrents. This fed three major rivers (renamed by the Spanish as
Los Angeles, San Gabriel, and Santa Ana), which drained into San
Pedro Bay. Winter monsoons and historic flood events, on average
every fifteen years, overwhelmed existing banks and tore new paths

through rock and soil. Even after changing course, the rivers left their mark. Floods deposited rich alluvium. Subterranean water moved beneath dry topsoil. Verdant corridors—of cattails and sedge, or oak and willow—ran to the coast, interrupting an otherwise arid prairie. The rivers always found the sea.

Here, at land's end, the watershed collided with the Pacific. It was a worthy match. Stream flows, wave action, tide, and currents—countervailing hydrologies—pressed against one another in contest, causing sediment to settle into mud and uplands. Marine particles came to rest as sandbars and beach. Eventually, a sandspit sheltered the interior bay. A coastal estuary, something neither fully continent nor ocean, took form. Southern California's watershed created six of these, all of comparable size. Together, they totaled perhaps eighteen thousand acres. The Los Angeles River formed the largest one, San Pedro's estuary, while flood events left the Ballona inlet, northwest on Santa Monica Bay. The Alamitos, Anaheim, and Bolsa estuaries developed from the San Gabriel River's various courses. Farther south, the Santa Ana River established Newport Bay.

By the 1850s, settlers and politicians imagined using such places to access Pacific trade. The US republic grasped for fortune in the western sea. But, before it could seize that imperative, it dissolved into Civil War instead.

Part Three
REBELLIONS

Secession and Lives Interrupted

O N APRIL 12, 1861, SECESSIONISTS FIRED ON A US military post in Charleston Harbor in South Carolina. The shelling of Fort Sumter culminated decades of growing political division over slavery. It also culminated decades of westward expansion. It was the West, after all, that made the slavery issue irreconcilable. The major battles of the Civil War would take place in the East, far from the Pacific Ocean. Nevertheless, the conflict would affect Western places like San Pedro Bay and the people connected to it.

George Davidson received news of the Fort Sumter attack while on furlough at New Harmony, Indiana. His life and career had reached a zenith. In October 1858, he had married Ellinor Owen Fauntleroy. She was the daughter of his deceased mentor, Robert Henry Fauntleroy, and granddaughter of the utopian reformer Robert Owen. George and Ellie passed their honeymoon on the voyage to California. He resumed his government duties. She assisted him at home and in the field. Davidson left surveys of the Canada border and San Pedro–Santa Barbara Channel to others. He now focused on geodetic positioning north of San Francisco to the Russian and "Wawlalla" (Gualala) Rivers. Settlers intruded on this rugged coast, renewing violence with Native peoples. That necessitated his work. The new assignment allowed him to stay close to Ellie and their son, born in November 1859. His pioneering studies

totaled sixty-six published volumes of astronomical and magnetic
research. He had contributed to every map or navigation chart of
the Pacific coast.[1]

But tragedy struck Davidson in the months leading up to the Civil
War. Bache found Congress too rancorous and reluctant to pay
for Coast Survey science. Davidson continued, despite a paucity of
funds. The work caused his neuralgia and rheumatism to return.
Then, in the spring of 1860, he was severely injured while setting signals on a mountain peak. While he recuperated, Edward Fauntleroy,
his nineteen-year-old aide and brother-in-law, died in his absence.
Distraught, George requested leave to the East. He and Ellie sailed
in November as the US awaited result of the presidential election, its
most contested one so far. The San Francisco Alta printed an appreciation of the departing thirty-five-year-old. It described Davidson
in distressing terms as "entirely broken down," an "invalid," and a
"mere wreck physically, of his former self." All were the product of
his government labors and unrelenting ambition.[2]

Upon reaching Washington, DC, George and Ellie learned that
the Southern states were to consider secession from the union. Wall
Street panicked, and politicians scrambled for a compromise. Given
the chance of war, Bache quietly secured the Coast Survey's vault of
maps. Davidson assisted until word arrived that Ellie's mother was
dying. They rushed to Indiana. Then, in April 1861—weeks after the
Fort Sumter battle, and just as Davidson prepared to resume duty—
his infant son died of injuries suffered in an accidental fall. Eleven
states had seceded by the time the stunned and grieving couple resettled at Philadelphia. So little of this world seemed certain. But war
did. How it would change the nation's history, and that of San Pedro
Bay, no one yet knew.[3]

Just as Davidson's family fell apart in 1861, the Civil War caused
his life's intellectual project to collapse. For a decade, he and the
Coast Survey had labored to unite the Pacific shore, from the 49th
parallel to the Mexican border. By opening the region to commerce,
Davidson's work promised to solidify a transcontinental republic

Ellinor Fauntleroy Davidson, with the child she and George would lose on the eve of the Civil War

and dissipate tensions between North and South. "*It knows no section,*" he wrote of the Coast Survey in 1859, "except to drive forward work *in them all.*"[4] This was Bache's design, a strategy to win greater funding from Congress. But it was sincere principle, too, a faith that science could achieve political order and remake the world.

However, the West overwhelmed hopeful intentions. Each advance of conquest, exploration, and settlement worsened the sectional conflict. Davidson privately blamed outgoing president James Buchanan (an "infernal old imbecile," in his words) for enabling secessionists "to play treason."[5] Yet he ignored an inconvenient fact: the Coast Survey's work—and his own painful efforts—contributed. The transcontinental map shattered because of those who dared to create it. Now, war left the task interrupted, perhaps forever.

DAVIDSON WAS NOT ALONE. Others had hastened the republic's undoing as participants in its territorial expansion. Among them were Edward Ord's cohort of army officers in the West. Without a transcontinental telegraph in place, news of Fort Sumter took twelve days to reach San Francisco by Pony Express. Word then spread to outlying stations. Individuals took stock of their loyalties, set aside oaths, and parted ways. Some paths would intersect. William T.

Sherman had left San Francisco five years before. He closed his struggling bank branch and set up another in New York. Then, the 1857
financial panic hit, destroying the company and his reputation. Sherman had yet to recover. He relocated to Rapides Parish, Louisiana,
to superintend a military school. The state's secession compelled
him home to Ohio. By July 1861, he had accepted reappointment to
the regular army. He led a brigade of US soldiers into the war's first
battle (and Union defeat) beside Virginia's Bull Run creek.

Like Sherman, the topographical engineers R. S. Williamson
and A. W. Whipple remained loyal to the government. So, too, did
their colleague, the former Washington governor Isaac Stevens. All
were veterans of the Pacific railroad surveys. Each would serve in
the Union Army. Whipple and Stevens would die for it. No engineer
served more gloriously—at least initially—than thirty-four-year-old
George McClellan, who in November 1861 took top command and
built that army into existence.

John Frémont anticipated a premier position as the war began.
In January 1861, he was en route to London to seek loans for his
troubled Rancho Mariposa. Frémont stopped to request a cabinet
post from President-elect Lincoln. In the end, Lincoln chose him to
command the army's Department of the West, centered in Missouri.
The state's railroads and rivers were indispensable to Union strategy.
Frémont used the Benton political machine to stabilize St. Louis. But
he lapsed into frontier habits, particularly his delusions of grandeur
and impunity. Without Lincoln's approval, he ordered the slaves of
Missouri secessionists to be freed, which alienated rebels and Unionists alike. (Both intended to maintain slavery.) His threat to kill captured rebels did likewise. Frémont had used the tactic before: early
in the US-Mexico War, he ordered execution of several Californio
noncombatants. His troops committed atrocities, as well, against
Klamath and Wintu Indian villages. But these were different times.
President Lincoln upbraided Frémont, then sent him to a new post,
where he failed miserably against Stonewall Jackson's Confederates.
Frémont resigned rather than accepting demotion. He sold Mari-

posa to cover massive financial debts. For the Frémonts, the Civil War began a decades-long fall. In old age, they would have little more than their pride and memories.[6]

George Davidson had none of Frémont's wealth or cachet to lose. Nonetheless, the war caused large parts of the Coast Survey to dissolve around him. Ambrose Powell Hill, an army officer who received Davidson's letters at the bureau's headquarters in Washington, DC, resigned to enter the Confederate military. Other staff left for the rebel states or had their work suspended. Some, too, returned to the Union military. James Alden, the hydrographic officer, had exited the Survey in 1860 to rejoin the US Navy. He would play a significant role in the blockade of the Confederacy and battles along the Gulf Coast. His intended replacement, David D. Porter, would emerge as one of the Union's greatest naval heroes, beginning with the Vicksburg campaign. Porter's boats skillfully covered the armies of William T. Sherman and Ulysses S. Grant.

Davidson had crossed paths with Grant in early 1854 at Fort Humboldt, California, just before the soldier ended his army career. Grant struggled to recover. By 1861, he had moved to Galena, Illinois, to join a leatherworks business set up by his father. It was no easy life. Grant asked the Union Army for reinstatement. His first inquiries—unlike Sherman's—failed. However, Congress enlarged the army after its defeat at Bull Run. In August 1861, Grant received a commission as brigadier general in the volunteer army. He took charge of troops at Cairo, Illinois, then at Paducah, Kentucky. With so many experienced and better-regarded officers, no one expected great things from Grant. By 1865, he would stand among the war's most surprising revelations.

JEFFERSON DAVIS'S REINVENTION garners less notice today, but was noteworthy then. He built his stature through the 1850s by a delicate mix of section, nation, and Western imperialism. He became a great patron of the army and Coast Survey, and perhaps the most well-regarded slaveholder in Washington society. He steadfastly committed to national institutions while Varina's parties convened the "most cultivated class" across section and party. After leaving the War Department in 1857, Davis reentered the Senate, where he clashed but also cooperated with Northern colleagues. He continued his support for Yankee science. In 1859, he fell terribly ill—blinded, in fact, by eye inflammation. Davis left his sickbed, nonetheless, to urge the Senate to pass appropriations for his "friend," Superintendent Bache. "I must go if it kills me," Jefferson told Varina. She agreed.[7]

The Mississippi senator kept careful distance from Southerners who talked too much of disunion. Yet he was no Unionist. He had spent much of the prior decade attacking the Compromise of 1850 and purging its supporters from the Democratic Party. In the 1860 presidential election, Davis worked to hold his national party together. But he did so with the intent of blocking Northern candidates: Stephen Douglas of the Democrats; William Seward, the Republicans' presumed nominee; and Lincoln, the ultimate nominee. The Mississippian spoke of Lincoln's win as a calamity. He told crowds it meant an attack on the South and the Constitution. Davis's nationalism was real, yet only so long as it served his sectionalism. Friendships broke on similar grounds. After 1860, Bache disappears from Varina's memoir of her husband's life.

Davis chose to make the republic's rupture his defining moment. He initially opposed secession and participated in negotiations to resolve the crisis. He pushed for concessions on slavery and its future in the West. By 1861, Davis had decided he could no longer balance section and nation. On January 10, he delivered his last speech to the Senate and called for an amicable separation. Then he departed. The family arrived at their Brierfield plantation. Days later, the Confederacy's constitutional convention selected Davis—ahead of several radical secessionists—as provisional president. Delegates reasoned that an avowed, accomplished nationalist would best lead their Southern republic and military. The former US patriot reemerged as

Southern separatist and rebel-in-chief. In doing so, Davis cut a path for others to follow.[8]

THE SECESSION CRISIS marked a reinvention, too, for Edward Ord. The moment caught him amid the complex ties he once used to navigate the California borderlands. The Ords were committed Unionists who disapproved of slavery, likely because of their Jesuit beliefs. But disapproval had degrees and left many doors open. During the US-Mexico War, Edward described the institution as an *unfortunate necessity* of labor, one perpetuated by consumers of cotton and food and no less by Northerners "who prate of freedom to the poor black."[9] His mixed views had drawn the officer in the 1850s to Southern patrons like Davis and William Gwin. Something changed, however, as he contemplated death in the Civil War. Edward wrote Molly that the "cause of Union" was the "cause of humanity." Slavery was an affront to God and a "great evil." It should be "discontinued wherever it is possible."

Still, the Ords remained tied to Maryland, a border slave state with a culture of moderation. For Edward, that often meant a divided mind. Despite strong feelings against slavery, he also thought attacks on it were unlawful. And he regarded with equal suspicion Southern ideologues, disunionists, "rabid abolitionists," and Republicans. In 1859, he transferred briefly to Fort Monroe, Virginia. From there, the army assigned him inland, along with thousands of others, to guard the trial and execution of the antislavery insurrectionist John Brown. Ord and his artillerymen occupied the Harpers Ferry arsenal, where they reported to Colonel Robert E. Lee. In Edward's telling, the army sent him, ahead of more ranking commanders, because he was not a Yankee. Molly still had Thompson relatives in Charles Town, where Brown was jailed and hanged. Two years later, Edward continued to worry about "Northern filibusters." Molly warned her husband against "abolition sentiments"—"They will surely do you no good what God has ordained."[10] He disagreed with her faith.

Even so, he complained of a "few abolitionists" in Congress who sought to mislead the nation into a war *against slavery* rather than against secession alone. "They are the same sort," he remarked, "as the men who started this rebellion in the South."

Ord knew such men—and women—more than he let on. As late as the summer of 1859, the Ord brothers had sought Gwin's help to find property and jobs. Edward, while stationed in Virginia, had a chance to catch up with the senator. Mary Gwin's face and manners reminded Ord of his own wife. He wrote Molly of the Gwins' good tidings and promise to visit when next in California. President Buchanan would host the Gwins at his summer retreat on Old Point Comfort. They—Molly and Edward—were invited. "I know," the soldier told his wife, "you will be a favorite."[11]

Molly's Southern ways had helped Edward reassure patrons and advance his career. But as the Civil War began, those ties became a liability. They would cause him consternation and, at times, regret. Edward had married into a slave-owning, political family, some members of which took an active part in the rebellion. He and Molly were distant from that in 1861, living at Benicia, California, with their four children. Rumors circulated nonetheless that they were Confederate sympathizers. Competition among officers fed the whispers, but some perhaps knew of Ord's past. He fretted over this, even after leaving to command Union volunteers in the East. He implored Molly to keep quiet and to mind the company she kept. Yet she persisted. In an 1864 letter, Edward labeled his wife a "secessionist"—no small charge. He feared others might uncover this secret and more. At his most despondent, Ord wrote of his marriage as "my unfortunate connection."[12] Yet he—and Molly—would prove surprisingly fortunate. Ord continued his talent for being in the right place at the right time and for having the right friends. Others he would let be forgotten.

For Ord, Davidson, and countless more, war became the singular divide between past and future, old lives and new. It served the same purpose for the United States and for San Pedro Bay. Among its many changes, the Civil War brought to the estuary precisely what Ord, Davis, and others had sought (and what Davidson and Bache had refused). San Pedro became a harbor, one that linked the Pacific coast to desert hinterlands. Building an inland empire was preliminary to developing a modern port. Using the bay's waters, the US would take hold of the Southwest's most forbidding spaces and prepare a growing city to do the same decades later. Wartime changes transpired in Davidson's and Ord's absence. Instead, two less-known figures, James Carleton and Phineas Banning, would lead events. Their story begins with the peculiarities of Southern California, and with regional rebellions that foreshadowed a national catastrophe.

Chapter 15

A REGION OF REBELS

T HE SECESSION CRISIS OF 1861 HAD MANY ROOTS. BUT
it was Western conquest that brought matters to a head. With
the Mexican war, the United States had acquired a massive territory
reaching to the Pacific Ocean. Many Americans hailed this as destiny.
But it also proved calamitous. The United States took on the West's
endemic disorder and its many peoples resistant to external authority. Anglo settlement added to the mix US political conflicts—that
over slavery most of all. By the late 1850s, the West convulsed with
instability. The tremors sharpened the disagreements in Washington, DC. They also troubled Southern California, which lay much
nearer, and more vulnerable, to the erupting chaos.

It was in some ways an old pattern. For three centuries, California and the Southwest had doomed the plans of European imperi-
alists. Most recently, the region exhausted the Mexican Republic's
effort to control its northern frontier. Vast spaces and difficult terrain left the region ungovernable. Towns and military forts were
insufficient. Independent Indians like the Comanche, Ute, Navajo,
and Apache controlled their homelands despite Mexican claims.
Empowered by environmental resources and horse/firearm technology, they evaded and repelled incursions. They isolated Hispanic
colonists, harried commerce, and seized goods and captives. Settlers at Los Angeles (established in 1781), along with those at Santa
Fe (1610), El Paso (1680), Goliad (1747), and Tucson (1775), found
themselves at the mercy of fearsome neighbors. From the vantage

of these *norteños*, Mexico scarcely existed, and mostly to demand tax payments.[1]

US expansion took advantage of Mexico's predicament. Yet it inherited much the same. The rapid influx of colonists and soldiers created a two-thousand-mile-long front of indigenous rebellion. This spanned during the 1850s from California's North Coast, to Puget Sound, to Utah, to the Rio Grande Valley, and west again to the Mojave Desert. Indian revolt flared along the Oregon coast, across the Willamette Valley, and far up the Columbia River. The partition to create Washington Territory in 1853 spread this unrest as far as the Kootenay River and Bitterroot Valley (today in western Montana). Settler governments in the Northwest moved ruthlessly against the tribes. The worst violence earned distinct names: the Yakima War, the Rogue River War, and the Coeur d'Alene War. Numerous other engagements went unnamed. They differed in size, but not in blood.[2]

Edward Ord, in charge of a 3rd Artillery company, saw action in several Northwest Indian wars. In May 1856, he wrote Molly from "Big Bend, Rogue River" at Oregon's southwest coast to convey the madness and misery. Settlers set off a cycle of ambush, reprisal, and atrocity. Then, Ord told with bitterness, they expected the army to relieve them. But there was no relief. The officer observed how "indistinct fights" and free-fire combat in shadowy forests had killed—contrary to politicians' denials—innocent, peaceful Indians.[3] Victims included women and children who "stood but little chance." He shocked Molly, then three months pregnant, with the aftermath: "Squalling" babies cried. Native women wept "decrepit" over their dead. Ord watched them gather corpses and anoint themselves with tar and ash. They mourned for a world lost. Yet the women also engaged in "vice," so he disdained, with unruly troops of the territorial militia. Both groups, he scribbled, were "unworthy" of army sacrifices. "The more of each other . . . they kill, the better for all honest men."

Ord's sympathies were with his country. But they contained surprises. He wrote a report, which he asked Molly to forward to the *New York Herald*, on a group of Indians who feigned surrender, only to murder US soldiers and then meet death themselves. Ord despised such treachery. Yet his mind opened to universal truths. Settler militia, he noted, were "not as brave" and "with a worse cause" than Native Americans who fought to defend their homes. No matter, he thought; their downfall was inevitable. He hoped the army might pressure them toward an easier peace by isolating tribes from the river salmon. He led his company in burning the village and food stores at Mackanootney, twelve miles upriver from the sea. Edward, meanwhile, promised to bring Molly some "Indian curiosities." He also snatched up real estate, even as he condemned settlers for their land lust.[4]

Molly read her husband's letters with rapt attention. But similar brutality and fatalism occurred closer to home. California had its share of Indian wars through the 1850s. Violence broke out as hopeful miners took to the hills while others prospected farms, ranches, and timber. Along the Redwood Coast, between the Russian River and Cape Mendocino and east over the mountains, bands of Pomo fought to scare off colonists. In the Shasta foothills and on the rim of the Sacramento Valley, the Wintu, Maidu, and Yana peoples defended their lands and streams. Settlers kidnapped Native women and children, as state laws condoned Indian servitude and human trafficking. Farther east, near Tule Lake, the Modoc killed several dozen emigrants in the Bloody Point Massacre of 1852. Militia, in return, killed forty Modoc during a false peace parley. Meanwhile, conflict shuddered down the Sierra Nevada and spilled onto desert trails. Paiute took up arms throughout their homelands, from the Truckee River south to Owens Valley and Walker Pass. At California's southeast corner, the Yuma, Mojave, Hualapai, and Western Apache did, too. They fought with neighboring Indians. They raided wagon trains and US Mail. They seized or killed captives in events like the Oatman Massacre.[5]

A sensational depiction of the Oatman Massacre,
printed in 1858

Desert violence forced the army to garrison Southern California's frontier below the Sierra Nevada. In 1851, it located Fort Yuma
near the confluence of the Colorado and Gila Rivers, seven miles
above the US-Mexico border. Emigrants, and later, the Butterfield
Overland Mail, used an adjacent road to cross between Los Angeles, Tucson, Santa Fe, and El Paso. North of Los Angeles, within
Grapevine Canyon, the army established Fort Tejon. Founded in
late 1854, it protected travelers on the trail from Stockton and the
San Joaquin Valley. Three hundred miles due east, the army in 1859
established Fort Mojave. It sat along the Colorado River, two hundred miles above Yuma, and guarded a wagon road to towns of the
upper Rio Grande. The US established Forts Yuma and Mojave in
response to uprisings among the namesake tribes. Fort Tejon had the
task of safeguarding peaceful Indians while subjugating others. Military posts began to set up the desert interior as California's field for
opportunity. Adventures and investors in San Francisco, Los Angeles, and San Diego looked to develop mines, ranches, beaver trap-
ping, wagon and riverboat service, and Indian trading operations.

Soon, that would take small parties into the Sonoran Desert, sparking new rounds of violence.

BY MID-DECADE, the US had made the West's endemic problems its own. It also carried new unrest to the region. The potential scale of crisis became clear in 1857, when relations collapsed between federal officials and the government of the Utah Territory. Ten years before, the Church of Jesus Christ of Latter-day Saints (LDS) had fled religious persecution in Missouri and Illinois to resettle near the Great Salt Lake. Their population enabled Congress to organize the territory as a concession for California's statehood. Connections between the regions increased as Mormon merchants and homesteaders moved into Southern California. But religious intolerance and mutual suspicion arrived, too. Within a few years, US and LDS leaders began a standoff over Utah's autonomy and selection of its appointed federal officials. To prevent an insurrection, the War Department diverted soldiers from the Bleeding Kansas conflict.

War in Utah seemed imminent. It grew more so in October 1857 when raiders, thought to be a mixed force of Mormon settlers and Indians, ambushed a California-bound wagon train. They killed 130 men, women, and children at the site once known as Vegas de Santa Clara, on the Old Spanish Trail. Most died after surrendering. Newspapers called it the Mountain Meadows Massacre. Some speculated that church leaders had ordered the attack. The 1853 murders of the Gunnison railway survey party, seventy miles to the north, remained unsolved. California's press stoked outrage with tales of LDS impunity, of bodies left unburied, to be torn by wolves, and of women and child survivors held in polygamy and slavery. Federal officials took note. In the spring of 1859, they dispatched a special cavalry unit from Fort Tejon. Its commander, James Carleton, had instructions to rebury the victims and investigate local complicity.[6]

The so-called "Mormon War" never started. Nonetheless, it was a sign that long-standing US political conflicts had found fertile ground. This included the sectional crisis over slavery. With settlers arriving in California from all parts of the republic, the state became a microcosm for the nation's political divides. Those hailing from the South continued to favor slavery, its westward expansion, and its sanction under law. Some even brought African American slaves, despite prohibitions in the state constitution. Other settlers, regardless of origin, grew more amenable to slavery as they faced California's high labor costs. This was especially true in the peripheral counties—around Los Angeles, San Bernardino, Santa Barbara, San Diego, and the upper Sacramento Valley. Here, the labor shortage was acute, money to pay wages did not exist, and an Indian and Hispanic underclass abounded. Many Southern Californians took up peonage and child apprenticeship, practices that dated to the Spanish period. They also contracted laborers from Latin America and China. Or they indentured penniless convicts from judges and sheriffs.

Despite being a free state by law, California internalized slavery both as practice and as its defining political issue. Democrats dominated the state government through much of the 1850s. When they did not, it was only because of their internal squabbles (between William Gwin and David Broderick, most of all). For a time, Southern colonists controlled the party, allowing them to place allies on the courts and in federal offices. Appointees included Edward Ord's brother and father-in-law, New Town investors Benjamin Wilson and J. Lancaster Brent, and Senator Gwin himself. They (and critics) called themselves "the Chivalry," or the Chivs. Some favored no limits on the spread of slavery, a view that triumphed in March 1857 with the US Supreme Court's *Dred Scott* decision. Others championed Stephen Douglas's doctrine of popular sovereignty, which allowed local voters to choose. Both positions undercut the Compromise of 1850, the very basis of California's statehood. Similar tendencies prevailed in Utah and New Mexico, which codified slavery of African Americans in 1852 and 1859, respectively.[7]

The expansion of proslavery politics through the Southwest intensified counterefforts by antislavery Californians. Together, the two sides drove dissension to new heights. But it was hardly a contest. A majority of California's voters rejected the Republican Party in two presidential elections. They did so in 1856 despite the Republicans' nomination of favorite son John Frémont. Four years later, Democrats won an even higher percentage of the presidential vote. Candidate John Breckinridge received pluralities in Southern California. Together with the party's second candidate, Stephen Douglas, Democrats would have won the state. Instead, Abraham Lincoln claimed California's four electoral college votes, but only because Democrats split their ballot with such enmity.

CALIFORNIA'S SECTIONAL CONFLICT spilled eastward into the desert and across the Mexican border. During the 1850s, the state was a staging ground for plots to expand US territorial claims and slaveholding. This, in one regard, continued old habits. The state originated, like Texas before it, in the opportunistic migration of Anglos to a Mexican province, followed by their rebellion (with the help of local Spanish speakers). By nature, Gold Rush-era migrants were freebooters. Many envisioned the American republic as perpetually growing. Californians led the effort to develop a proslavery Territory of Arizona in the Gadsden Purchase. Senator Gwin supported this, along with acquisition of the Hawaiian Islands and the Northwest coast to Alaska. Others hoped to take more of Mexico. In the fall of 1853, filibuster William Walker sailed from San Francisco with a small army to conquer Sonora and Baja California. Newspapers followed his unsuccessful six-month campaign (and subsequent trial in federal court). Henry A. Crabb, California legislator and witness in Walker's trial, attempted his own capture of Sonora in 1856. By this time, Walker had left to build a proslavery Central

American republic. Both would-be conquerors died. Yet their ideas lingered. "Let Sonora, and as much of Mexico as may be necessary be acquired by fair and legitimate means," the *Los Angeles Star* proclaimed. Editors denounced the filibusters, but only because they had failed, thus delaying the desired result.[8]

Conquest and rebellions haunted California. Its settler imperatives, combined with Indian resistance, sent disorder far into the interior. The desert became its hinterland, one of dreams, but nightmares as well. Unrest might spread into the state. Or homegrown chaos might explode outward to consume the larger region. Both fears would shape how Californians understood 1861 and the early moments of the Civil War. Some, with keen eyes and ears, sensed the foreshocks. James Carleton spent the 1850s learning to make war in the Southwest. Phineas Banning, meanwhile, learned that the coming fight would be impossible without the humble movement of his mules and wagons.

Chapter 16

BANNING SEES AN OPPORTUNITY

I N SEPTEMBER 1858, A CROWD GATHERED AT THE SHORES of New Town. They were guests of Phineas Banning, the young proprietor of a freight and landing business. He called the audience to witness an auspicious moment, or so he hoped. They eyed a distant column of smoke belched by a tugboat as it towed barges carrying cargo and passengers. The flotilla approached from across the bay and up the channel to Banning's wharf. The *Los Angeles Star* recalled that "all hands, the ladies included, assisted in hauling the hawser" (the ropes used to tether vessels beside a dock).[1] The arrival set off celebrations, music, speeches, and toasts, which lasted into the next day. It hailed the entry of commercial shipping inside the estuary.

A more sober group might have noted the event's prematurity. Nature, not industry, prevailed over mudflats and marsh. The estuary's shallows were barely passable, even at high tide. Banning hid this bothersome fact by loading his barges with boxes that were empty or half-full. By this ruse, they floated high in the water. Meanwhile, an endless plain of grasses, ciénegas, and sparse trees spread inland. The small pueblo of Los Angeles, with four thousand inhabitants and not a single bank, lay almost invisible, twenty-five miles off. Undeveloped land and sea dwarfed New Town's dozen structures. It was no place for commerce.

Nevertheless, newspapers played up the milestone. Banning's wharf was a glimpse of the future in a region eager to leave the past behind. It marked an improvement over the difficulties of loading and unloading ship cargo—lighterage, in technical terms—through the shorebreak at old San Pedro. Editors called for a county road to move goods between the wharf and Los Angeles. Proponents scored an additional stroke of luck. One week after Banning's demonstration, a hurricane passed over Southern California—one of only two known instances in which a tropical cyclone reached the area from lower latitudes. The October 1858 storm damaged ships and piers along the open shore. But New Town's wharf and the few vessels *inside* the estuary stood unscathed. Protected by the sandspit of Rattlesnake Island, they rode out high winds and heavy surf.[2]

The lessons of that day had begun to form in Banning's mind long before. His association with San Pedro Bay started in 1851 by chance and good fortune, precisely what so many migrants sought in Gold Rush California. A merchant in Philadelphia sent him as escort to a shipment of goods. Its purchasers, David Douglass and William T. B. Sanford, enticed him to stay with a job offer. Banning worked as their clerk, learned to drive stagecoaches, and soon became a manager. After one year, he bought their wagon operation. Banning remained tied to Sanford, whose sister Rebecca he would marry in November 1854. Through business, he met local landowners and visiting federal officers like George Davidson and Edward Ord.[3]

Banning was not an original member of the New Town syndicate. These were Sanford, J. Lancaster Brent, and Benjamin Wilson, along with pharmacist John Downey and Henry R. Myles, a drugstore proprietor and Wells Fargo & Company agent. By 1857, Manuel Domínguez had joined them as a coinvestor. But, given his role as wagon provider, Banning was involved in the scheme from the start. As of late 1854, his enterprise included forty wagons and fifteen coaches. Soon, his horses and mules pulled to mining camps at the Kern River, near Walker Pass, and east to the Colorado River. Banning also attempted a line to the Salt Lake. He won a US contract to supply Fort Tejon, the new army post one hundred miles (and fourteen hours) north of San Pedro on the San Fernando Road.[4]

*Phineas Banning—master
of spectacle, army freight
contracts, and San Pedro
Bay shipping*

In 1858, when Banning officially gained his stake in New Town, he brought what no original member could. He provided the means to connect ship with wagon, the sea with far-off inland points. The venture speculated in these geographic connections as much as in land. Syndicate members traded up the coast to San Francisco. Wilson, Myles, and Sanford each had tried to establish wagon lines. But none wished to reenter the market, which by mid-decade became crowded and competitive. The investors required a partner willing to shoulder this risk. In exchange, they offered New Town's most strategic parcel: thirty-five acres adjacent to the estuary's channel. This became known as "Banning's landing."[5] Here, the entrepreneur built his wharf. Banning also held shares in the development's larger, undivided (that is, land held in common) portion.

NEW TOWN NOW HAD transportation. But it required more. Investors used their political connections to secure a wagon-grade road to

Los Angeles, paid for by county subsidy. Domínguez, who recognized the benefits, allowed the road to cross his estate. That was the easy half. The syndicate also required federal endorsement. Without a government chart of the estuary, merchants and insurers could not be confident that cargoes would safely reach New Town, or that it was a harbor at all. And the Coast Survey resisted such a request. Superintendent Bache and George Davidson distrusted speculators as a rule. Davidson participated in the study of San Pedro Bay in 1852. By mid-decade, he classified its outer waters as roadstead—that is, partially protected anchorage. He utterly dismissed New Town's so-called harbor. Davidson cautioned Congress and sea voyagers, who ultimately chose where to drop anchor, that the bay had significant demerits.[6]

Even as Banning debuted his wharf to the public, New Town's backers struggled to win over the Coast Survey. In November, they placed an editorial in the *Los Angeles Star* that accused the USCS of neglect. The piece attacked the Survey's Channel Islands work as a waste of money. It criticized Davidson by name. *Star* publisher Henry Hamilton was a voice for boosters and Chivalry Democrats of the Brent-Wilson camp. Most interesting about the editorial was a minor detail. Authors proposed a federal harbor improvement, and provided their own design: a breakwater made of wooden boxes filled with sand. Although they claimed no engineering expertise, they were confident the breakwater would open the estuary's entrance and be a "work of easy accomplishment." At the time, the plan's origin remained a mystery. A decade later, when it reemerged, the answer would become clearer.[7]

A copy of the *Star* editorial reached Bache in Washington, DC. He, in turn, sent word to Davidson at the field station on "Table Mountain" (today's west peak of Mount Tamalpais), north of San Francisco. Here, Davidson braved cold temperatures and high winds as he worked to position San Francisco, Point Reyes, Bolinas Bay, and the north Farallon Islands. The scientist answered Bache by deriding the *Star* as a provincial "curiosity" and those who placed the letter as "petty and rival interests about San Pedro."[8] The authors withheld their names. Yet Davidson knew who they were. He promised to "place the matter right."

From his rarefied mountain, Davidson penned a rejoinder to the *Star*, which the paper published in February 1859. He disputed the charge of neglect with a litany of USCS accomplishments. And he repeated his doubts about the estuary ("lagoon," he wrote, in quotation marks) and the dangers that rendered it "impracticable."[9] Above all, he offered a spirited defense of government science against the corrupting tide of self-interest. "*No public work on this coast is conducted with as much economy as the Coast Survey*," he exclaimed. "But there is not an officer in any department who does not know the extortionate demands made upon him, when acting in an official capacity." As for private citizens who hoped to capture taxpayer money, Davidson decried, they "openly boast of their determination to make as much as possible out of the Government. They consider it the legitimate subject of plunder, and chuckle when successful." Davidson saw purity in the Survey's quest for knowledge, and, no doubt, in his years of monastic suffering. He grasped the big picture of the Pacific coast, which the government had to weigh and balance in fairness. New Town, after all, was but one scheme among many.

The scientist believed he had settled the dispute. But before his letter made it to press, he had the high ground pulled out from under him. The *Star* published Davidson's words. Alongside, however, it printed news that Superintendent Bache *already* had approved a new chart of San Pedro Bay. The paper followed this with an anonymous letter—by someone aware of Bache's decision—in praise of the growing harbor at New Town, or "Banningville." Somehow, Bache had a change of mind. He instructed James Alden and the steamer *Active* to make the soundings that summer.[10]

WHAT ULTIMATELY CHANGED the Coast Survey's stance? Banning's September 1858 publicity stunt, for all its overstatement, certainly helped. Yet he built up another means of persuasion: War Department and military demand. And this weighed heavily on the superintendent. By 1858, Banning was the leading army freight contractor in Los Angeles. His mules and teamsters pulled cargoes from old San Pedro to the outlying posts of Southern California: Fort Yuma and Fort Tejon. In 1859, Banning began to supply Fort Mojave. His service gained a reputation for speed and reliability. Even when not, he offered "liberal terms" like lighterage, extra wagons and animals, and hauling of firewood or water at no added charge.[11] Banning earned the confidence of army commanders: John W. Davidson and James Carleton at Fort Tejon; Winfield Scott Hancock, assistant quartermaster at Fort Tejon and later at Los Angeles; and Thomas Swords, the Pacific Department quartermaster in San Francisco, among others. Banning hosted officers when they passed through. He allowed troops to camp on his seaside property, which was temperate and free of mosquitoes. He treated soldiers to refreshments. He became popular among the uniformed set.

US flags and federal dollars: Banning found his formula, one that no government scientist could refute. Of course, his supply contracts would have been far less without the rebellions inland. Native Americans, Mormon refugees, filibusters, and separatists roiled the West. Banning understood that the tumult increased military presence and kept his wagons rolling. Settlers and Indian reservations would follow. All required the freight lines and wharves of enterprising patriots. His business model stood upon an assured foundation, bloody and idealistic at the same time.

By this calculation, Banning took the lead in shaping the estuary's history. Like Davidson, he saw the big picture. But it was a different vision, in which science—and perhaps government and the army, too—operated at the behest of private profit. He learned to wield this vision from and for New Town. It was his reputation that retained army contracts. Whenever new ones became available, military officers recommended Banning by name. There is evidence that some officers gave him preference, in breach of regulations. Banning's winning bid to supply Fort Mojave was, in fact, the highest bid, as one officer later marveled. By 1860, Banning lost the contracts for Forts Yuma and Mojave to George A. Johnson & Company (predecessor to the Colorado Steam Navigation Company), which offered slow, though cost-efficient steamboat delivery. War soon enabled him to recoup the loss.[12]

Bolstered by the army's needs, Banning deflected criticism of New Town and its quixotic seaport. At the very moment Davidson mailed his letter to shame speculators, Banning played host to the 6th Infantry and 3rd Artillery. When soldiers arrived in the desert weeks later, they named their temporary base "Camp Banning."[13] Davidson, steadfast but nearly alone on Mount Tamalpais, had neither the means nor the inclination to counter this.

The USS *Active* arrived on May 25, 1859, to sound San Pedro's outer waters and those within the estuary's channel. To celebrate, Banning devised another spectacle. On June 9, he led New Town's associates, plus fifty guests, on a pleasure cruise to Catalina Island. Banning's steamboat, the *Comet*, paddled from the wharf, scraped

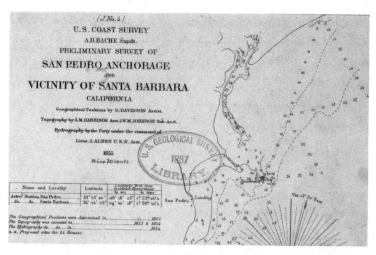

A Coast Survey chart of the San Pedro anchorage, 1855. It does not depict the estuary as a harbor, but Banning and his allies would change this.

over the sandbar, and crossed into the deep. Here, passengers boarded the *Active* for a sixty-mile voyage to the island and back again. Alden's crew served them complimentary breakfast and dinner. It was a sign of Banning's knack for enlisting government to his purposes.[14]

The Coast Survey issued its new chart of San Pedro Bay—New Town included—in September 1859. The *Star*'s editors announced the news on the one-year anniversary of Banning's first, publicity-seeking shipment. They credited several members of the Coast Survey, but not Davidson, even though the map relied on his geodesy. By this time, Banning had expanded his waterfront to include a storage house, yards, and an 1,800-foot railroad (pulled by mules). Observers noted busy smokestacks and, with exaggeration, predicted that the town would possess "all the necessaries": grocery, hotel, saloon, blacksmith, bowling, and billiards. By 1861, Banning's fleet had twenty-three freight wagons, seven stagecoaches, five lighterage boats, three small steamers, two hundred mules, and one hundred horses, with up to 150 employees. His teams pulled to California's Great Valley, the Sierra Nevada, and deserts beyond. A new road eased travel to Los Angeles. Lumber arrived at his warehouse from the Humboldt and Oregon coasts.[15]

Against powerful doubters, but with more potent friends, Banning made the region's commerce his domain. He would not bring a fully-loaded, seafaring vessel up to his wharf until January 1861—two years after he had declared New Town open for business. But here again, Banning's timing was opportune. The Civil War would bring unprecedented government dollars to the bay and its new master.

Chapter 17

BECOMING THE DESERT SCOURGE

F ROM THE COAST, WAGON WHEELS CUT TRAILS TO THE
interior. This, too, was not yet a place for commerce. Instead,
horse-bound riders and violence permeated California's desert
periphery. James Carleton of the US Cavalry inhabited this troubled place and conformed to its punishing landscape. He fought its
various peoples. He learned their unconventional modes of warfare.
He studied the environment as weapon and shield. Southwestern
chaos—nature and human—became his enemy and his tutor. And
he vowed to bring it to order.

Of Carleton's several legacies, he is remembered most for leading the army's removal and internment of the Navajo between 1863
and 1868. But that episode marked the last of his six decades on the
frontier. Carleton's life began in Maine. Here, boundary disputes
had lingered since the War of 1812. During that war, a British offensive forced the Carleton family from their home on Moose Island,
near the Bay of Fundy. James was born in 1814, just as the conflict
ceased. Life was trying in the far Northeast. Nature, for much of
the year, was a cold mess of mud and ice. Atlantic storms pummeled
Maine's rocky headlands and pine-fir forest. The international border remained in flux two decades later.[1]

As a young man, Carleton hoped to escape from this bleakness.
He dreamed of succeeding as a writer in the New England tradition.
He practiced his craft and the language of Yankee Protestantism,
which lionized labor, thrift, self-reliance, and reform. Yet the fron-

tier's instability did not let him go. Unsettled claims nearly renewed
war with Britain. Carleton joined the militia and discovered his apti-
tude for military life. Superiors recommended him to the regular
army. In the end, Carleton left home to train as a cavalry officer.
That did not curb his intellectual bent, which continued in search
of an outlet.

IN THE WEST beyond the Mississippi, Carleton found new purpose.
Here, the army patrolled an internal border with Native Americans.
It guarded settlers, missionaries, and traders and proudly supported
US expansion. It reclaimed captives and African American slaves
"stolen" by the Comanche. It also monitored reservation tribes, such
as the Cherokee and Sauk, recently removed from the East. These
indigenous refugees became pioneers in an unfamiliar land. Their
plows, in some locales, first broke the prairie soil. Each group faced
assault from multiple corners. The cavalry rode in between, even as it
favored settlers overall. Carleton entered this bewildering zone while
training as an officer at Fort Gibson, on the Arkansas River (in what
is now Oklahoma), and Fort Leavenworth, on the Missouri River
(today's Kansas). He saw nobility in the work, particularly in protection of reservation Indians—now considered federal wards. This,
he believed, made it possible to uplift them into peaceable, Christian
farmers. The prerequisite was their defeat. Carleton would pursue
both goals, no matter the context. His reform and military impulses
became one.

The Mexican War shifted Carleton's mission to the Southwest.
He fought notably at the Battle of Buena Vista in February 1847.
This surprise US victory would open Coahuila and Mexico's north
to occupation. The Mexican army, led by Antonio López de Santa
Anna, had entrapped its smaller enemy and shattered its left flank.
Carleton's company of dragoons—the name given to light cavalry—
raced to secure the breach. They rode over ridges and ravines to shield
a counteroffensive by the Mississippi Rifles. At the head of this well-

US dragoons in Mexico. Carleton's daring in battle alongside Jefferson Davis would earn him a reputation by the 1850s.

armed infantry was a planter, congressman, and recommissioned army colonel, Jefferson Davis. Despite taking heavy casualties, their units stabilized US positions, allowing artillery fire to tear through the Mexican lines. Santa Anna had reserves enough to continue the battle. But, unexpectedly, he withdrew during the night.[2]

The US triumph at Buena Vista came days after George Washington's birthday, a coincidence that elicited strong passions. According to legend, generals Zachary Taylor and John Ellis Wool wept in each other's arms. They believed they had remade the fragile American republic, the founders' experiment, into a permanent nation. That glory came at the expense of a sister republic. But in the moment, mythmaking prevailed. Carleton and Davis emerged as heroes, along with their commanders. Newspapers praised the pair in the same column. Superiors and subordinates did in the same breath. Taylor honored Davis and his troops with a parade near Monterrey before sending them home. Davis, who had been wounded by a musket ball, returned to acclaim in Mississippi and then in Washington, DC.

Carleton, meanwhile, hitched his fortunes to the memory of Buena Vista. The cavalryman came away with a double promotion in rank and the insight that small mounted units could fight over difficult terrain and devastate the enemy. He would apply that lesson through the remainder of his career. But first, he wrote a 250-page history of the battle, which Harper & Brothers published in 1848. The work displayed self-taught literary skill on par with that of the best West Point–educated officers. Carleton understated his own exploits, but praised Davis's as "gallant," "glorious," and "pure gold."[3] The book added to the Mississippi senator's reputation. It also burnished the name of Zachary Taylor, whom voters soon elected as US president. Carleton had impressed himself on suddenly influential figures.

While Davis and Taylor embarked on political careers, Carleton brought his experience to annexed New Mexico. He spent the years between 1851 and 1856 stationed at Albuquerque or outlying forts. Here, the army moved to halt Indian raids on civilians and pastoral tribes. Carleton blamed the raids for numerous ills: poverty, vice, lawlessness, and morally degraded soldiers. In his mind, raiding blocked Anglo settlement and kept the territory in a backward state. He crusaded against Native marauders, criminals, and drunken troops alike.

Carleton became known as an expert on desert tactics. He noted the horse warfare practiced by the Navajo, Comanche, Jicarilla Apache, and others. They struck in small groups with quickness and stealth, then evaporated into the terrain. The same environment, or rather an insufficient knowledge of it, caused US forces to stumble. Distances, extreme heat, and an absence of water or shelter—most especially shade and forage grass—limited the reach of dragoon horses. The landscape became Carleton's additional adversary. It had harrowing place-names like the Llano Estacado ("staked plains"), Sangre de Cristo ("blood of Christ"), and Jornada del Muerto ("journey of the dead"). To combat twin enemies, he hired indigenous scouts and trackers. He sought help from veteran traders like Kit Carson. And Carleton pondered logistics—that is, the challenge of supplying forces over ground that gave so little.

It was not long before army brass and politicians came calling. In 1856, Secretary of War Davis asked Carleton to participate in ambitious reforms of the military. He sent the officers Richard Delafield, Alfred Mordecai, and George B. McClellan to study the Crimean War, which had begun three years earlier. The commissioners were broadly knowledgeable. Yet Davis recruited additional experts, individuals he sought to advance or who offered specific skill sets. Edward Ord was encouraged to travel to Russia, perhaps to observe artillery. He declined. Carleton received an invitation to review the Delafield-Mordecai-McClellan reports. He accepted and relocated to Pennsylvania with his wife, Sophie, and their children. The slaughter in Crimea showed the need to rethink cavalry and its uses. New battlefield weapons, like accurate rifles and cannon, proved immensely destructive to horses and riders who, by tradition, charged an enemy's lines with sabers, lances, and pistols. The reports would advise discontinuing the use of cavalry as an attacking force in pitched (or set-piece) battle.[4]

The West was the exception. Davis saw it as the one place where cavalry attacks could continue if innovated. Carleton's input was critical. With great interest, he reviewed McClellan's observations on the Cossacks—the semi-autonomous, horse-based bands of the Caspian steppe. The Russian Empire used them as elite cavalry, providing specialized training, modern weaponry, and Christian education. Carleton believed the US might enlist Native Americans in similar fashion. Davis, who served as a frontier dragoon in the 1830s, agreed. The war secretary's interest in Western rails and slavery required taming the Southwest as quickly as possible. It required, too, proxies drawn from subject peoples as much as from army officers and spoils seekers.[5]

After a year's sabbatical, Carleton returned to the desert. Now, Southern California's frontier became his base. Davis had left the

War Department by this time. Yet his influence lingered in projects like the Camel Corps, which imported from Ottoman Egypt seventy-five animals. The army sent some camels and two handlers to Fort Tejon. They worked alongside dragoons to improve long-distance supply, including to Camp Cady, a forward post along the Mojave–Salt Lake Road (near today's Barstow). Carleton served at both stations. He became fascinated by the Tejon Reservation, with its mixed tribal population, farms, and livestock, as a model of reform. Residents, he pondered, might someday provide Cossack-type units.[6]

Notions of uplift, which were humane in Carleton's mind, coexisted with ideas that were cruel and cutting. Few Native groups gave up old ways voluntarily. In practice, the army existed to subjugate tribes, while utilitarians like Carleton proclaimed that they had averted a greater evil. The officer's attitude toward Mormon settlers showed similar conflation. In early 1859, Carleton led an operation to investigate the Mountain Meadows Massacre. The troopers' presence heightened tensions between the US and Latter-day Saints. That was fine by Carleton. He made it known that he came with no olive branch. At the site of the massacre, the officer ordered troops to erect a wooden cross. It bore the inscription, *Vengeance is mine; I will repay, Saith the Lord.*[7]

Carleton chose the Biblical (and Book of Mormon) verse as a warning to the murderers. Yet the words were a statement, too, about the region's contested power, its bloody history, and Carleton's conviction that he was a seer of order to come. The Southwest claimed by the United States remained a place of perilous consequence for any aspiring nation and empire. To change this required a soldier willing to match peril against peril. Carleton was the purifying scourge. Some enemies, however, were more vulnerable than others. Mormon suspects, whom Carleton blamed, remained at large and shielded—so he believed—by Utah's government. But independent Indians, whom the officer continued to view as the crux of trouble, had no protection beyond their own. His dragoons punished the tribes between southern Utah, the Colorado River, and Cajon Pass.

It would not be the last time he used a fight among Anglos as pretext to bring destruction to indigenous homelands.[8]

WHEN THE CIVIL WAR ERUPTED, Carleton was in the midst of an Indian campaign. Secession and disunion intruded into his world and immediately diminished it. The nation's fate no longer seemed to lie in the desert. Instead, the impasse centered on Washington, DC, and a few harbors, cities, and transportation corridors. That caused a drawdown of federal resources in the West, at least initially. During the spring of 1861, the War Department shut down Southern California's frontier installations, including Fort Tejon and Fort Mojave. It prepared to withdraw troops from Fort Yuma as well. Officers were needed in the East to train the mass of volunteers. Army regulars marched to Benicia, Los Angeles, and San Diego, where they waited to ship out for the Atlantic coast. It was likely that Carleton would return to New England, where he had not lived for two decades.[9]

The desert was abandoned. But not for long. Secessionist feelings took hold in the Southwest. Following the exit of Texas on February 1, armed groups in New Mexico seized plazas and public buildings. Settlers living in the Gadsden Purchase, between the lower Rio Grande and Tucson, organized themselves into a breakaway rebel territory. The majority Hispano and Indian populace remained indifferent. They had lived under US rule for barely a dozen years, and before that under Mexican authority for two dozen. In their view, flags and political claims shifted like the wind. Allegiances were neither rushed into nor revealed too soon.

Not every New Mexican understood that wisdom. Several army commanders surrendered their federal property, firearms, and supplies. Others joined the rebellion. One was Louisiana-born Henry H. Sibley. As of 1861, he led New Mexico's 1st Dragoons, Carleton's former command. But Sibley resigned, evaded arrest, and escaped to Texas to take charge of rebel units. He briefly detoured to Richmond,

Virginia, where he counseled with Confederate president Davis. As a sign of things to come, Sibley renamed his contingent the Army of New Mexico. He sent word to officers that they would march "on to San Francisco."[10] Their anticipated path ran through Southern California.

A Southwest invasion began in July 1861, when a regiment of three hundred Texans crossed north from El Paso into New Mexico and up the Rio Grande. They advanced quickly through the small, dispersed towns and captured US installations, except for Fort Craig, a lonely outpost west of the river. The remnant Union Army, which had stood as a triumphant force of occupation only months before, now found itself trapped, dispirited, and running out of time. On August 1, Confederates pronounced the lower Rio Grande Valley below the 34th parallel to be the Territory of Arizona. Sibley arrived and took command of the invasion in December. Before moving on the Pacific, he first aimed to capture northern New Mexico's wealthy towns and the gold mines of Colorado's front range. Sibley sent a single company 250 miles west to capture Tucson, hub of the Sonoran Desert and Santa Rita-Patagonia mines.

Past conquerors faced a new invading force. Fifty-two-year-old

Tucson had been part of the US for only eight years when Confederates captured it in 1862.

Kit Carson stood among the ironies. As a young man, he had fled an apprenticeship in Missouri and taken to the Santa Fe Trail in what was then Mexico's north. In further violation of borders, he guided Frémont's expeditions across the Rocky Mountains and Sierra Nevada. Carson then fought in the US-Mexico War. He returned to the Southwest in the 1850s to work as an Indian agent and scout. He became proficient in Spanish, plus a number of indigenous dialects.

Carson lived as a vanguard of United States imperialism. Yet 1861 reversed that role. Early that year, he witnessed secessionists—fellow Anglo pioneers and Southerners like him—attempting to pull down the US flag at Taos. Carson dispersed the mob, reportedly by threatening to shoot. Even so, Confederates advanced up the Rio Grande. And this caused Carson to look upon former friends as enemies, and vice versa. Such were the cords of memory in an unpredictable land. He raised a Unionist regiment, the 1st New Mexico Volunteers, mostly made up of Hispanos and Pueblo Indians. The multiethnic force later regrouped with army regulars at Fort Craig, under the command of Kentucky-born Edward R. S. Canby.[11]

Reversal occurred, too, on the opposite side of the desert. Southern Californians knew the interior was an open and defiant space. Fort Yuma, at the Colorado River, provided the only military obstacle to a Confederate march on the Pacific. With supreme irony, Unionists put their faith in other barriers: hostile tribes and the severe landscape itself. Wild rumors reached them of Confederate victory in New Mexico and of invaders amassing on the state border. Many feared that secessionists already had infiltrated. Los Angeles's proslavery, Southern immigrants emerged as the foremost concern. Chivalry Democrats of the Brent-Wilson circle had prevailed in recent elections. The Republican Party and Lincoln government remained unpopular. The *Los Angeles Star* printed seditious articles, while Confederate sympathizers gathered openly in Main Street saloons. In May, would-be rebels raised a separatist Bear Flag and paraded through "the Monty," a settlement fifteen miles east. The *Daily Alta* labeled it the most disloyal community in the state. Like the people of New Mexico, Angelenos watched their neighbors closely. They saw old political leanings in different, ominous shades.[12]

The attack on Fort Sumter made secessionists more brazen. They formed militias, which drilled in public and threatened to march on the pueblo. Some citizens renounced their allegiance to the United States. They might have noticed that the Stars and Stripes were absent from Los Angeles's plaza and courthouse. (When news of Fort Sumter arrived on April 24, someone wisely lowered the banner before secessionists could capture it.) All of this set off talk of impending insurrection. Local quartermaster Winfield Scott Hancock stayed up through the night to guard the army's adobe storehouse, located at Third and Main Streets. His wife, Almira (who considered herself a "pretty good shot" with a Derringer), did, too.[13] At last, federal soldiers arrived on May 14. The contingent of fifty dragoons rode in from Fort Tejon, led by James Carleton. It was the first garrisoning of Los Angeles since the US-Mexico War.

Angelenos believed that a new war of conquest had begun. They were correct, but only in part—war had never ceased. Restless imperial designs had long ago trespassed into the Southwest and remained unfulfilled. Stalemate connected patchwork conflicts, past and present, and favored those who understood. Carleton certainly did. And he found himself in an opportune position to shape the region's Unionist defense. He had battles new and old yet to fight.

Chapter 18

UNION AND DISUNION IN LOS ANGELES

THE LOS ANGELES UNION CLUB HELD ITS FIRST RALLY on May 25, 1861. James Carleton and Winfield Scott Hancock of the US Army led a small crowd in patriotic fêtes. Carleton had arrived with his troopers ten days prior. Reinforcements from Fort Tejon and Fort Mojave followed. Their presence steadied Unionist civilians enough to reclaim public space and to hoist the US flag once again over the old Spanish plaza. To present the colors, the Union Club chose a well-liked business owner from San Pedro Bay. He was a Democrat with known Southern ties and associates. Phineas Banning stepped to the flagpole and into the Southwest's Civil War.[1]

Unionists hoped their show of loyalty would turn the tide against disunion. Instead, secessionists gained momentum. Three more states—Virginia, North Carolina, and Tennessee—would join the Confederacy by July. The federal army's defeat at Bull Run that month stunned Unionists to the point of despair. Officials at Washington, DC, contemplated evacuation. Newspapers pronounced the end of the American republic. Attention now focused on the upper tier of slave states, Maryland and Kentucky most of all, which had yet to declare allegiances. Indecision in these "border states"—so-called for their location between North and South—reflected divided views. Compromisers and fighters, Unionists and disunionists, all vied for influence. As Banning and Carleton stood at the plaza, the direst stages of the secession crisis had yet to begin.

The same was true in Los Angeles. The national emergency

compounded the region's troubles as border state and borderlands. Much like in Kentucky or trans-Allegheny Virginia (soon to be West Virginia), Californians split their loyalties. Communities became torn by the "demon of suspicion," as San Francisco's pro-Union *Daily Alta* put it.[2] Authorities maneuvered to keep order ahead of state elections in September. But each time the national government faltered, secessionists grew bolder. At the same time, discontents surged forward from a history of conquest. Southern California faced potential rebels far beyond avowed Confederates—the usual suspects in border slave states—or even the Chivs, who dominated politics. Unionists' fears soon encompassed an overwhelming share of the Southwest's population, and especially its Hispanic, Native American, and Mormon peoples. These groups might join with the Confederacy. Or they might cooperate in insurrection and then go their separate ways.

FEAR AND ANXIETY had their origin in the region's paradox. Conquest had established US claims to Southern California. But it had left behind turmoil and disaffection. The *Alta* marveled at how, around Los Angeles, "foreigners" and "native Californians" continued to outnumber "Americans." The pueblo remained an indigenous place where Gabrieleño-Tongva survivors comprised one-tenth of the population. Mexican and Indian workers were essential to the surrounding ranches and vineyards. They labored under various forms of coercion as the city closed communal lands and demolished adobe districts along with the nearby Indian village. Visitors felt the oppressive mood. Edward Ord wrote (with some sympathy) that the area's Mexicans, "Sonorians," and Indians suffered on the cusp of rebellion. Amid the secession crisis, many wondered if that cusp would hold.[3]

Yet the question ran further. As racial and ethnic antagonisms reshaped civic politics, Californios grew estranged from Anglo elites and began to challenge their authority. The pueblo was plagued by violent crime through the mid-1850s. Settler vigilantes targeted Mexican, mestizo, and Indian defendants, while mobs on both sides threatened revenge. Ranchers blamed Anglo Democrats for this. Their grievances, more so than the issue of slavery, split Los Angeles and its ruling class by 1861. But Californios had further reasons to be angry. Most had yet to receive federal land patents. Drought, between 1855 and 1858, destroyed their grasslands and cattle. Their independence and influence slipped while Americans waxed about the blessings of liberty. Rather than grappling with complex social problems, Anglos resented the "inimical feelings" they ascribed to the entire "Spanish race."[4]

Most Southern California Unionists, the small number of Republicans especially, ignored this dispute. Instead, they looked to the historical record. Most ranchers—like Domínguez—had joined the Democratic Party and its Southern-leaning Chivalry faction. Together, they made for an invincible coalition. Los Angeles's multiethnic group of Democrats supported protection for "domestic institutions," a legal euphemism for slavery. They endorsed the *Dred Scott* decision and carried the region in the 1860 presidential election. Domínguez's adobe served as a precinct ballot site and reliably proffered votes.[5]

Did that mean Californios would take up the cause of secession? In a sense, they already had. The Pico Act, named for the assemblyman and rancher Andrés Pico, aimed to secede the lower counties to create a new US territory of "Colorado," perhaps open to slavery. The state legislature approved the bill in 1859 and forwarded it to Congress. Furthermore, ranchers like Pico had rebelled before. In 1846, they rose up and bloodied American troops. Based on presumed facts, the *Alta* reported in June 1861 that "Mexicans are holding secret meetings at every hamlet and rancho," waiting (by implication) for the opportunity to revolt.[6] Suspicions peaked after a wave of horse thefts. Californios claimed the animals were stolen. Unionists whispered that ranchers had supplied them to secessionists across the Colorado River. They slandered some Californios by name. Domínguez escaped public accusation, but not suspicion. For anxious, outnumbered Unionists, the region's history—its grievances and engrained distrust—gave meaning to a Civil War centered thousands of miles away.[7]

Similarly, Unionists saw local Mormons as borderlands rebels-in-waiting. Antagonism lingered from the 1857 Utah War that had never begun. Newspapers fanned hostility with tales of LDS secrecy, despotism, kidnapping, murder, and hatred of "Gentiles" (non-LDS), but polygamy most of all. They expected Mormons, as "sworn foes to the United States government," to seize any opportunity to throw off federal authority.[8] Concerns centered on San Bernardino, sixty miles (and six and a half hours by wagon) east of Los Angeles. The settlement, where Mormons comprised two-thirds of the population, occupied a strategic site near Cajon and San Gorgonio Passes. Several years before, at the height of the Utah crisis, newspapers accused San Bernardino's Mormons of plotting civil war and inciting Indian violence. This logic returned in 1861. Residents were said to insult the US flag, intimidate loyal citizens, and cheer the name of Jefferson Davis. Reportedly, they harbored secessionist fugitives and enlisted outlaws and nearby Spanish-speaking settlers in the cause.

Fears about disloyalty at San Bernardino signaled further alarm about the great desert beyond. The inland village of 1,500 persons served as gateway to the Salt Lake Road, as well as Yuma and the Colorado River. Adjacent to the latter was the Mexican state of Sonora, considered by Unionists a den of instability. By the 1860s, it, like California, had gold, agriculture, and settlements. Through its port at Guaymas, it provided prime access to the Pacific Ocean. US expansionists, slaveholders especially, coveted Sonora. American adventurers flocked there. Yet, even without them, the province epitomized the spirit of insurgency. Sonorans fought a civil war for two decades ending in 1861. (French military intervention reignited conflict the next year.) Unrest sent refugees to the mines of California and New Mexico, where they gained a fearsome reputation with

prospector's pan, frying skillet, horse, and pistol. Joaquin Murrieta was *Sonorense*. Other gold rush bandits were said to be. Anglos
dubbed Los Angeles's toughest Spanish-speaking district "Sonora
Town," regardless of the residents' place of birth. Confederates were
already present in Sonora, seeking alliance and material support.[9]

Unionists looked out from Los Angeles and imagined desert enemies all around. "We have open and avowed secessionists and southern sympathizers," the New Town investor Matthew Keller wrote
to army commanders, "who exercise most political influence with
the native population" (meaning Spanish speakers). "We are surrounded," he continued, "by barbarous and hostile Indian tribes,
that may at any moment be excited against us and the Government
by rebels or marauding Mormons."[10] And "we are threatened with
rebellion across the plains" by Confederate Texans. Keller's con-
glomerate of enemies, villains of past and present, summed up the
region's sectional and imperial predicaments. A motley group of
rebels—real and imagined—threatened to upend manifest destiny
and smash the American republic to pieces.

KELLER'S WORRIES may have come as a revelation to some, but
not to James Carleton. He had fought against the desert for a decade
already. In July 1861, as Unionists panicked, the army sent him to
San Bernardino to assess the danger. Not only did he advise that
troops garrison the inland settlement, but he argued that the whole
of Southern California was under threat and that the US could not
afford to withdraw its forces. His report found a receptive audience with General Edwin V. Sumner. Sumner was new to the Pacific
Department and knew little about local society, but the sixty-four-
year-old veteran understood risk. He had commanded troops in
Kansas during 1856 in the effort to halt its civil war. He was a second
cousin of the famed solon, Charles Sumner, beaten nearly to death
in the Senate for condemning slaveholders. And he served in Abraham Lincoln's security escort, which guarded against assassins on

the president-elect's rail journey to Washington, DC. Most important, Sumner had known Carleton since 1839, when the elder officer commanded the Cavalry School of Practice at Carlisle, Pennsylvania, and from their assignments in New Mexico. He trusted Carleton's judgment.[11]

Sumner and Carleton agreed that the Union Army should begin a massive buildup of force around Los Angeles. That marked a wholesale shift. At the start of 1861, the military had virtually no presence within a hundred miles of the pueblo. As outlying posts shut down, army regulars prepared to depart California entirely. Most did. Edward Ord, promoted to brigadier general, sailed from San Francisco in October. Sumner had planned to send Carleton as well. No longer. He organized a new military district of Southern California, with Carleton as its second-in-command.

The district's immediate task was to raise and deploy a force of volunteer soldiers. Units assembled near Lake Merritt, in Oakland. Once ready, they shipped to San Pedro Bay and inland to several encampments. Most important were Camp Latham, on rancho lands near Ballona Creek (and present-day Los Angeles International Air-

James Carleton, during the Civil War—ready to fight enemies old and new

port), and Camp Oak Grove (later called Camp Wright and relo-cated to the ranch of Unionist Juan José Warner) on the Yuma Road northeast of San Diego. By the first week of October, 1,200 troops were drilling at Camp Latham. By early 1862, two thousand soldiers occupied Los Angeles and San Bernardino, counties with a civilian population of seventeen thousand persons.[12]

The greater task would be to outfit the soldiers. The army had to source supplies on an unprecedented scale, far above the region's capacity to produce. It was a national problem. The Quartermaster Department in Washington, DC, cobbled together a system of private contractors. This proved more difficult on the Pacific coast, however. Distances, low population, and the absence of factories or railroads made a system overseen from the East impossible. In response, Quartermaster General Montgomery Meigs allowed the Pacific Department and Southern California District to tackle their own supply needs. As they did, San Pedro Bay became receiving point for a large volume of goods: uniforms and boots, plus rifles, tents, mess kits, bayonets, knapsacks, and bedrolls. The district also required massive amounts of food. One shipment in October 1861 consisted of forty thousand rations. At Fort Yuma alone, military animals consumed thousands of pounds of barley each day.[13]

Failures troubled the buildup from the start. Recruits drilled without firearms and uniforms. Without tents and blankets, they shivered and grew rheumatic. Cavalry—Carleton's forte—suffered particular shortages. Mounted troops required specialty items, like breechloading Sharps carbines, pistols, and horse tack. Most saddles in Southern California were Mexican-style *charros*, designed for roping cattle, not for combat. Officers found the region's Spanish-breed ponies too small, "unbroken," and easily spooked by gunfire. Faster, sturdier "American stock" Morgan or quarter horses were in short supply.[14]

Transportation itself was insufficient. The Pacific Department faced a shortage of mules that was entirely self-inflicted. Early in

1861, it needed to repay bondholders, who panicked as the secession crisis spiraled. To raise cash, the department announced a fire
sale on account of "the Government having more Mules in Southern
California . . . than is considered necessary."[15] In March, five weeks
before the shelling of Fort Sumter, the Los Angeles depot auctioned
off two hundred animals. Months later, when desperately in need,
the army no longer had them. And speculators asked "exorbitant
prices." Carleton and his district superior, George Wright, groused
in anger. Their only solution was to source the animals elsewhere
and add them to the list of cargoes shipped into San Pedro Bay.

Supply challenges mounted as the district contemplated the next
stage of its strategy. After completing the buildup and pacifying
Southern California, Carleton and Wright planned for a counteroffensive into the desert as far as the Rio Grande Valley. It was a distance of 860 miles, beyond anything required of troops in the East.
No large infantry had ever attempted a desert crossing. Carleton
thought their boots would disintegrate from scorching sand and
salts. Yet a more basic obstacle existed. An army of several thousand
soldiers would exhaust surface water, game animals, and forage
grass. It would exceed the landscape's capacity to support life. The
only solution was to land even larger cargoes at San Pedro and move
these inland by wagon. But the district owned no shipping facilities.
Worse, it had only fifteen wagons as of early October.[16]

The army would need to hire a private freight service. Phineas
Banning knew it and was auditioning already. In September 1861,
the first batch of California Volunteers began to reach Southern California. At old San Pedro, that would involve wading through tum-
bling waves, then hauling gear up the steep bluffs. But officer Joseph
Rodman West reported his fortunate experience at New Town.
"The command in disembarking," he wrote, "is most particularly
indebted to Mr. Banning."[17] He complimented the freight operator
for being "prompt, efficient," and "untiring," he explained, "to render every facility at his disposal for the comfort of the men and offi-

cers." One month later, Banning loaned West twenty-two wagons to ease the march to Warner's Ranch.

As the army searched out local partners, Banning had no equal. He owned wagons and a wharf plus his own fleet of tugboats and small steamers. He kept warehouses as well as a saddlery, metal forge, grain and feed mill, lumberyard, sawmill, and slaughterhouse. He enhanced his credentials by joining the Republican Party. When Carleton arrived in October, he declared Banning the "only person" capable of supplying the army's needs. His endorsement came at a critical moment. Days before, the War Department had recalled Edwin Sumner to Washington, DC. George Wright assumed command of the Pacific Department. Carleton gained full authority over the Southern California District and its supply contracts. Within weeks, Banning received his first of many. Soon, his wagons plied the roads between the estuary and destinations inland.[18]

Banning's indispensability reshaped the army's footprint and priorities. On his recommendation, the district selected New Town for a supply depot, barracks, and headquarters. It located the installation one mile back from the water, on lots donated by Banning and Benjamin Wilson. Banning's firm would build the facilities. He put his wharf and bayfront land at the army's disposal as well. Here, the Quartermaster Department constructed a storehouse known as San Pedro Depot. In January 1862, the army named the upland facility, Camp Drum, after the Pacific Department's assistant adjutant general. That officer would sign off on all shipments. It would be a test of his fiduciary responsibility to refuse requests from his namesake post.

Drum tried, at times. In April 1862, he arrived at New Town for an inspection. He found things suitable, but noted some irksome waste. The depot possessed no mules of its own. "Consequently," Drum wrote, "transportation for all stores landed on the wharf has to be hired to transport them to the warehouses, a distance of 200 or 300 yards."[19] In the first month alone, this cost the government $725—"enough almost," Drum figured, for the depot to have purchased the mules. It did not, of course. The animals were rented instead from the nearest source. Marginal cost showed Banning's skill at bending military practice to entrepreneurial gain.

UNIONISM IN CALIFORNIA was a courageous political stance. Yet it served as business strategy as well. Government spending, and the promise of inland commerce, provided powerful incentives, as changes amid New Town would show. At first, the syndicate seemed to lean toward secession. It certainly was not favorable to the Lincoln government. Within the group, only Banning gave full devotion to the Union. He abandoned the Democratic Party and his former Chiv allies at the *Los Angeles Star.* Yet Banning's army contracts soon brought most of his fellow investors into the patriotic fold. They, in turn, put him forward as the development's public face, its visionary. In 1862, syndicate members renamed it Wilmington, after Banning's birthplace in the Unionist border state of Delaware. They also subdivided the 2,400-acre parcel into city blocks, town lots, and twenty-four farm lots. Investors like Benjamin Wilson and John Downey, who by circumstance had become state governor in 1860, remained

New Town, renamed Wilmington, as subdivided in 1862.
Banning's reservation and wharf sit on the estuary waterfront.

Democrats and dubious of Lincoln and the Republicans. Nonetheless, they gladly accepted federal money and tempered their criticism to match.[20]

Like Banning, other California entrepreneurs would find lucrative opportunities through their Union loyalty. They included four Sacramento businessmen who would later redefine the estuary story. One of them, Leland Stanford, replaced Downey as governor at the start of 1862. He parlayed his office and Republican Party ties to benefit one of his investments, the fledgling Central Pacific Railroad. That summer, President Lincoln and allies committed, in the Pacific Railway Act, to fund the Central Pacific as a portion of the transcontinental railroad. The US government would pour massive amounts of cash and credit into Stanford's company to build eastward across the Sierra Nevada. Within ten years, Stanford and the Central Pacific associates would find their way to Wilmington before entering the Southwest.

Union and disunion had great power to change the history of San Pedro Bay, as well as that of the surrounding region. It advantaged some, while it disadvantaged others. The latter proved true for several persons previously tied to the estuary. Edwin Sumner had grown so convinced of the secessionist threat to Southern California that he sought to impose martial law and make preemptive arrests. He pursued this policy even as he departed aboard the SS *Orizaba*, where he arrested fellow passengers suspected of intent to aid the rebellion. These included J. Lancaster Brent, the Wilmington investor, and former Senator William Gwin. US officials worried that Southern politicians would travel abroad to seek diplomatic recognition for the Confederacy. The most famous incident would begin four days after Sumner made his arrests, when the US Navy intercepted two Confederate agents on the RMS *Trent* bound from Cuba to Britain. Apparently, Gwin was a suspect, too. During the voyage, Sumner perhaps noted a foreigner seeking the senator's ear. It was Mikhail Bakunin, the revolutionary Russian dissident. The white-haired General Sumner already feared a variety of rebels: Southerners, Chivalry Democrats, Spanish speakers, Mormons, and Native Americans. He would not stand for the American Union, a mere ten years older than he, to have enemies around the world.[21]

The republic's fate hinged on geographic connections—whether built by diplomacy, maps, or military supply. As a habitual empire builder, Gwin linked many places: from Mississippi plantations to California's gold mines, and thence to Washington, DC, and across the oceans. Among these places, Gwin once had designs on the San Pedro estuary, where, late in 1861, Union troops and cargoes amassed. But the estuary, with its potential imperial ties, was no longer his. It belonged instead to the army's desert scourge and his patriotic wagon purveyor. Together and at odds, the two prepared to strike out.

Chapter 19

WEAPONIZING
THE LANDSCAPE

As the Union Army concentrated near Los Angeles in late 1861, Carleton planned for an invasion across the Colorado River. The counteroffensive would retake areas lost to rebels and restore overland mail routes. If successful, it would open a western front against the Confederacy. Yet the challenge of providing for soldiers became Carleton's main preoccupation—so much so that he and the army entered into battle against the Southwest landscape as much as they did against its human belligerents. San Pedro's estuary made both battles possible. It became the entry and departure point for soldiers, wagons, and mules bound for the interior. Through Carleton's management and Banning's freighting, the Civil War connected two paradoxical landscapes: deserts and wetlands. Unionists employed these in tandem, as they altered supply and scarcity hundreds of miles inland.

Although he had tackled shortages at Los Angeles previously, Carleton faced more extensive want in the interior. Three desert zones punctuate the expanse to the Rio Grande: the Mojave, between Southern California's mountains and Colorado River; the Sonoran, east of the Gila River bend; and the Chihuahuan, which runs farther from Arizona's Chiricahua Mountains. Together, this great desert posed an existential threat. California Volunteers could not count on even the most basic of resources. Low rainfall hindered the growth of sheltering trees and exposed troops and animals to sunburn, heat exhaustion, and sandstorms. Surface water was absent for much of the year. One veteran recalled wells so low that they filled merely a pint cup at a time.[1]

The absence of water raised particular difficulty for army livestock. Wild grasses and legume plants might be available during a desert spring. Otherwise, they dried out. Working animals have intense caloric demands. If not met, the animals break down and become burdens to a military, much like wounded soldiers or enemy captives. The planned offensive would be only as robust as its horses, mules, sheep, and cattle. As a consequence, Carleton's army would carry the large share of its water, fuel, and animal feed. Forces in the East faced no such task.

To provide for the column, Carleton planned its offensive as a series of leaps between human-made oases. Supplies had to collect at desert drop points *before* large numbers of troops or animals could move. On each day—or rather, each sunrise, since units often marched at night—they would reach sites where shade, water, and materiel awaited. This included boxes of rations and subsistence, forage piles, six-gallon water kegs and six-hundred-gallon rolling water tanks, ammunition, and medical supplies. Staff organized a complex schedule of advance or in-time deliveries. Carleton planned to have 150 army wagons, each with a six-mule team. Using these and Banning's fleet jointly, the district distributed goods along initial legs of the route. Prodigious amounts of freight shipped into Wilmington, across Banning's wharf, and onto the backs of animals and men. As Carleton later found—perhaps as he anticipated—caching supplies in a hostile landscape guaranteed wider conflict.[2]

By March 1862, the Union Army had built up sufficient numbers between Camp Drum and Fort Yuma to begin its advance. Time was of the essence. Rumors circulated that Unionist New Mexico verged on defeat. In fact, US troops remained in the fight, but besieged at Fort Craig. Their predicament worsened early in the month, when Sibley's Confederates captured Valverde Crossing on the Rio Grande. That cut the fort's supply line from Albuquerque and Santa Fe, which Sibley soon occupied. On February 28, rebel cavalry claimed Tucson. The town's five hundred residents, just one-third of the prewar population, surrendered without a fight. Rebel scouts began to reconnoiter the Gila River Valley.

Despite the urgent situation, Carleton's force was unable to move. Southern California saw some of the heaviest rains in its recorded history that winter. Roads washed out. Streams and canyons flooded. Mud and quicksand plagued the trails. Teamsters quit. Horses and cows drowned by the hundreds in semi-liquid earth. The Colorado River, swollen by Rocky Mountain snowmelt, over-ran its banks at Fort Yuma. The deluge swept away grazing herds, plus a wealth of hay stockpiled for the campaign. The river rose so high that it cut an additional channel and, for a time, stranded the fort on an island. The same storms took down the San Francisco–Los Angeles telegraph, disrupting communication at a critical time of planning. Calamity postponed the army's advance by one long, worrisome month.[3]

WINTER'S END brought Carleton some unforeseen advantages. When units finally marched in April 1862, they entered a desert transformed. Surface water, wells, and grasses were plentiful—as if the San Pedro estuary and Colorado delta had spread over the land-scape. The spring bloom also included supplies. Army freighters and Banning's crews endured difficult conditions to place cargoes at the targeted sites. By May, 230 wagons, 1,450 mules, and 2,350 troopers had taken to the road.[4]

As the soldiers advanced, old worlds gave way. For pastoral tribes of the interior, this brought new commerce. The Digueño provided the army with hay cut from San Diego's Cuyamaca Mountains. The Cocopah assisted at fords that the tribe operated on the Colorado River. Toward the Gila River, Carleton set up alliances with the Maricopa, Pima, and Papago to provide food and information. The former cultivated rich floodplains midway up the river. This was the most reliable source in the Sonoran Desert for forage and grain. To trade for the crops, Union wagons arrived bearing five thousand pounds of gifts, as well as ten thousand yards of calico, manta, flannel, and dyed cotton cloth prized by indigenous people.[5]

In exchange, the tribes asked for protection from the Apache. Indians long had preyed upon other Indians in the Southwest. Carleton knew to use this to broker relations. But who gained ultimately? Whereas the army couched its transactions in equal terms, one-sided expropriation lurked beneath. Trade eroded Indians' self-sufficiency and created a wedge by which to pry them off ancestral lands. River tribes who allied with the Union soon found themselves resigned to reservations, annuities, and federal wardship. Carleton had no intention otherwise.[6]

Similarly, the Union Army revived trade networks straddling the US-Mexico boundary. For centuries, commerce had connected settlers at Ures (established in 1644), Sonoyta (1693), Magdalena de Kino (1700), Tubac (1752), Altar (1775), Santa Cruz (1775), and Tucson (1775), as well as indigenous communities at Los Algodones, Yuma, Las Calabasas, San Xavier del Bac, Maricopa Wells, and Pima Villages. By the 1850s, mining camps at the Gila River, New Mexico's Piños Altos, the Patagonia mines near Tucson, and Sonora's Sierra Madre entered the economy.

That confronted Carleton with a border dilemma. The army's defensive concerns required it to interdict migrants and goods. At the same time, it needed free-flowing commerce. The commander tried to balance competing demands. In October 1861, the Southern California District declared the state and international lines at the Colorado River closed. Soldiers confiscated boats and controlled crossing points. They cut back grasses and thickets to gain a clear view of the riverbank. They recruited ship captains, Native Americans, and beaver trappers to surveil for suspicious activity. Any person seeking to cross had to sign an oath of allegiance and carry a passport issued by the army. It was the first true attempt at border control in the region.

But the army violated the sanctity of borders as well. In November 1861, US troops trespassed thirty miles into Mexico to seize a river ford called Gonzalez Ferry. They destroyed property owned by a Mexican citizen. And they would return periodically to send spies through the Tinajas Altas mountains. Carleton attempted to source goods, beef cattle especially, in Sonora and open a wagon line from Guaymas or Libertad-Port Lobos. Mexican authorities refused to give a formal concession.[7]

In the end, the army received only piecemeal shipments hauled illegally across the border. The Union capture of Tucson, without a fight on May 20, established it as a hub for wagons moving in all directions. Carleton arrived two weeks later and set up a supply depot. He also declared the founding of the US Territory of Arizona and a desert military district, based initially among Tucson's adobes. (Southern California's district remained headquartered at Wilmington.) As both the territory's governor and commander, he prohibited sedition, mandated the oath of allegiance, and replaced jury trials with military tribunals. He criminalized vagrancy and leased out

The mountains of Arizona and Southern California inhibited army supply and kept cargoes moving to and from Wilmington.

peons to Arizona's multiethnic landowners. These were ominous signs of his mastery over the region.[8]

Even so, supply needs kept the army yoked to San Pedro Bay. Upon reaching Tucson, Carleton tried to cut back on high-cost deliveries from Wilmington, including those under Banning's contract. But demands were too great, and alternative routes lacking. San Diego offered only a seasonal source for supplies because of mountainous terrain inland. The Colorado River delta enabled ships to reach Fort Yuma, a far cheaper mode of delivery. However, ships were subject to delay, accident, and impoundment by customhouses along Mexico's coast. Forage and subsistence easily ruined aboard vessels. And neglect at Fort Yuma's storehouse caused spoilage and waste. Despite being cost-effective, the Colorado River became infeasible for time-sensitive or perishable deliveries. In the end, Banning's wagons proved as intractable as the supply problems they were enlisted to solve.[9]

CARLETON'S COLUMN made haste to reach New Mexico. What they did not know was that the territory's Unionists had already won a decisive victory. At the Battle of Glorieta Pass, fought east of Santa Fe on March 28, units of Colorado and New Mexico volunteers defeated Sibley's rebels. More important, they burned Sibley's meager supply caravan and killed hundreds of captured horses. As a result, Carleton's troops would encounter scant Confederate resistance. Instead, war in the Southwest transformed into a new type of conflict.[10]

The army would wield the desert's scarcity and estuary's bounty against any and all opponents. The first to suffer was Sibley's force, which was now without wagons, animals, food, and equipment. Sibley never built (or was not allowed) supply chains due to the distance from populated areas of East Texas. Instead, his troops had subsisted off former federal depots and tribute seized from fields, yards, kitchens, and storehouses. As the once-proud conquerors retreated, they found little left to take. "The Southern soldiers here are retiring," a resident of Mesilla, New Mexico, wrote in June 1862, "The reason is that they have consumed and destroyed everything. . . . The people here are with their eyes open toward the North, in the hope of being relieved from the devastation of these locusts."[11] However, the condition of Sibley's command indicated that war had devastated locusts and residents alike.

By comparison, Carleton's troops moved with lines of plenty. They joined the offensive of Colorado and New Mexico regiments, and by late summer had pushed two hundred miles into West Texas. Like links in a chain, cargoes from Fort Yuma and San Pedro Bay (and ships departing San Francisco) followed to replenish storehouses. When necessities—or even niceties like tobacco—ran short, officers obtained these express from depots extending back to Camp Drum. Union supply trains were so efficient that California soldiers marched without knapsacks, which quickened their pace and lifted morale. Perhaps most remarkable, the army's surgeon reported that not one soldier died as a result of want, weather conditions, or physical strain. The same officer took this as a sign that the "desert had been conquered."[12]

Carleton, of course, had his own measure of conquest based on memory and long-standing feuds. Here again, the estuary supply line would make all the difference. The presence of thousands of California soldiers, plus their many thousands more animals, quickly depleted the desert environment. The army became a vortex siphoning material from nature and local populations. As units moved toward Fort Yuma, Carleton ordered officers to disperse all Native Americans from water sources, so they would not exhaust what his column needed. The practice returned after troops left the Gila River. This alone set the Union force on a collision course with independent tribes, most especially the area's Apache.[13]

Yet bountiful supply trains, which traversed Arizona and New Mexico during the hot summer of 1862, guaranteed a clash of unprecedented fury. Wagons presented a logical target for Indians seeking to block army incursions and secure their own needs. Groups like the Apache, Navajo, and Mojave long supplemented their agriculture and commerce by raiding. The power to take was the basis of their independence. Carleton had spent years trying to suppress this activity. As he prepared the Union offensive against the Confederates, he intended to punish the tribes for past deeds, and those to come. He drew up preemptive expeditions against the Western Apache and Navajo, as well as against the Mojave, Owens Valley Paiute, and Kern River peoples to be waged in his absence. As he preferred it, there would be no negotiation or treaties.[14]

"Unconditional surrender," a phrase popularized by Ulysses Grant, became the Unionist objective in the Civil War. So it would be, too, in the Southwest. Carleton commenced his campaign against local Apache immediately after arriving at the upper Gila. This, he calculated, would win over ex-secessionists, pastoral tribes, Anglo and Hispano colonists, and the Sonoran government. It was the desert's version of national reunion. "If the *Tontos* are hostile," Carleton instructed his cavalry officer Edwin A. Rigg, "shoot or hang every one."[15] There was no if. And Carleton knew it. By July, cavalry reported harassment from Apache and Navajo. Troops found fresh graves and dried-out, blackened corpses. Miners in New Mexico's Piños Altos pleaded for protection.

The desert's scourge had returned. Carleton resumed his war against Indian rebels that secession had interrupted, only now, the Civil War provided him greater wherewithal. A lifeline of external supply allowed the army to field superior numbers. Troops reoccupied abandoned installations and added more posts and garrisons. They did so at key places of sustenance and transit: mountain crossroads, streams, artesian springs, forage pastures, shady groves, and bottomlands suited to crops. These were vital sites in a landscape of choking limitation. Tribes resisted. But in the zero-sum game of the desert, each loss (or draw) pushed indigenous societies into crisis. And that spelled more violence.[16]

By a barbarous circle, ecological warfare became an end in itself.

When the California Column arrived at the Rio Grande in August 1862, it gained access to a rich supply line from across the Plains, the Mississippi River, and Northeastern railroads. It no longer needed to take from the desert. Nonetheless, Carleton continued the campaign of expropriation, all the more because of the abundance now in hand. He sent officers Joseph R. West and Kit Carson against the Mescalero Apache that fall. They had license to confiscate or kill livestock, execute tribal leaders, and destroy food and shelter. (All sides took scalps.) Next, West pursued the Mimbreño. This included the arrest of chief Mangas Coloradas during parley, followed by his torture and killing. Days later, California troops attacked the tribe's camp and executed captives who had surrendered under white flag. Carleton wrote the officer to affirm the measures taken: "I do not look forward to any peace with them, except what we must command. . . . Entire subjugation, or destruction of all the men, are the alternatives."[17]

Meanwhile, Carleton initiated his offensive against the Navajo. In late 1863, he ordered Carson with four hundred troops, including a contingent of California Volunteers, to enforce starvation and

APACHE CRUCIFIED.

"A strange and ghastly sight." By war's end, the desert was littered with signs of barbarity, including this one—a warning left by Maricopa Indians to their Apache enemies.

exposure during midwinter. Carson later recounted that his soldiers destroyed three thousand peach trees, confiscated 1,200 sheep, and passed an entire day burning corn. Starting in January 1864, groups of Navajo capitulated. That began the Long Walk of three-quarters of the people to Bosque Redondo, a new military reservation Carleton established on the Pecos River. He conceived this as a model farm and reformatory, drawn from observations of California's Fort Tejon and the Civilized Tribes of Indian Territory. As he prevailed on the desert, his thoughts turned to sage reflection. "They have fought us gallantly," he wrote to the War Department. "But when, at length, they found it was their destiny . . . to give way to the insatiable progress of our race, they threw down their arms, and, as brave men entitled to our admiration and respect, have come to us in confidence of our magnanimity."[18] Congress, Carleton urged, had an obligation to "repay" the tribes with support. That, of course, meant supporting his program.

DESPITE CARLETON'S TALK, the desert remained finite in its hospitality and stubborn to human conceits. Here, lofty ideas too easily withered into ignoble devices. The Civil War in the East, with its language of ideals and universals, did not change this. On a spring day in 1864, Edwin A. Rigg's detachment of California Volunteers entered southeastern Arizona's Tularosa Valley. Miles south of the upper Gila and near the ephemeral Rio de Suaz, they established a post named Fort Goodwin. Troops and civilians who journeyed the Tularosa had suffered furious attacks. Now, the Union Army claimed the site, along with a nearby water source and mountain trails. Surrounding canyons held springs, lush grasses, shallow ciénegas, and well-tended fields of corn and wheat. This was a refuge for the Chiricahua Apache and their horses, to which they must return. "The valley is the most beautiful one I have seen in Arizona," Rigg pronounced. "It has evidently been a place of great resort for the Indians."[19]

As they had done elsewhere, California's Union soldiers demolished food stores and structures. They reclaimed stolen US property such as guns, blankets, and animals. To prevent theft, troops had orders to shoot any mule or horse that lagged. The military hoped that Fort Goodwin might allow a supply road from the Sonoran coast and bring in settlers. "This valley will soon be thickly populated," Rigg predicted. "The locating of Fort Goodwin has settled the reign of the murdering Apaches who have held it so long," "Their race is nearly run," he added. "Extinction is only a question of time."

As he thoughtfully scratched the words into paper, Rigg's boots and equipage—his soldiers', too—bore the dust of the desert. Beneath that, hid faint traces of estuary mud.

Chapter 20

STARS FALL AND RISE AT WAR'S END

For nineteenth-century Americans, the idea of extinction clarified changes lived through or made by their own hand. By some veiled mechanism of time, they said, the old lost its grip and passed away to usher in the next age. Extinction narratives eliminated the need to explain. They obscured connections and cleaved past from present. They provided a way to remember and forget. That was their appeal.

So it was that when the Civil War drew to a close in the winter of 1865, Unionists meditated on the loss of antebellum worlds. They pronounced that the Union had been saved, free government vindicated, and slavery abolished by the Almighty's will. That same victory had meaning, too, in the Southwest and on the Pacific coast. Here, parallel expansion of the US's southern and northern sections proved incompatible. A slaveholding nation and empire could not coexist with its free-labor, free-soil variant. War eliminated the losing half. Southern nationalism no longer had a place in the American West. It took new form by sanctifying the Confederacy's defeat and resisting Reconstruction.[1]

At the same time, the Civil War neutralized other Southwest rebellions. US soldiers already had overwhelmed many indigenous groups. They would continue to try against the remaining holdouts. Utah's loyalty to the Union dissipated fears of armed Mormon rebellion. California's Chivalry Democrats, meanwhile, foundered in their attempts to legalize slavery. They then emigrated, relented, or

took up new political banners. Free labor triumphed when California's legislature ratified the Thirteenth Amendment. When the same
chambers rejected the Fourteenth and Fifteenth Amendments, it was
not the Chivs, but a free-labor coalition that cast decisive votes.[2]

At San Pedro Bay, war produced one difference above all. In place
of rival imperialisms and Western insurgencies, a more coherent
US expansion found root, one that formed in partnership among
Union officers and patriotic entrepreneurs and in the movement of
boots, horse hooves, wagons, and steam vessels. San Pedro's estuary
became a harbor. And its reach extended to a vast, arid interior soon
to fill with settlers and markets. That made it the inception point
for imperial projects to come. Phineas Banning and James Carleton
represented that possibility. The two found prominence as defenders
of the Union and harbingers of change. Nevertheless, they would
fare quite differently. That was because of their individual differences, but for another reason, too. In the race to claim the interior
and coastal waters, private enterprise had far more to gain and fewer
limits than did the government or military, at least in the short run.

PROFIT HAD TAKEN HOLD of the estuary. The Union Army
pumped funds and supplies through Banning's facilities at Wilmington for the duration of the war. From there, shipments fanned
out to inland points. Banning contracted for a share of this business.
Yet, even where he did not, his investments benefited. Camp Drum
became the preferred entry for army horses bound for Arizona. It
was home to the Camel Corps and "Turk" expressmen like Hadj Ali
and Abdel Kadir. The quartermaster depot at Wilmington supplied
frontier posts: Fort Mojave, Fort Tejon, and Fort Independence, a
new station in the Owens Valley. Shipments originating at the estuary supported posts in parts of Arizona, too. Wagons hobbled to
Fort Whipple (named for the officer and Pacific railroad surveyor
killed at Chancellorsville), near the territory's capital of Prescott.[3]

Banning's services eased the flow of government deliveries deep into the desert. In fact, Carleton's removal of military wagons from Los Angeles to Tucson and Santa Fe only increased the army's reliance on Banning in the rear. Either by direct shipments or by relieving traffic elsewhere, Banning enabled army offensives against the Paiute, Mojave, Chiricahua, Yavapai, and others. The umbilical cord that maintained posts like Fort Goodwin in the Tularosa Valley led back to his account books. Military inspectors took notice. For the time being, they reviewed Camp Drum and the Wilmington Depot positively. Officers who dared not did not last long.[4]

But besides army freight, Banning ferried emigrants who followed on the soldiers' heels. A correspondent for the *San Francisco Alta* reported the buzz at Wilmington in late 1863, as travelers waited to reach Arizona's untapped opportunities. The estuary became a transit point for settler colonists, and the profits redounded to Banning and company. His freight lines thickened to serve boomtowns at Frazier, near Tejon; the Kern River and Walker Pass; Panamint and Cerro Gordo, near the Owens Valley; El Dorado Canyon and La Paz, on the Colorado River, and 150 miles farther, to the Walker mines near Prescott. By the close of the war, the entrepreneur's wagons rolled to Tucson, the Salt Lake, and the Rio Grande on roads cleared of secessionists and Native peoples alike. His coastal trade ranged north past San Francisco and south into Mexican waters. Banning gained access to copper camps on the Arizona-Sonora frontier that a competitor had monopolized before.[5]

Wealth accumulated rapidly. Steady shipping and passenger arrivals lifted property values in Wilmington. Speculators scrambled for choice lots, and construction multiplied. As of December 1864, the town had sixty dwellings; numerous hotels, saloons, and boarding rooms; several merchandise stores; and a post office. The road from the wharf to the army depot provided a thoroughfare named Canal Street. The town in between became a center for business. One mile upland, at Drum Barracks (renamed from Camp Drum), the afternoon sea breeze snapped a large Stars and Stripes into view. The federal banner unfurled from an unusually tall pine pole that US

WILMINGTON.

The changing estuary. Wilmington in the late 1860s, with
Banning's steamboat and Camp Drum at rear.

Surveyor Henry Hancock had sourced from Mount San Bernardino. It was Hancock's associate, Henry Washington, who surveyed from that peak in 1853 to set off the Los Angeles real estate rush.[6]

It was apt that the US flag dominated the skyline. Association with the Union was what had distinguished both Wilmington and Banning. Their resulting prosperity was not undeserved. Southern Californians' ambivalence created high risk for those who did business with the unpopular Lincoln government. But risk heightened the rewards. Benjamin Wilson estimated that the US spent no less than $1 million on construction within the subdivision. This was an astounding infusion of money to a place so deficient previously that cattle hides and promissory notes were exchanged as currency. Greenbacks accumulated in one set of hands most of all. The army put the total value of Banning's wartime contracts at $200,000. In 1865, the freight man reported the highest taxable income in Los Angeles County, at $20,000, well above the rest. He built a three-story, twenty-three-room home, complete with a staff of servants. He reinvested in machinery and a growing pool of laborers to speed development. He extended his wharf to two hundred feet, long enough to exchange cargoes at low tide.[7]

In addition to profits, Banning gained a moral victory and unrivaled stature in Los Angeles politics. He had supported the Union cause from the start. Now, the cause was vindicated. That death and sacrifice fell harder on others mattered little, since most Californians were far removed from battle. Locally, Banning's contributions had no equal. Admirers and opponents credited him with building support for Lincoln's reelection and for the president's National Union Party. The Chivalry group that once dominated Los Angeles was "dead," as a correspondent explained to its former leader and now-paroled Confederate, J. Lancaster Brent. Brent's erstwhile partner was the reason. "Through his influence," the writer told of Banning, "the county went anti-Democratic for the first time."[8] Political shifts ran even deeper. Los Angeles Democrats denounced the Southern insurrection by the summer 1863 and endorsed the Union war effort. (They refused to condone the emancipation policy.) With their help, Banning won election to the state senate. He would be remembered for the rest of his life as a "staunch Republican," "ardent Abolitionist," or simply as "General" for his nominal commission in the militia. All had a grain of truth, but made for a handy reinvention of his past—and Los Angeles's as well. Southern California must have seemed unrecognizable to Brent. He never returned.

Banning's remembrance contained one further omission. The Southwest Civil War lingered on as an Indian war. And he took every advantage of conflict-opportunity. This ungenteel fact surfaced at times. The *Wilmington Journal* newspaper, recently launched by Banning to promote Lincoln's reelection, closely monitored the inland frontier. It ran stories demonizing Native American resistance and advocated for high military spending. Banning printed his weekly on a press the army had seized from the *Los Angeles Star*, his former promoter. Of course, the entrepreneur no longer needed the *Star*. His business was embedded in a world of desert violence, as well as union and freedom reborn.[9]

THAT DESERT'S MASTER was James Carleton. Between 1863 and 1865, Carleton—like Banning—stood at the zenith. The officer brought to the Southwest a new order that he had long ago conceived. He transformed the landscape from a citadel of insurgents into a blunt instrument for their subjugation. But his repute did not wear as well as Banning's. New Mexico's merchants (like those in Los Angeles before) accused him of fraud and, worse, of rewarding rebel sympathizers. Others complained of his autocratic ways, confiscation of property (without due process), and suppression of political opponents. Some Unionists said Carleton perpetuated the territory's peonage system, a form of virtual slavery. Others were angry when the army moved to eliminate this system. US settler societies could be a knot of compulsions.[10]

Nothing polarized New Mexico as much as Carleton's Indian wars and reservations. Armed conquest had its champions, particularly among those who stood to gain land, mineral rights, or tribal annuity contracts. But skeptics, including Native American leaders, argued that the same policy exacerbated raiding, resistance, and reprisal. Carleton, they muttered, had become the obstacle to New Mexico's progress. It was an irony he found baffling. Although the War Department initially refrained from upbraiding him, Carleton became a lightning rod for humanitarian concerns and charges of profiteering. It did not help that his showpiece, the Bosque Redondo reservation, proved a costly failure and a prison of suffering and iniquity.[11]

Evidence mounted, and opponents gathered strength. They won election to the territorial government, which caused more critics to emerge. Meanwhile, internees defied Carleton's authority and left Bosque Redondo. The officer ordered them apprehended, but escapes continued. Indians were starving and dying as they tried in vain to farm the Pecos. The national press covered the tragedy, along with the millions of dollars sunk. Congress and the army launched investigations. In 1865, Senator James Doolittle, a Republican, traveled West to gather testimony about the massacre of Indians at Sand Creek. One eyewitness told how Colorado Volunteers—the Unionist heroes of Glorieta Pass—had engaged in "indiscriminate slaughter of men, women, and children."[12]

Doolittle's trail led south from Sand Creek to Santa Fe and Carleton. The officer cooperated with investigators and remained firm in his sense of right. He ascribed the indigenous population decline—both off and on reservations—to the effects of war, alcohol, epidemic diseases, and syphilis, but also to "causes which the Almighty originates."[13] The officer hoped Doolittle, as chair of the joint House-Senate special committee on Indian affairs, would advocate greater funding. Instead, the War Department cut Carleton's independent command by folding it into the army's Pacific Division, and then the Department of the Missouri. In fall 1866, Carleton transferred to a cavalry command in Texas. Unlike other Civil War commanders of comparable success, he returned to the regular army with only a one-grade promotion, to lieutenant colonel. That meant field duty

Carleton's civil war forced reservations on tribal people of the desert, such as the Coyotero Apache.

and sore hours in the saddle. Conflict persisted on the Southern Plains for several years, but in the charge of others.

Carleton died of pneumonia in 1873, but not before one additional rebuke. William T. Sherman, now the army's second-highest general, arrived in New Mexico in May 1868. He was part of a commission appointed by Congress to reopen negotiations with Indians. The effort, based in revitalized Reconstruction politics, lasted several years. It became known as the Peace Policy, once Sherman's commander and friend Ulysses Grant took office as president. At Bosque Redondo, Sherman conferred with Navajo leaders. He condemned their sufferings and replaced Carleton's unilateral terms with a new treaty that allowed the tribe a reservation on their ancestral lands. They were free to return home. From his exile in Texas, Carleton saw his model reservation closed and his methods forsworn as a relic of misguided times. Those methods would return. Peace could not withstand ongoing Native insurgency, escalating violence, or the temptation of settlers and bureaucrats.

In spite of Carleton's new order, the desert held surprising capacity for those with eyes to find it. Only after long-distance railroads entered the region would the army overwhelm the last indigenous rebels. By that time, industrial corporations had eclipsed small entrepreneurs like Banning. And tactics had shifted from Carleton's forte in light cavalry. Even still, the Apache evaded capture. During 1882, they battled troops under command of a Civil War veteran, Adna Romanza Chaffee, at the Battle of Big Dry Wash, near Mogollon Rim. Three decades later, Chaffee would administer the construction of the Port of Los Angeles and watch over San Pedro's estuary in its final days.[14]

HENCE, the story of the Civil War in the desert returns to its coastal source. For all of the panic and counterstrategy precipitated in early 1862, the Confederate threat to Southern California vanished into the mirages. This has relegated San Pedro Bay to insignificance within conventional accounts of the war. Not so, however. The estuary's story reached across the continent and revealed itself in one of the war's most potent moments. Edward Ord—former Coast Survey officer, artillery captain, Democratic spoils seeker, and enterprising agent of land speculators and Southern politicians—traveled east in the autumn of 1861 to take charge of Union regiments in Virginia. He served as a division commander and, like so many others, did so with little notice. He was wounded in October 1862 in defense of Corinth, Mississippi. The next year, he transferred to the Army of the Tennessee to serve under Sherman and Grant during the final weeks of their famed Vicksburg campaign. It was a path blazed by Sherman. Grant had emerged from obscurity by scoring two victories in the Union's first major offensive. Sherman, who struggled with command before, asked to transfer to Grant's staff. Ord's opportunity came in mid-1863, when Grant fired a corps commander who had tried to undermine him. As Ord had learned in California, someone else's misstep could be his gain.[15]

But he had made some missteps, too. His connection to secessionist in-laws, the Thompson family, and to wife Molly caused Ord continual melancholy and worry. Edward knew of the rumors. Some even appeared in newspapers. By 1864, Ord concluded that he was destined to "lose by my wife's opinions."[16] He lamented in a letter, "I have had this fact of my wife's family politics thrown in my teeth by Generals' wives." He nearly resigned his commission. A year later, Molly received a loud beratement from Mary Todd Lincoln during the president's review of troops. (Mary was furious that Molly briefly rode her horse alongside the president's and ahead of the first lady's carriage. But she likely knew the rumors, too.) "Women's views," Edward observed, "are helping to make men cut each other's throats both practically and politically." For those climbing the army career ladder, husbands and wives were partners for better and worse.

Fortunately for Ord, his star rose with those of his superiors and by the faults of others. In 1864, Grant brought him, along with Philip Sheridan, to the Virginia theater to defeat Robert E. Lee's armies. Ord was wounded a second time. When he recovered, Grant put him in charge of the entire Army of the James, including its sizable African American units. Once again, Ord replaced a general (the colorful Benjamin Butler) who had lost Grant's confidence. As he climbed the ranks, Ord's prewar reputation lingered among Southern officers and politicians. When Confederate peace commissioners approached Union lines in January 1865 to negotiate with Lincoln, they passed through Ord's district. When rebel general James Longstreet inquired about Lee and Grant meeting to discuss peace terms, Longstreet did so with Ord. Ord briefly provided a conduit between two armies, two governments, and two commanders-in-chief.[17]

He fought superbly, as well. Ord took a leading role in the Siege of Petersburg and Grant's final offensive. By spring, his force occupied trenches on the Union far left, where they pinned down A. P. Hill's rebel lines. This enabled Sheridan's Army of the Shenandoah to flank farther left and capture the last railroad supplying Confederate defenders. At dawn on April 2, Ord's troops joined in a massive frontal assault. Army of the James units were among the first to break through and overtake Petersburg's outer defenses, including the key artillery post, Fort Gregg. After nine months of stalemate, Lee's resistance collapsed. The Army of the James's XXV Corps, an African American force made up mostly of former slaves, soon occupied Petersburg. Other members of XXV Corps became the first Union units to enter Richmond. Subsequently, Ord's command captured roads and bridges to keep Lee's tattered group from escaping.[18]

Accolades abounded, as Ord proved to be in the right place at the right time and capable of rising to the occasion. Like Sherman and Sheridan, he featured prominently in commemorative artwork. Ole Peter Hansen Balling's *Grant and His Generals* (see page 171), which today hangs oversized in the National Portrait Gallery, shows Ord not in front, but directly to the rear of Grant and Sherman, on a dark horse just below and right of the federal flag. Balling received permission to sketch the likenesses in person. Ord was distinguished by prematurely gray hair, which worried him while he was an unmarried man, but now made him instantly recognizable.

After Petersburg and Richmond fell, Ord stood among Grant's inner circle. He was present at the McLean House in Appomattox Courthouse on April 9, when Lee surrendered and effectively ended the war. Following the signing of terms, Ord showed he had lost neither old habits nor his keen eye for speculative property. He lingered among the McLeans long enough to negotiate the purchase of the marble-capped table upon which Lee signed the surrender document. Today, the table resides at the Chicago History Museum.[19]

Ord claimed one more piece of real estate. Grant temporarily assigned him to Richmond, the former Confederate capital, to oversee its occupation. Soon after, Edward and his Virginia-born wife moved into the Confederacy's executive mansion. Their former patron, Jefferson Davis, had lived here only weeks before as the one and only president of a separatist slaveholder nation. Once a respected war hero and public servant, Davis now was reviled in both North *and* South. He was, too, a fugitive hunted by the government he had sworn to protect—as soldier, senator, and secretary of war—but in the end betrayed. (Union troops would apprehend Davis on May 10.) By the time Ord rode into Richmond, the letters he exchanged with Davis about Pacific railroads and San Pedro had vanished. Family legend holds that Sherman ordered these destroyed when his troops swept across Mississippi in 1863. William Gwin's correspondence perhaps met a similar end? Ink and pulp washed down the river like so much blood drawn by lash and sword.[20]

As the Civil War ended, Edward Ord no longer had to worry about his past. On a sunny April day, he and Molly left behind their complicated life in California, as well as her once-intemperate views.

*Edward and Molly
Ord pose in victory,
with the Appomattox
table and Jefferson
Davis's former home, in
Richmond, Virginia.*

In the process, they concealed a more complex history of the American nation and a far-off estuary. The Ords came to Richmond among the conquerors. To mark the turn, the couple posed before the camera of Mathew Brady, seated and solemn on the Davises' former porch, with the Appomattox table placed conspicuously behind them. The image tells how remembrance and forgetting often are one and the same.

Mapping the Disappeared

THE CIVIL WAR BROUGHT PARCELS, PROPERTY, AND cargo to the estuary. But nature remained. It passed without record, except for one document: a Coast Survey map commissioned by George Davidson in 1870 and published two years later. This showed topography at several stages of tide. Details give the clearest image of the ecosystem that once was.

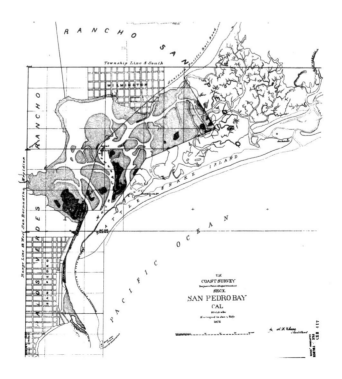

As with other coastal lagoons, a mosaic of surfaces comprised San Pedro's inner bay. Waters from land and ocean built up varied soil composites and four distinct habitats. The core was a large mudflat. At high tide, water subsumed it and created a field of movement for small fish, mollusks, echinoderms, and crustaceans. The shallows, those with eelgrass especially, offered safety to eggs and juveniles. But the flats were a place of risk, too. Low water exposed them to air. Creatures most threatened by predation or drying out burrowed below. The more resilient—aquatic snails and crabs especially—took the opportunity to move about. Yet both they and the burrowers could become food for invading gulls and shorebirds. Mudflat occupants were easy pickings. They persisted out of sheer number.

The estuary's second zone was its marsh. Although this sat above the waterline, soils here were inundated by salt. That would kill most plants, but the hardiest survivors adapted and flourished. Specialized halophyte vegetation took root along the bay's north and east limits. In lower intertidal areas, dense cordgrass (genus *Spartina*) prevailed. With hollow stems that circulate air, cordgrass can tolerate being submerged to half its height. In higher areas, leafless pickleweed (genus *Salicornia*) spread. Like all succulents, it stores fresh water in cell tissue and can expel salts. Pickleweed does so by growing in fleshy, segmented stems, like a chain of tiny gherkins. After concentrating toxins, the stem tip falls off. Its sacrifice enables the remainder to survive. Marsh plants are evolutionary anomalies, heirs to remarkable invention.

Elsewhere, sand stretches formed wherever waves and the prevailing currents deposited ocean sediment. This third ecological zone included the four-mile-long barrier named Rattlesnake Island (later Terminal Island). As is true today, beach and dunes life had to survive a gamut of conditions: wet or dry, salty or not, cold or extreme heat. Crabs did particularly well, as did snakes presumably washed down river. Creeping sea lavender, sagewort, and saltgrass grew with taproots long enough to find fresh water. Birds nested in colonies. Meanwhile, millions of microscopic creatures lived between the grains.

The sand regions were a surprisingly functional ecosystem. They also sheltered calm, open water inside the bay. The size of this fourth zone altered constantly. Twice a day, with the rising tide, the sea entered. For several hours, it fractured the contents into a collection of islands and shoals. During higher tides, water opened the bay to large fish and marine mammals. Then it reversed and gave way to dry land. Only channels of several inches to several feet deep remained at ebb tide. Aquatic life took refuge here, or in isolated pools, until the sea returned. The process repeated ad infinitum.

By its nature, the estuary defied the parcels, property, and cargoes that settlers imposed. Alkali soils rendered the bay unusable for agriculture. Shifting sandbars were a hazard to boats and threatened to cut off Wilmington's commerce. Through much of the bay, no more than a skiff could navigate. Late nineteenth-century humans saw the estuary for what they wanted. They ignored its wondrous tenacity. That would strand the 1872 map as a lost moment out of time.

Part Four
CAPITAL

"IT COSTS MONEY TO FIX THINGS."—*C. P. Huntington.*
As it is Plain that Most of Our Congressmen Are for Sale, They Might as Well Display Their Prices Prominently.

Extravagant Costs and
Huge Opprobrium

I N D E C E M B E R 1 8 6 8, G E O R G E M. W H E E L E R S T E P P E D
ashore at Wilmington. The officer came to study logistics so that
the army might at last overpower the Native rebels of the interior.
It was, in some ways, a greater challenge than on the Plains. In the
Southwest desert, there were no large Indian encampments to attack
or bison herds to kill. No railroads existed to move troops and
equipment. Wheeler would travel on horseback, 1,700 miles counted
by wheel odometer, to observe thirteen forts and their supply chains.
He would recommend some for closure or relocation, all with an eye
to concentrating military force. The journey would inspire his later,
famed explorations of the West beyond the 100th meridian.[1]

Wheeler's arrival marked the army's return to California in the
wake of the Civil War. Here, it confronted the lingering problems
of US expansion. The inland region remained vast and inhospita-
ble. Apache raids continued. Disputes over Mexican-era property
claims dragged on. The sea offshore hid uncharted dangers. These
challenges were heightened because of the added cost. Local entrepreneurs angled for land, contracts, and tribal annuities. Settlers
provoked Indians to fight, then reveled in the opportunity to service
or sell to federal troops. Government agencies were too small and
incapable to do otherwise. They needed civilian proxies, and that
was their dilemma. They had to pay up.

Old challenges thus entangled with a new problem of the age: the problem of capital. Wartime spending stimulated the growth of banks, finance, and corporations. It created railroad, ship, and wagon companies. It entrenched rapacious profiteers. War built up the private sector and made it even more of a partner in frontier colonization than before. The partnership was fraught. It produced waste, expense, and corruption that infected government projects and commandeered public money. Private frauds worsened the state's fiscal burden. Through the postwar decades, they shook the economy and undermined the trust of citizens. They deprived soldiers and reservation Indians and reignited conflict. Frauds weakened every effort to consolidate the West while, perversely, they set off demands that government sink still greater sums.[2]

But war also empowered government, the military especially, and its sense of upholding the national interest. Confident from its victory in the Civil War, the army would attempt the first reform of capital in the West. It aimed to draw down the spiraling cost of rebellions. And it looked to the coast—and beyond, into Pacific waters. Problems existed in a thousand places. Yet, for the expanse Wheeler traveled, auditors traced the source to the same estuary-harbor where he arrived. The army had scrutinized Wilmington Depot, Drum Barracks, and Phineas Banning before. But urgency outshone the criticism and kept shipments moving. After the war, scrutiny came back like a rising tide. It did so slowly, at first, because waste was so ubiquitous. Quartermaster General Montgomery Meigs had made Union soldiers the best-equipped and best-fed in history. Yet his supply system was a nineteenth-century operation, decentralized and semi-regulated at best. Officers hastily awarded contracts while opportunists scrambled to grab their share. Banning's wharf and wagons were one small piece of the problem. Nonetheless, military accountants were on their way.

The army reconsidered investments at San Pedro Bay amid the daunting task of Reconstruction continent-wide, in the South as well as in the West and on the Pacific coast. The United States could scarcely afford the massive force that won the Civil War. The $1.8 billion expense nearly shattered the North's economy. That the Confederacy's broke far sooner was of little consolation once hostilities ended. No one knew how much spending the Union could tolerate. Many triumphs in history turned pyrrhic when the bills came due. To avoid this, Congress drew down its one million troops and sold off surplus goods and animals. By late 1866, it allowed for only fifty-four thousand soldiers, roughly half to serve in the occupied South. Most of the other half shifted to the Plains to dampen Indian unrest and protect emigrants, the US Mail, and railroad crews. In all theaters, the government determined to employ and outfit soldiers with frugality.[3]

The challenge of making the numbers work fell to the quartermaster general. Meigs had allies, however, including the Pacific Division's new commander, Henry Halleck. Both had proven themselves master bureaucrats during the Civil War. When expenses set off alarms, the army sent Halleck from Washington, DC, to San Francisco. By the end of 1866, the division (with its subsidiaries, the Department of California and Department of the Columbia) supported 3,400 soldiers in a territory that stretched from the tidewater through Arizona, Nevada, and Idaho. This same area consumed a disproportionate share of the federal treasury. Expenditures here were higher per unit and per capita than in any other administrative region, far beyond anything seen in the battlefronts of the East or by European militaries. Costs were most extravagant in the District of Southern California (including Arizona and lower Nevada), the area supplied by San Pedro Bay and the Colorado River. This showed no prospect of decreasing. Swaths of the Southwest remained in Indian rebellion. Pockets of Northwest backcountry, like the Modoc and Nez Perce homelands, did too.

Meigs's office and the Pacific Division functioned as autonomous parts, albeit within an army that had grown more coherent than ever before. Even so, they united against inefficiency. They mandated the review of future contracts to prevent "favoritism" or "collusions."[4]

Additionally, the division ordered a retroactive audit of its wartime contracts. Investigators' findings were eye-popping. Feeding troops in Arizona, they found, cost five times the already-high amount to do so in San Francisco. Hay to feed animals cost $84 per ton by the time it reached Fort Yuma, versus $7.50 per ton on arrival at the frontier post of Fort Riley, Kansas. The division had built desert posts with no regard for maintenance. Many were placed ineffectively. Troops and horses suffered in miserable heat and did more idle drudgery than soldiering.

Inspectors blamed the extravagance on aridity, distance, and difficult transportation. Yet they returned again and again to the human variable: contractors and their delivery routes had distorted the military's footprint and bloated its budgets. This made for a strange postscript. The Civil War in the East exacted a shocking price in blood. Operations in the Southwest exacted a severe monetary toll. Meigs saw a parallel between the two. Late in the war, Confederate partisans killed his army engineer son behind Union lines in the Shenandoah Valley. The body lay overnight in the rain and mud before federals could recover it. Meigs regarded rebels, including

Fort Yuma—entry into Arizona for supplies, animals, soldiers, and travelers. By the mid-1860s, it was busy with activity.

his brother Henry and former friends Robert E. Lee and Jefferson Davis (who wrote John Rodgers Meigs's recommendation to West Point) as traitors and murderers. The quartermaster thought little better of profiteers who enriched themselves off the sacrifice and grief of others. As news from the Pacific reached Washington, his outrage boiled.

At the close of 1866, Meigs dispatched his senior investigator, James Fowler Rusling, to search out the centers of waste. Rusling had just concluded his study of desecrated and poorly kept Union gravesites in the South. The work compelled Congress to establish a system of battlefield cemeteries under federal watch. Now he would take on another troubling war legacy. Meigs instructed him to assess all depots and supply routes in the Far West. The inspector traveled from Fort Leavenworth, up the Platte River to Denver, then to Salt Lake City and Boise, and down the Columbia River to the coast.[5]

Rusling visited San Pedro Bay in early 1867 and penned a scathing report. At this point, most supplies bypassed the estuary and moved to Arizona by way of the sea and Colorado River. (Troops still landed at the harbor before marching inland.) That did not stop Rusling from assailing Wilmington's facilities. They were of no significance, he disclosed, "except to Mr. Phineas Banning."[6] "I must say," the officer added, "I admire his genius and audacity in getting Drum Barracks to become what it is; namely, a huge opprobrium and scandal to the [Quartermaster's Department]."

Rusling's incredulity focused on the estuary, which, he explained, "is really no harbor at all." In descriptions similar to those by George Davidson and the Coast Survey, Rusling observed that the channel to Wilmington was impassable except at high tide. Even then, tugboats traveled single file and could not turn around if they met with distress. The lagoon was "nothing but a little crooked gut of the sea" where the US government had squandered a million dollars—all for the benefit of Banning's contracts. Judgment fell swiftly. The inspector recommended that the army shut down its operation and move it to San Diego.

With the army decided, it seemed certain that San Pedro would cease to be a federally sponsored harbor. When Wheeler arrived in 1868, he came to determine exactly how the military's desert footprint would change. The base at Wilmington was of no interest. Wheeler was so assured of its closure that he paused only long enough to select a horse and pack up equipment. However, the problem of capital was too unwieldy, too tempting, too bewildering. It outgrew reformers' grasp and cloaked itself in dazzling technology and the allure of self-made success. Transcontinental railroad corporations, exemplars of the problem, would quickly capture the public dime and imagination. Officials—the martyred Lincoln himself—had deemed them a necessity worth their every offense. Most of all, they were new and bursting with creative energy. They were not yet the obvious danger they soon became.

The army's attempt to curb and clean up expenditures would fail, both at San Pedro Bay and in the West generally. By the time of Wheeler's visit, the estuary's promoters had made calculated shifts. Banning conceded the loss of army money and quickly prospected the railroad. Hence, the tides of Reconstruction ran in two ways at once. In spite of reformers' efforts, nascent corporations emerged far and away as the prime force in Western development. They would dominate the estuary, confounding observers and critics alike. The US and its people would continue to pay extravagant costs.

For the moment, that is. The success of capital ultimately proved its own undoing at San Pedro Bay by a series of events that led back to Washington, DC. Until then, it was a rare few who kept the reformers' faith. But it was an even rarer few who kept the flow of money and cargoes coming.

Chapter 21

BANNING RIDES
THE TURNING TIDE

A S OF 1868, ARMY INSPECTORS AND BANNING'S WAG-
ons continued to ford the desert by horse and mule. The same
was true for settlers, at least the lucky ones who had the choice not
to walk. While they engaged in stoic hardship, a revolution gathered
nearby. It grew quietly at first, on drafting tables, within company
directorates, and inside courts and legislatures. Then it took ear-
splitting form as railroads, shipping lines, and other soot-spewing
craft. Steam motive power was the single greatest force for transformation during the nineteenth century, and California and the Far
West would experience its effects as a sudden jolt. Of the twenty-nine
thousand miles of railroad recorded in the 1860 census, only twenty-
two existed in the region. Masted sailing ships or hybrid vessels still
outnumbered steamships along the coast. During the next decades,
however, steam engines overran land and sea, stoked by daredevil
entrepreneurs, corporate innovators, and government handouts as
much as by coal or wood.[1]

Among its many effects, steam power intensified both the promise and problem of capital. It produced unprecedented wealth, but
inequality, too. It made and destroyed fortunes. It opened up opportunities for those who allied with its explosive energy; even more
so for those who learned to wield volatility and harm to further
advantage. Several such visionaries would shape the estuary story.
Banning would be the first. Just as the army ended its investments
at Wilmington, he attached himself to rail technology and its mas-

ters. Still, this was a precarious trade. The Western railroad industry struggled. It was unclear which companies would prevail, and therefore with whom to align. There was a great risk not only of losing money, but of making enemies. Banning bet well. Others tried, stumbled, and never recovered.

Like many entrepreneurs who entered the railroad game, Banning's embrace of the new rested upon his success with the old. During the war, Banning lent his personal army of animals, tugboats, and wagons to the Union government. Patriotic ties gained him wealth and prestige. He ran successfully for the state senate in 1865. As a candidate, he called for government subsidies to build Southern California's first railroad. The proposal set him apart from most other wagoners, who opposed railroads as a threat to their businesses. Banning was smart enough to know that an iron road running from his waterfront to Los Angeles threatened only his rivals. The county wagon road he championed ten years before had served no differently.

Plans for a track between San Pedro Bay and the city had circulated since the 1850s. California's legislature considered several bills to allow for a line, yet none had passed. Banning and fellow Wilmington investors Benjamin Wilson and John Downey used their political connections to revive the issue. Merchants, ranchers, and vineyardists lined up in support. The project connected scattered pools of investment capital, which were growing because of war-era commerce and rising property values.

Progress on the first transcontinental line added urgency. By late 1864, the Central Pacific Railroad (CPRR) started to build from Sacramento, up the Sierra, toward Donner-Truckee Pass. Leland Stanford and partners Collis Huntington, Mark Hopkins, and Charles Crocker now pushed the line through forest and granite. Federal policy helped, inflating their stock value with land grants and cash transfers. Under the terms of the 1862 Pacific Railway Act, the CPRR would connect with the Union Pacific Railroad, which was being built concurrently from Omaha, Nebraska. In Washington,

DC, officials took a keen interest in watching track mileage accumulate. They worried far less about the companies' financial misdeeds.[2]

The transcontinental railroad's construction—its builders' ballooning wealth especially—set off a mad dash to start up railroads that might connect as branches. The most ambitious hoped to develop into long-distance trunk lines and to profit from federal support. One aspiring scheme was the Southern Pacific Railroad (SPRR), established in San Francisco, but planned to be built in segments (a portion through Missouri by John Frémont's syndicate). Organizers aimed to lay track through Southern California, then east via the 32nd parallel. The destruction of slavery ensured that Jefferson Davis's preferred route now had broad support. But the SPRR group could not get the project started. In 1868, the CPRR's Stanford, Huntington, Hopkins, and Crocker acquired the scheme. Even so, they were hardly the "Big Four," as they later came to be known. They were simply the "Pacific Associates," one avaricious ring among many. And their company was not quite the monopoly power it would be a decade later. Instead, it was over-leveraged and vulnerable, a financial house of cards. Speculators knew it. Throughout California, they sought franchises for small railroads that might displace the transcontinental, or else force it into a painful buyout. For aspiring challengers, there was no need to build much. Incorporation papers or the deed to a strategic piece of land might be enough to compel the Associates to make a deal.[3]

Like other upstarts, Banning and his circle entered the railroad rush. They set up their project, the Los Angeles & San Pedro Railroad (LASPRR), and found big money willing to join them. After entering the state senate, Banning introduced bills to subsidize construction. By the spring of 1868, he had secured $225,000 from city and county voters. Their subscription covered half the cost. The LASPRR raised the remainder through a joint-stock company. Banning purchased the majority of shares. San Francisco investor Henry B. Tichenor bought in as well. During the 1850s, he had built a timber empire along Mendocino County's Navarro River and up

the coast to Oregon and Puget Sound. Tichenor & Company's products soon traded throughout the Atlantic and Pacific worlds. His
involvement with the LASPRR gave it a reliable source of wood at
a discount.[4]

But Tichenor offered more. He provided Banning a lesson in the
cutthroat tactics and high rewards of Bay Area capitalism. In the
early 1850s, the lumber mogul purchased waterfront lands on San
Francisco's Mission Bay for $2,700. The parcels were far from the
city's current waterfront, so who would want them? The railroad
boom changed this. The Central Pacific, fearing it might lose the site
to a challenger, purchased Tichenor's property in 1868 for $250,000.
His windfall caused buyers to snatch up tidelands around San Francisco and Alameda, by fraud if necessary. Banning, too, followed
Tichenor's example. As he devised the LASPRR, he applied for state
title to 626 acres of mudflats abutting Wilmington, a move that
would set up intense property conflicts to come.[5]

With his wartime earnings, as well as backing from Tichenor and
taxpayers, Banning's railroad became one of the earliest upstart proj-

*Banning's wharf in 1870—terminus for Los Angeles's first
railroad. Marsh and mudflats appear in the foreground.*

ects to open. On January 12, 1869, work crews connected the several miles of completed track to Banning's wharf. Two days later, steam tugs ferried cargo across the mouth of the estuary, up three miles of sinuous channels, and alongside the landing. Laborers transferred the freight onto rail cars. The event hinted at the growing convergence of steam transportation on land and water. And yet, the size of Los Angeles's first locomotive, named *San Gabriel,* understated the event. A photograph shows a diminutive "donkey" engine—something one might ride today at a children's amusement park. The machine belied Banning's large ambitions. Mindful of the Southern Pacific, he jumped to incorporate a rail line to the Gila River in Arizona. He would beat Stanford et al. to the desert, and then negotiate on his terms.[6]

Time was of the essence. On May 10, Central Pacific and Union Pacific crews met at Utah's Promontory Point to connect the first transcontinental through line. Stanford was on hand to drive the final spike. Although a large man, the executive had long ago traded hard labor for fine suits and a calf-leather chair. He raised the hammer dramatically above his head, lost his aim (and perhaps his balance), and missed. A second, more heedful swing hit the mark. The Associates then cast their eyes toward Los Angeles.[7]

Unfortunately for Banning, he failed to win a subsidy for his Gila River railroad. That proved fortunate in the end. The Associates did not easily forgive or forget. Banning would find that they made better patrons than rivals. Besides, public money had many avenues. And corporate clientage came with advantage and far fewer risks. Also lucky for him, the Associates finally took notice of California's estuaries. One recent oversight had been embarrassing to the company—like Stanford's miss with the hammer—and almost fatal.

BANNING WAS NOT alone in gaming the Central Pacific. The region was astir with aspirants, all looking to enrich themselves by rail technology. At the same time that Banning launched the

LASPRR, San Francisco investors woke up to the opportunity of "terminal railroads." All sought profit in a plain fact: The Central Pacific constructed its rails eastward from Sacramento, a river town sixty miles inland. The Pacific Railway Act incentivized building through the interior, but did little to encourage the CPRR to lay track in the opposite direction, west from Sacramento to the ocean. As early as 1866, rival ventures had organized to build these short, though indispensable, lines. Stanford and company were distracted and mired with their Chinese workforce in blasting through the High Sierra. Meanwhile, others seized the best routes to places where rail and ship might connect. If successful, they might block the CPRR from ever reaching the sea or the Golden Gate city, the largest in the Far West. Then they could starve the line of freight.[8]

Together with San Francisco financiers, who had spurned the CPRR's initial stock offering, the terminal syndicates plotted to steal the transcontinental out from under the Associates. The one that came nearest to doing so was William Birelie Hyde's Terminal Central Pacific Railroad (TCP). Hyde was a graduate of the prestigious Rensselaer Polytechnic Institute of New York. As a young civil engineer, he had the big dreams and precocious talent of the industrial age. Opportunity seemed to fall to him from the sky. After serving in the Civil War, Hyde returned west, and in 1864 found work in the Comstock silver mines of Nevada. He then joined the government survey of the Yosemite Valley, which President Lincoln had designated as a nature preserve. The expedition included Frederick Law Olmsted, who became the nation's leading landscape architect, as well as Hyde's friend Clarence King, later the head of the US Geological Survey. In 1866, Hyde took part in the Western Union Telegraph Company's exploration of Russian Alaska.[9]

On his return to California, he took a job as chief engineer of the Oakland-based TCP. The company sought a concession to Goat Island and surrounding tidelands. Also named Yerba Buena Island, the site today provides the midpoint for the San Francisco–Oakland Bay Bridge. The TCP planned a similar structure: a railroad trestle extending over water to reach the island. There, it would develop facilities to exchange ocean freight. The project required permission from the state (for the tidelands) and from the federal government (for the island and trestle crossing the bay). It also had to to deny the same to others, especially the Pacific Associates.[10]

Army engineers in San Francisco rejected the TCP's proposal, so the company sent Hyde to lobby in Washington, DC. In November 1867, he was in the capital, peddling the project to legislators and ad-interim secretary of war Ulysses Grant. Of Washington, Hyde wrote to his wife, Marietta "Mette" Butler, "I am dealing with all the elements that win battles in Congress. Greed, Ambition, Lust for Money and Power, Self-Interest, everything by which men's minds are influenced."[11] With some reluctance, he doled out promises and company stock in exchange for support. If he did not follow the tide of corruption, others surely would. The CPRR's Collis Huntington had arrived in the capital to do the same. Hyde remained long enough to attend impeachment hearings against President Andrew Johnson, then hurried home.

Back in California, the engineer-lobbyist found an epic battle unfolding. The CPRR had launched a flank attack on the terminal upstarts, his in particular. Somehow, Hyde and his backers emerged with the upper hand. In April 1868, the state granted mudflats to the TCP. But when the War Department withheld access to Goat Island, the TCP's tidelands became useless. The company collapsed like so many others. The CPRR had survived its most viable challenger thus far. And Hyde had proved a worthy opponent. Huntington decided to hire him before a competitor did.[12]

Hyde joined his former rivals just as the Associates prepared to construct the Southern Pacific. In November 1869, they sent him to Los Angeles to scout locations and raise support for a new subsidy. Among the assets, Hyde noticed Banning and his small railroad running between wharf and city. With a future trunk line through Southern California, the LASPRR would make an ideal terminal railroad. In December, Hyde would broker a deal with Banning to purchase his Wilmington landing, adjacent tidelands, and majority stake in the road. In the meantime, Banning helped Hyde prepare for the county subsidy vote. Initial signs were promising.[13]

Hyde spent the remainder of his visit surveying possible routes to the Colorado River. During the excursion, he stayed at the citrus ranch of Wilmington investor Benjamin Wilson. He promised Marietta that, with wealth to come, they would live on a ranch of their own. He was just twenty-seven years old. Success and power seemed his. But Hyde wanted more. He departed just before Christmas, traveling north by wagon roads that he hoped to make obsolete. Change promised great profits to those in the know. And Hyde had several years' head start. It would take that long for track crews to reach the outskirts of Los Angeles. He planned to return then to reap the rewards.[14]

Hyde's optimism heightened when the Army Engineers and Congress talked of harbor improvements at San Pedro Bay. In California, harbors would determine the railroad economy's winners and losers. Hyde and Banning had failed to rise to the top, but they were allies with the four who had. That would not prove security enough for either man—one especially. As it would do to the mule and wagon, the railroad could discard alliances seemingly overnight.

Chapter 22

PAST PLANS AND
UNDYING HABITS

T HE YEAR BEFORE HYDE SECURED WILMINGTON FOR
the Southern Pacific, another visitor looked over the bay. R.
S. Williamson arrived in his role as the army's lead coastal engineer.
But he had been here already. In 1853, he commanded the Pacific
railroad expedition through Southern California. War Secretary Jefferson Davis approved the route. Congress refused, however. Sectional politics kept Williamson's effort from amounting to anything
more than voluminous maps and papers. The rejection deprived him
of the acclaim he deserved. By 1868, things had changed. The engineer returned while Davis was self-exiled to Canada (and later Britain) after two years in a military prison. The Union victory removed
the political stain from the 32nd parallel route, and a Republican-
dominated Congress resurrected the project. Investors, tempted by
government incentives, pumped money into construction schemes.
They also gained in their ability to steer federal money. Williamson
returned to San Pedro for both reasons.[1]

The army was different as well. It had long advocated for rails
across the West. Now, commanders aggressively pushed development, even allying with corporations and railroad interests. This
seemed, at the outset, consistent with Civil War idealism and
attempts to root out private influence from the contracting system.
Locomotives promised rationality and scale. They signaled evolv-
ing military know-how. To consolidate advances, the army in 1863

combined its former Topographical Engineers and Army Engineers into a unified Corps of Engineers. In 1867, it established a special board at San Francisco to manage its Pacific coast projects. The office answered to the War Department and the chief of engineers, yet it was semi-independent, operating alongside the army's Pacific Division and Department of California.[2]

Physical reconstruction of the continent would proceed in parallel with political reconstruction of the former Confederacy. Over the next decade, the Army Engineers planned improvements to Southern harbors, the Mississippi River, and tributaries. The Pacific Engineers would do likewise across the West's coastal and inland waters. Staff cleared channels and reinforced shorelines. They built lighthouses, defensive batteries, forts, and navigation improvements. In doing so, they laid the foundation for a new kind of American expansion, extraterritorial and maritime in quality. The work led back to familiar places. This was true for Williamson as well as for others. They would take up unfinished projects, relics that gained new meaning in the era of capital.

San Francisco, 1870. Bustling with commerce, but without connection to the transcontinental railroad. Entrepreneurs and army engineers looked to change that.

Among the San Francisco cohort of officers, Williamson had the most experience with California and Oregon. Tuberculosis caused his departure from the 1853–54 Pacific railroad expedition and forced him to take continual leave thereafter. The West's dry, clean air was best to stave off consumption. Nonetheless, Williamson felt compelled to serve in the Civil War. He became chief engineer for the Army of the Potomac, led by friend and former colleague George McClellan. Williamson's health deteriorated nearly at the pace that McClellan's relationship with President Lincoln did. To save his life, the army sent him back to the Pacific, where he became head topographical engineer. He stepped aside in 1867, when a more ranking officer, Barton Stone Alexander, arrived. Williamson then took charge of lighthouse and navigation projects.[3]

Alexander signaled a new importance for the Pacific coast engineering office. He had sterling credentials, having served in the US-Mexico War, in the Civil War as engineer for the Army of the Potomac, and later as chief engineer for the defenses of Washington, DC. Yet Alexander was renowned for two marvels he constructed during the 1850s: the Smithsonian Institution's "Castle," and the Minot's Ledge Lighthouse, built outside of Boston Harbor and considered an impossible project. Both structures still stand today.

The Pacific Engineers would work to promote settlement and commerce, tasks that put them at cross-purposes with reform-minded wings of the army. Alexander found it within his duty to advocate for merchants, politicians, and resource speculators who sought to gain by federal projects. In the late 1860s, he endorsed petitions for government levees and drainage gates to be built throughout the Sacramento-San Joaquin Delta. This was a boon to the San Francisco-based Tide Land Reclamation Company, which owned 120,000 acres of marsh. The company—with $12 million in capital, steam plow machines, and two hundred Chinese workers— promised to remake the delta into farms, homes, and profits. Alexander loaned his reputation to the company's proposal and assured investors it would pay a dividend. Whether he approved or not, the corporation featured his words in promotional literature. Did the interests of capital make for sound engineering and serve the national interest? The Pacific Engineers tended to think so. But not always. The office remained part of an army still reeling against cost and corruption. Every potential harbor and river town had its promoters, trying to finagle an appropriation. Alexander disappointed many.[4]

ENGINEERS, like quartermasters and Coast Survey officers, faced the difficult task of discerning a line between state and economy, between enterprise and graft, that did not yet exist. Hence, it was unclear how Alexander would respond when Wilmington investors demanded improvements to San Pedro Bay. In 1868, Phineas Banning and others petitioned for government harbor works. Locals (with Senator Gwin) had asked the same of the War Department and Secretary Davis in 1855. Davis, in turn, had deferred to the Coast Survey. This time, the Wilmington group put the request to Alexander while State Senator Benjamin Wilson introduced a bill of support. Timing seemed against them. The request coincided with the Quartermaster Department's investigation of Camp Drum and its dismissal of the estuary-harbor as a scandal. Yet Banning and associates gained ground. They cited the imminent completion of the Los Angeles & San Pedro Railroad, as well as the Central Pacific transcontinental. Alexander overlooked the speculative nature of the request and approved a survey.[5]

So, too, did Alexander's fellow commander and ostensible boss, Edward Ord, the newly appointed chief of the Department of California. Ord returned to the Pacific with rank and influence. During April of 1865, he had led Union-occupied Richmond, feeding and pacifying locals. In the confused hours after President Lincoln's assassination, it was Ord who counseled Grant against a mass arrest of city leaders. Weeks later, when Robert E. Lee first sought a pardon, he consulted with Ord. From Richmond, Grant moved him through a series of important posts: first, the Department of the Columbia, where settler conflicts with Indians raged. Luckily for Edward and Molly, Grant immediately rescinded the order. The *Brother Jonathan*, which ferried Ord's replacement, George Wright, struck a reef and sank off Northern California. Two hundred people perished, including General and Mrs. Wright and Anson G. Henry, friend and physician to the Lincolns. Grant sent Ord instead to the Department of the Ohio (later renamed Department of the Lakes).[6]

When Congress reinstituted military governments in the South, Ord commanded the Fourth District, headquartered at Vicksburg, Mississippi. After a local newspaper incited violence against Union soldiers, Ord had the editor, William McCardle, jailed and tried by an army tribunal. That spurred a key Supreme Court decision that sustained federal authority to pursue Reconstruction. Nevertheless, in early 1868, Ord asked to be reassigned. Back in California, he no longer had to settle for underpaid fieldwork or political spoils. Financial opportunity came to him, as it did to many Union generals. He perhaps considered his life's fortunes when, in September, he and Alexander dispatched R. S. Williamson to San Pedro Bay.[7]

The engineer had instructions to study the estuary and devise works to deepen its entrance. It was no small matter. Government infrastructure projects were not yet ubiquitous, as they later would become in the New Deal and World War II eras. The study also contradicted Williamson's prior judgment. Fifteen years earlier, he had written that San Pedro had "no harbor." Yet, in his report of February 1869, he concluded that harbor improvement was possible. This would serve the needs of Los Angeles and its Southwest wagon trade. It would do the same for anticipated railroad freight. Oddly, Williamson based his decision on the same evidence that Quartermaster inspectors saw as inefficiency and corruption. He even appended an endorsement letter from Banning. But there was an added twist. An Engineers project required significant new congressional spending. Williamson's report ran counter to those of army reformers who wished to abandon the estuary and cut off the flow of money.[8]

That was a hurdle that his commanders, Alexander and Ord, would need to clear. They did so with notable effort and by some mysterious turns. First, Williamson's findings detoured from the normal chain of command. In early 1870, Alexander endorsed the project. At that point, the decision should have gone to the chief of engineers and the War Department in Washington. But, before they could weigh in, Alexander's approval became public knowledge and was misunderstood—or misrepresented—as final. Benjamin Wilson's bill asking Congress to fund the San Pedro harbor works had stalled in the state senate. Representatives from the competing port cities of San Francisco and San Diego were the cause. To counter them, Wilson disclosed a letter Alexander had written in support of the project. The engineer claimed Wilson had made this revelation without permission. But whether his letter was off the record or not, Alexander nonetheless was willing to communicate internal army information to a legislator and land speculator. And Wilson was too cunning to keep useful secrets. Publication of the letter tipped the debate. The state senate approved Wilson's resolution and called on Congress to finance construction.[9]

Next, Alexander modified the specifications to make them more palatable to his money-minded superiors. Williamson recommended a jetty made of granite blocks. He placed this structure perpendicular to the western tip of the bay's sandspit, Rattlesnake Island, and extending to a squat promontory known as Deadman's Island. The design utilized a simple principle of hydraulics that army engineers had tested on the Mississippi River. It would constrict the tide's ebb flow, a daily river dissipated and shifting in nature, to a narrow channel. The total volume of water, estimated at 257 million cubic feet, would scour away the entrance bar as it flowed out to sea. But Alexander revised the design to a cheaper structure of wooden cribs, to be filled with sand and small rocks, secured between wooden piles. Wood was far less durable than stone when immersed in seawater; in fact, shipworms rapidly devoured it. The project's construction engineer later would be baffled at how flimsy and vulnerable it was to an errant ship, vandals, or wood thieves. Plus, timber was not standard material for army coastal improvements. Alexander knew this. He had championed the superiority of stone in his construction of Minot's Ledge Lighthouse, where even iron had failed. But he knew also that cost estimates matter, especially at Wilmington, a place the army deemed so dubious.[10]

PERHAPS THE OPINION of another swayed Alexander from his best instincts? The wood-crib design had originated a decade before, with the Wilmington developers. It appears in the November 1858 *Los Angeles Star* editorial, which defended the estuary's so-called port against criticism from George Davidson and the Coast Survey. But, given that the Wilmington investors were not engineers, they would have needed someone else to draft such a design. It is likely that the plan originated with their hired surveyor, Edward Ord. The jetty might have been the "breakwater" he suggested to Pacific railroad surveyors in 1854.[11]

In 1870, Ord was Alexander and Williamson's superior, charged with signing off on the project. They would have known of Ord's prior work at San Pedro. His name appeared on existing Coast Survey maps. But his private ties to Wilmington's syndicate were less known. Ord was good at holding his secrets, even as he continued to worry over them. He wrote to William T. Sherman of his efforts not to appear "bought" by the region's politicians and speculators.[12] Military contractors posed a particular menace. They used newspapers, so Ord alleged, to enflame the Apache wars, tighten their grip on federal money, and "keep the troops in their hands." Banning's *Wilmington Journal* certainly tried. Yet Ord's skepticism had to contend with his long-standing belief that the estuary was destined to be a harbor. How, exactly, this shaped his thoughts is unknown. The switch in materials might have been his. Or, perhaps Wilmington's investors revived the idea, knowing it would cross his desk? Ord disclosed no conflict of interest. (He did recuse himself from investigations of Camp Drum because his brother James Lycurgus Ord had supplied the post with horses.)

In any case, both Ord *and* Alexander reviewed Williamson's report. And the former agreed when Alexander predicted that a wood-crib jetty would increase depth at the entrance bar to eighteen feet. Congress agreed as well. In 1871, legislators appropriated an initial $200,000. Banning and Huntington, with their pending agreement to bring the Southern Pacific to the estuary, lobbied for the funds. Construction began the next year. Because of problems with contractors, saltwater damage, and cost overruns, the wood jetty would take a decade to complete. It produced only six feet of depth. Shipping continued to grow regardless, fed by more talk of transcontinental railroads.[13]

This time, there would be no inspectors to decry the jetty's ineffectiveness. Those voices had moved on. James F. Rusling, the army's harshest critic of Wilmington, left to practice law in New Jersey. Henry Halleck, who had campaigned to clean up the Pacific Division, transferred away and was in failing health. Williamson's declined as well. Although only forty-six years old in 1871, he retired from field duty and abandoned his life's work. Another officer, George H. Mendell, assumed charge of coastal improvements. (Mendell reported the jetty's shortcomings and revised the design back to stone riprap.)

Ord also moved on. By the end of 1871, he had become commander of the Department of the Platte, headquartered at Omaha Barracks. Molly arrived with their seven children, including a baby who died the next summer. Despite personal loss, the move was a step up for the Ords. The middle Plains was more of a center of industry and culture than California, and much less foreign. Edward promised Molly better climate and house servants—"white" rather than Chinese.[14] It also remained a hotspot for Indian conflict, as the transcontinental railroad set off destructive changes. In that context,

Ord sent out the first detailed military survey of the Yellowstone region. Quietly, he continued to shape facets of modern America.

Proponents and doubters of the estuary-harbor each made their way forward. But not all. Williamson lived just long enough to see his railroad route through Southern California vindicated. He died at San Francisco in November 1882, his lungs consumed and no longer able to breathe the ocean air. Which fate was worse? Elsewhere, government reformers would live to see their past words proven incorrect or simply ignored.

Chapter 23

RAILROAD SWINDLES

M UCH HAD CHANGED BY THE TIME GEORGE AND Ellinor Davidson resettled at San Francisco in 1868. Wealth recast the rough-and-ready town they once knew. Mansions and hotels spread like weeds and choked out the clapboard modesty of the Gold Rush. There was even an opera theater. The scientist enjoyed refinement. But, in this case, he knew its cause and worried. He complained to his brother of "gigantic railroad swindles" that now characterized the West.[1] Government had financed the frauds, building up private fortunes while denying the Coast Survey much-needed funds.

Davidson continued to distrust grifters and speculators. More so because he felt their temptation. Nearly everywhere, Americans succumbed to railroads, their opulent chiefs, and heady promises. Some guarded against these enticements. Yet, most would be overwhelmed by the times. The railroad economy was a vortex that pulled people in—or else cast them aside. Initially, Davidson refused. He spent months updating his mariner's guide, the *Coast Pilot*. Its revised 1869 edition continued to dismiss San Pedro's estuary as narrow and impassable. Subsequent events forced him to relent, however. The Coast Survey and its science would either follow the money or wither away. For self-preservation, Davidson made his compromise. He was not alone.[2]

That arc started years earlier and far away. The Civil War halted the Survey's routine work, but gave it unprecedented stature, too. Since the 1850s, the bureau had amassed data on Southern harbors

*The Palace Hotel, 1875—a sign of industrial
wealth in San Francisco*

and rivers, like the Potomac and James in Virginia and the Savannah at the Georgia–South Carolina boundary. Maps offered military advantage—that is, if Unionists could secure them before Confederate spies did. Superintendent Bache led a hurried effort to gather items from libraries, desks, and bookshelves and place them in a safe repository. Meanwhile, he strengthened the hand of scientists at the capital. He helped found the US Sanitary Commission and the National Academy of Sciences. The former set standards for military camps, hospitals, and nurses. The latter advised Congress.[3]

Nevertheless, Bache struggled with the betrayal of sectional union, especially by some of his closest friends. The rebellion's near success left him in despair, precipitating his physical collapse—likely a stroke as well. By the autumn of 1864, the fifty-eight-year-old was an invalid, dependent on his wife's care and sinking away in spirit. "It is very hard for a man like him to do nothing," Mrs. Bache confided to Ellie Davidson.[4] Officials sent him on paid leave to Europe. When the war ended, George Davidson brought him home. Former students shed tears at their mentor's condition: "Instead of the intellectual giant, I found one whose mind was shattered," a visitor recalled. Bache remained superintendent until he died in February 1867. Washington newspapers listed the cause as "softening" of the brain. Others said the Civil War killed him.

Without Bache, the Coast Survey's prospects began to dim. President Andrew Johnson named Benjamin Peirce as superintendent. Peirce embodied the science of the era, which included his affinity for scientific racism. He was a world-renowned Harvard mathematician, director of the college's observatory, and a member of Bache's Lazzaroni. At first, he revitalized the Survey. His initiatives included sending Davidson to Alaska in 1867 with the instruction that "Americans must control the Pacific."[5] That sentiment was widespread in the capital. It bridged partisan debates over Reconstruction. Davidson's work helped persuade the naysayers. He presented his 140-page report on Alaska to Secretary of State William H. Seward. Congress finalized the purchase the next summer.

Although Peirce brought academic repute, the Coast Survey lost its former political standing. The new superintendent had few of Bache's connections in Washington, and fewer among the military. Bache was well liked around the capital. Peirce preferred life in Massachusetts. Politicians were also different now. Their attention drifted toward lobbyists and investment schemes. The capital's salons and scientific clubs winnowed. Most of all, Peirce lacked a strategy to keep his agency central to the US's pursuit of empire. Instead, the army and navy intruded on Coast Survey turf.

Davidson grappled with the changing situation. Peirce assigned him to San Francisco to head all Survey operations but hydrography. Fiscal constraints in an age of excess angered Davidson. He grumbled more when railroad corporations linked both ends of the continent far ahead of his attempt to do so with geodesy. Confidence men in fine suits received the nation's praise. Davidson still had his youthful ambitions, but he could no longer keep up with them. He was middle-aged now, and a father. Rheumatism, neuralgia, and eye injury kept him from fieldwork. Administrative duties diverted his time. He often sent out crews while he stayed in the city to compute results. Relaunching studies of the Washington straits and San Pedro–Santa Barbara Channel became a priority.[6]

For someone as expert as Davidson, numbers and shapes could elicit joy, but also frustration, even alarm. Since the mid-1850s, he had been aware of discrepancies in the channel study. However, in January 1870, Davidson noticed startling errors. These appeared as mislocations when he compared recent data against that taken before. Locally, the errors were small—in the range of one to five miles. But, because geodesy uses compounding calculations—its large-scale triangles are assemblages of small triangles—mislocations cascaded upward. Davidson found that the Southern California errors threw the entire map of the Pacific shore off position. They affected three of the four vertices used to position the Washington Territory; two of the four used to locate San Francisco. Most disconcerting, mislocations clouded the US–Canada border and threatened an international incident.[7]

To correct the mistakes, Davidson returned to San Pedro. He spent several months there in 1870 and 1871. The work required him to retrace old steps. He measured from various stations, including the one Edward Ord had located near the Domínguez home. Along the water, he searched for wooden scantlings Ord had placed to mark his sightlines. Once Davidson reconstructed the small triangles, he reconnected them north to Monterey, San Francisco, and Seattle. Without fanfare, he stabilized the republic's map and its budding hemispheric designs.[8]

Davidson's stay at San Pedro coincided with approval of the jetty project by the Pacific Engineers and Congress. In private, he objected to the jetty and disputed its design. He met with Barton S. Alexander to challenge the officer's rosy predictions. Davidson criticized railroad influence on this and other scientific matters. He alleged to Peirce that the Southern Pacific had threatened to send its lobbyists to block Coast Survey funding. At the same time, the chance of gaining dollars from railroad projects proved irresistible. In 1870, he complied with Alexander's request to supply hydrographic charts of San Pedro Bay. Finding these wanting, Davidson ordered his field crew to make a complete survey of the estuary, the first of its kind. This resulted in the 1872 USCS chart, which shows the bay in detailed relief.[9]

Davidson found that pure science could not win against the railroad age. Some accommodation, however, might sustain it enough to continue as a relevant voice. He advised Peirce to pursue projects agreeable to the Army Engineers and railroad companies. Among the suggestions, Davidson's views on San Pedro shifted. He wrote from Domínguez Hill about the "success" of Banning's railway and the "rapid selling" of real estate along the line.[10] (Davidson owned a home near LA's Elysian Park, as well as properties in San Francisco. He was no speculator, but he was no fool, either.) The estuary, he told Peirce, had become "the port of Los Angeles." He likewise proposed work in Alaska, Mexico, and Panama. All had scientific value. They also had constituents in politics and the private sector. Davidson's grandest idea was for a survey (later known as the 39th parallel arc) to extend geodetic lines eastward across the continent. The project relied upon the transcontinental railroad and telegraph. It revealed where and why the scientist would adapt to the times.[11]

So it was with San Pedro. While never his harbor of choice, the estuary altered Davidson's cartography, and ultimately the shapes he chose to see.

RAILROAD SWINDLES came in many forms. Some made casualties of insiders. Such was the case for William Hyde. The engineer, admirer of orange groves, and doting husband had succumbed to the promise of wealth years before. After his own schemes faltered, Hyde joined the Southern Pacific (SP), which sent him to squeeze taxpayer money from Los Angeles. He returned there in June 1872 to procure the essential piece, a city-county subsidy. He found victory less assured than three years before, however. Rival interests stirred up opposition. Critics had heard about the SP's bullying and extortion of towns in the San Joaquin Valley. If that did not make Hyde's job difficult enough, the company embroiled itself in yet another controversy. After dismissing its Central Pacific workforce upon the line's completion, the SP received blame for high unemployment and for having imported low-wage Chinese laborers. Statewide, an anti-railroad political movement began, led by the Workingmen's Party. The Chinese became the easier target. Los Angeles had experienced a bloody anti-Chinese riot eight months before Hyde's return. In a space of several years, the company traded away its reputation as the herald of a better future. It now seemed a manipulator of politics, robber of public funds, and cause for social unrest.[12]

That was not what Hyde had signed up for. But his bosses had their ways, and he had no choice but to please them. He doled out corporate money to shore up the SP's eroding position. Voters narrowly approved the subsidy in November. Still, executives were displeased. They refused to admit that their own deeds were at fault. Instead, they blamed Hyde. This influenced what would happen next.[13]

Once voters agreed to pay for the trunk line, Hyde turned to Wilmington. The SP needed a harbor to bring in construction materials. Hyde envisioned something bigger. In January 1873, he proudly sent Collis Huntington an engineering plan that would transform the estuary into the SP's permanent rail–water terminal. "The final desire," Hyde wrote elsewhere, "must be to bring the ship and car together."[14] He proposed that the War Department pay SP subsidiaries to build (and thereafter control) the waterfront. He predicted that machines could complete the job within one year—well before government had time to reconsider. Hyde even suggested that the SP take charge of the federal jetty, construction of which had stalled. "Ownership and practical control of that little harbor," he counseled Huntington, "will be one of the very handsomest pieces of property that the Company can own."

Hyde glimpsed the future. Huntington did not. As the SP's black-bearded wizard of finance, he did not take advice from just anyone. Plus, the company faced severe problems brought on by its reckless growth and constant battle against competitors. As banks collapsed and economic crisis spread, the SP teetered on insolvency. And Hyde was asking to spend more money! Huntington rejected the estuary plan and opted for a limited strategy. The SP finalized its buyout of Banning's railroad, wharf, and warehouses. (Banning's operation would conduct all lighterage and freight handling.) This cemented an alliance, formal and informal, that would last five decades to the estuary's demise. However, once construction reached other supply sites, the SP lost interest in Wilmington. A connection to San Francisco opened in August 1876, the Colorado River one year later, and New Orleans in early 1883. Later, Huntington would relocate shipping operations to deepwater piers at West San Pedro and Santa Monica.

Among swindles of the age, some were self-inflicted and slow to appear. Neither the Southern Pacific nor Banning's company developed the estuary to the extent that Hyde envisioned in 1873. They would not devise a comparable plan until too late. Rather than truly integrating ship and rail, the SP disabled ocean transportation in order to divert freight over land. This was profitable. But by passing up Hyde's advice, Huntington missed his best opportunity to take charge of San Pedro Bay.[15]

Instead, Huntington decided that Hyde's usefulness, much like the estuary's, was temporary. Company leaders refused the engineer-lobbyist's requests for promotion and stripped him of the power to make decisions. They named stations along the rail line (the route of today's Interstate 10) for Banning, D. D. Colton, and other underlings, but not for Hyde. They grew rich; he did not. Following numerous slights, Hyde resigned in frustration. However, his luck of being in the right place at the right time had gone. Lack of work, due to the economic depression, forced him to beg for his old job. He received only intermittent positions. While scouting mines in the Idaho Territory in 1882, Hyde sickened from erisypelas, a flesh-eating bacterial infection. He died at age forty—far from the fortune that once seemed so close at hand.[16]

PROMOTION OF THE estuary-harbor caught elements of the US Army off guard. They too would need to choose which swindles to accommodate and which to cry foul over. Contradictions abounded. In 1871, while jetty construction progressed, the Quartermaster Department completed its closure of Drum Barracks. Ownership of the land reverted to Banning and Benjamin Wilson. The two asked to buy the camp's buildings at a bargain price. The army refused and held a public auction. Banning and Wilson won every bid.

The end of Drum Barracks did not quite extricate the army. Instead, it became caught in litigation over the flume that supplied the base with water. In 1869, Banning filed suit against Manuel Domínguez to gain both the wooden aqueduct and the rancho land on which it stood. At coastal Southern California, as in the West generally, water equaled agriculture, real estate, and big profits. Rails and harbors multiplied the potential gain. Domínguez knew this and insisted the flume was his. It was the first sign of future conflict between the families. Banning's firm had built the structure, but the army had paid for it. He expected to gain private water at public cost. And why not? Despite losing the barracks, Banning had a successful record of persuading commanders. This time, they fought him. On advice from General of the Army William Sherman, Edward Ord dispatched an attorney who claimed the flume as federal property. Litigation continued for six years until Banning lost in court.[17]

The ongoing scuffle over Drum Barracks caught the attention of Quartermaster General Montgomery Meigs. At the close of 1871, he prepared for his first visit to the Pacific coast and thought to see Wilmington for himself. The quartermaster traveled west on the Union Pacific–Central Pacific line. The UP's financial crimes were legend by now. That did not blunt Meigs's enthusiasm for rails. When heavy snows delayed the train for three days at Cheyenne, Wyoming, Meigs pondered how a railroad across the Southwest would do better. For some reason, he missed seeing that the anticipated desert line was the cause of his ongoing headache at San Pedro Bay.[18]

A greater object obstructed his view. As he traveled, Meigs enthused about how transcontinental railroads would open the Pacific to US prospects. That was his visit's purpose. He reached San Francisco by mid-January of 1872 and marveled to wife Louisa, "the city is American and yet foreign unlike anything else."[19] He shopped for robes of Alaskan lynx. He visited the Cliff House—a landmark restaurant— and took pleasure in watching sea lions and pelicans below. On the twenty-third, Meigs attended a banquet at the Grand Hotel for Japan's delegation. Minister Iwakura Tomomi and General Yamada Akiyoshi, leaders of the Meiji Restoration, had arrived with a group of men and women to study US institutions, schools, and railroads. Meigs joined them, and his fellow generals Barton Alexander and John Schofield, along with a set of the city's grandees. Together, they cheered ties between the countries with no idea of where that would lead in seventy years, just one lifetime. Instead, they toasted "civilization," friendship between the "Oriental" and "Caucasian race," and the two nations' shared battle to unify internally.

There was more at work. The Grant administration pursued a number of programs in the Pacific besides those in the South and West. It sought to annex British Columbia. And it soon sent Alexander on a mission to the Hawaiian Kingdom. There, he secured US military rights to Pearl Harbor in exchange for a free trade agreement. Attempts to acquire a similar concession at Pago Pago, in the Samoan islands, failed for now. American politicians had long talked up expansion into the Pacific, but army leaders like Grant, Meigs, and Alexander were different. They were builders. They knew how to wield supply chains, coal stations, and raw military power. Fresh off victory in war, they stepped into the world with methodical confidence.[20]

*New foreign relations in the Pacific: members of the Japanese
delegation at San Francisco, 1872*

Civil War, Western railroads, and Pacific designs: Meigs had all
of this on his mind while he steamed to Wilmington. He sized up the
troublesome harbor, then rode Banning's railroad to Los Angeles.
The mix of adobe and Victorian buildings fascinated him. Meigs
returned east from San Diego by stagecoach. Back in Washington,
the quartermaster attended another dinner, this time to honor the
army's desert explorers John Wesley Powell and George M. Wheeler,
whose exploits would bring more rails and settlers.[21]

US expansion was alive with renewed vitality, the steam engine
its beating heart. Like a good soldier, Meigs lined up behind. That
did not keep him from reiterating that the decade-long Camp Drum
debacle was an "absolute and shameful waste."[22] Somehow, the two
thoughts coexisted. That put Meigs in a further conundrum. Given
that San Pedro Bay was a place of past scandal and future investment, *who now would listen to his complaint?* The country was half
deep in an age of corruption. Meigs tendered his report to an exemplar of the times, War Secretary William Belknap. Belknap used his
office to encourage fraud while he and his wife collected kickbacks.
He was so obscene in shorting frontier forts and tribal annuities that
George Armstrong Custer eagerly testified about the crimes to Congress weeks before his death at the Little Bighorn. Belknap resigned

in disgrace, damaging President Grant's policies on Reconstruction
and Indian affairs.

Meigs grew old and retired. A younger generation of officers
remained. They included George Mendell, the engineer supervising the San Pedro jetty and all harbor improvements on the coast.
Initially, Mendell worked to assuage private interests. He soon lost
patience. San Francisco's land speculators, so he warned his army
superiors, jeopardized the "dignity of the Government and hence its
ultimate interest."[23] Private claims at San Pedro's estuary, he wrote,
would do the same. Mendell's words echoed those of prior dissi-
dents. They hinted that the tide might switch against the railroad
and its ilk. But not yet. For time being, the scientist-reformer stood
in the wilderness of industrial America.

Chapter 24

AN OLD RANCHERO DEFIES THE TIMES

B Y 1869, IT WAS A TIMEWORN AMERICAN ADAGE THAT one person's misfortune is another's opportunity. Still, in an era when wealth might be suddenly made or lost and when people often were not what they seemed, sorting the two fates could be difficult. The *Daily Alta*'s Richard L. Ogden tried anyhow. Word had reached the newspaper that Los Angeles's old cattle ranches were bankrupt, devastated by floods and drought. Abel Stearns, the Yankee Californio and largest landholder, saw his 200,000 acres reduced to dust, bones, and debt. Rumors circulated that the Coronels, Sepúlvedas, and Picos faced ruin, too. They were desperate to sell. San Francisco investors looked to buy.[1]

Ogden arrived at Wilmington on one of Banning's vessels. The new railroad completed his journey to Los Angeles. Weeks later, the journalist sent in his story. He described a society in transition, as one nation and race—in his telling—surpassed its predecessor. The place remained shabby, Ogden thought. But he found its "Spanish plumage" giving way to a "pinfeathering state of Americanism."[2] Steam engines, Anglo property, and the ranchers' downfall gave signs of progress. Yet Ogden struggled to make sense of an anomaly. One holdout defied his narrative. One name was absent from the list of failing Californios. That was "Old Don Domínguez," he wrote. "There he sits in lonely grandeur, defying alike taxes, Yankees, and compound interest—nothing can induce him to sell." Ogden presented his readers with a relic: the defeated figure in denial of the times. The last man standing. The epitome of misfortune.

Ogden was only half correct. And this made him entirely wrong. His commitment to a narrative of US conquest and white supremacy disguised opportunity and to whom it truly belonged. Amid the new economy and the failure of others, the Domínguez family remained adaptive. With Anglo son-in-law George Carson as manager, Domínguez shifted his stock to include sheep, which proved quite resilient in drought. He acquired lots in Wilmington. He allowed Banning's railroad a seven-mile easement and prepared to sell town parcels along the route. The family would exploit its groundwater and, later, petroleum rights. Most important, they avoided any duress to sell. If there was to be a railroad and harbor, the clan Domínguez determined to wait and get the better price. Ogden, in his ham-handed ways, missed seeing that they shared the same calculus as the *Alta*'s readers. The heirs of Spanish conquest remained positioned to take the windfall wanted by others. Their patriarch was no relic. Quite the opposite.[3]

A modern rancho. Contrary to newspaper accounts,
the Domínguez family continued to be successful
in the era of capital.

Events would prove Ogden wrong in one more way. He celebrated Anglo enterprise as the moment's defining force. Yet at San Pedro Bay, the Domínguez family held the upper hand through a combination of acumen and luck—that most celebrated of Yankee virtues. And they did so precisely because of US actions. By holding off large sales for decades beyond their neighbors, the family enabled legacies of Southwest conquest to combine with growing Pacific imperialism. This potent mix would reshape the bay's development. But first, it came together in a legal contest, over a government matter, as the family sought to restore their estuary claim. In doing so, they demonstrated indisputable stakes in the future—no matter what Ogden or anyone else thought.

As Los Angeles grew, so too did the Domínguez family's advantage. Construction of the Southern Pacific trunk line increased land values. By the late 1880s, the Atchison, Topeka & Santa Fe arrived and the Salt Lake Railroad approached. The county's population of 15,300 doubled during the 1870s, then quadrupled during the 1880s to 115,000. Total assessed value of property rose accordingly. Lumber for new buildings crowded the ships inbound at San Pedro Bay.[4]

The family itself changed also. Manuel died in late 1882. His wife, Engracia, followed months later. Their passing put the twenty-four-thousand-acre rancho in the hands of the next generation. Like their parents before them, the Domínguez heirs would adjust to the times, protecting their equity and future profits. They started with a problem, however. Manuel's and Engracia's wills each named the other as sole beneficiary. This put the estate in probate for the county court to distribute. Six daughters, aged between thirty-five and fifty-five years, stood to inherit property, livestock, and other assets. Only one of them, Victoria, was married. Two, including the eldest, Ana Josefa, were widowed. The court appointed Victoria's husband, George Carson, to administer the process. That reflected the gender and racial assumptions of the time. Yet the sisters knew to use these to their benefit. Carson was keen, politically connected, and loyal. He served as their public figurehead to the end of his life.[5]

Following probate, the Domínguez women requested a partition of the rancho. That commenced a second, more important court process, which required detailed surveys and maps. The family hired George Hansen, an esteemed civil engineer, for the job. Hansen was born in 1824 to a family of Baltic or Scandinavian heritage in Fiume (now Rijeka), a port city on the Austrian Empire's Dalmatian coast. The Hansens were sojourners. And George maintained the tradition. He sailed around Cape Horn, lived briefly in Peru, and arrived in California by 1850. He even ventured to Japan circa 1871, quite early in Americans' ability to do so, and wrote to George Davidson of the experience. (Davidson would visit Japan in 1874, then continue around the globe.) Hansen was fluent—in fact, apt to recite poetry or philosophy—in Spanish, German, French, and English. More than any other character of this story, he personified California's global borderlands and the international mobility that the Port of Los Angeles represents today.[6]

As an engineer, Hansen regularly found work at Rancho San Pedro and throughout the wider region. In 1855, he placed the rancho's outer property line in relation to the public lands survey. The work enabled Domínguez to confirm the grant under US law. Hansen then was elected city-county surveyor, an office he held off and on into the 1880s. He helped develop Los Angeles's private water utility and was considered the local expert decades before anyone had heard of William Mulholland. Hansen also laid out the colony of "Annaheim," set up by German-speaking vineyardists and famous today for its Disney amusement parks. His mark was everywhere that real estate or irrigation existed. He liked to boast that, in an age of industrial unrest, his work fostered stable property and law. To the degree that this was true, his Rancho San Pedro partition would prove the exception.[7]

Hansen and crew began working the Domínguez Ranch in June 1884. They cataloged its pieces into "bottom land," "highland," "lake farm," "prairie," and "dunes," which the surveyor bundled equitably for each heir. By the spring of 1885, the court had allotted eight tracts, seventy-four parcels, and water rights among the Domínguez sisters. To promote future development, they set aside three hundred acres for county roads. Draft maps of the partitioned estate set off excitement. Newspaper advertisements enticed buyers to a coming real estate rush.[8]

Yet the partition remained incomplete. Hansen needed to segregate a ninth "Estuary Tract" that included mudflats, marsh, and Rattlesnake Island. Due to its potential as harbor frontage, the Domínguez sisters agreed to leave the 1,760 acres in undivided, equal shares. Hansen's task was to determine its boundary. And this hinged in part on the line between the rancho and Wilmington. Elsewhere, it followed the line of so-called navigable waters. Both originated in Edward Ord's 1855 survey and map of the inner harbor. However, Hansen could not locate Ord's key landmark. The officer had chosen a spring called Los Barriles, known thereafter as "Ord's wells," which lay on scrubby uplands west of Wilmington. The wooden barrels that once identified the site had vanished. Documents from Ord's survey were gone, too. Through interviews—of local clerks, old-timers, and Indian ranch hands—Hansen finally tracked down a map. He credited the find in his diary to the "help of Juan Toro."[9]

Even with the map, Hansen could not resolve the tract's boundaries. Los Barriles had served as the indispensable reference point for all property claims derived from the rancho, Wilmington, and the US reservation for navigable waters. The number of surveys multiplied after Wilmington's subdivision in the mid-1860s. The Southern Pacific's arrival set off a second round, as Banning and the railroad snatched up tideland claims. These various surveys disagreed on the exact location of Los Barrilles. Ord's process muddled things further.

He executed a meander survey, a technique used to chart the perimeter of lakes or other inland water features. The methodology was all about convenience. Meanders were done on foot or horseback without getting wet. They avoided difficult astronomical measurements. Ord recorded his meander line relative to the start and end point. That landmark's discrepancy thus threw off the entire line.[10]

WAS HANSEN SURPRISED? It was not the first time he had found fault with one of Ord's maps. In 1869, while city surveyor, Hansen disputed the location of "Ord's Rocks," a landmark in the soldier's Gold Rush-era plat of Los Angeles pueblo. Apparently, Ord's sense of precision sufficed then, although no longer. But to take issue with Major General Ord was a daunting task, given his reputation as a Civil War hero. All the more so in 1884.[11]

Ord never again lived in California following his transfer to the Department of the Platte. In 1875, he moved to the Department of Texas, another front in the US war against Indian rebels. While in command, Ord made a contribution to international law and rules of engagement when he permitted US troops to pursue Native American riders into Mexico. His career hit its own limit soon after. Upon Ulysses Grant's exit from the White House, President Rutherford Hayes declared an end to Reconstruction and chose other generals to promote. According to rumors, Hayes forced Ord to retire in 1880 to clear an upward path for Nelson Miles. Triumphs against the Comanche, Cheyenne, Lakota, and Nez Perce (later, too, Geronimo's Apache) had earned Miles great acclaim. General of the Army Sherman, Ord's frequent advocate, could not object. Miles had married Sherman's niece.[12]

In retirement, Ord chased the wealth that he had dreamed of as a younger man and that now seemed so easy. He ventured to Mexico City under special passport as a liaison for American interests, including the Southern Pacific and Standard Oil. He set up a company, the Mexican Southern Railroad, with backing from Grant, Collis

Huntington, and Mexican presidents Manuel González and Porfirio Díaz. Edward married off his daughter Bertie to general Jerónimo Treviño. Borderlands strategies—marriage and business—remained a basis of partnership just as in 1850s California. Dangers lingered, too. In 1883, Ord steamed from Veracruz to New York to gather investors. It was the reverse route taken by US invaders four decades before. He contracted yellow fever and died in Havana on July 22, at age sixty-four. Neither Edward nor Molly (who lived a decade more) wrote memoirs, so details of their past died with them. Sometime after her husband's passing, Molly penned a letter to son Eddy Jr., worried over the family's stake in the Mexican railway and "powerful men" who might steal it. "Write to me soon, and remember me," she closed, "I am so lonely."[13] Edward Sr. received full honors upon his burial at Arlington Cemetery. Patriotic elegies failed to note how Ord's pursuit of frontier opportunity finally had done him in.

Hansen began the Rancho San Pedro's partition only months later. After studying the estuary, he returned to the Domínguez heirs with startling news. By accepting Ord's map and the discrepant surveys linked to Los Barriles, the US government had altered the Rancho San Pedro in violation of its own laws. Hansen reasoned that the family could claim back eight hundred acres (or seventy-six percent) of the navigation reserve that Ord had expropriated. The moderate amount of land belied its significance. The estuary had become valuable, acre per acre the most valuable portion of the estate. Lands gained their value as a direct result of federal spending and two powerful recipients: the army and the Southern Pacific Company. The sisters must have appreciated the irony. They would reassert their family legacy as historic justice.[14]

ACROSS THE BAY SHORE, another family thought much the same. Phineas Banning had established his stakes at the estuary entirely atop Ord's work. He filed the majority of tideland claims, joined by adult sons William, Joseph Brent, and Hancock. By a web of con-

veyances and deeds, the Banning claims passed under sway of the
Southern Pacific. Like the railroad's political influence, the connection between the family and the corporation grew more pervasive as
it became less visible. In the early 1880s, Banning's company leased
back its wharf and yards—an arrangement hidden and made to look
like a sale.[15] Yet it continued to function as the holding company for
the SP's real estate. This became clearer during a lawsuit filed against
Banning by another speculator-developer. Maneuvers hint at a test
case. The suit coincided with a new state constitution and changed
laws regarding tidelands property. Banning certainly understood the
case was about the Southern Pacific. "Certain parties are endeavor-
ing to steal some of our tide lands," he wrote to a SP representative.
He did not name the parties, but he ensured that "the interests of the
railroad are and will be fully protected."[16]

Read in retrospect, Banning's words predict how his family's fortunes would rise and fall with the railroad's. They also provide a
fitting coda to his life. Banning professed independence, but the old

*Undated image of "Indian workmen" hired by Banning "to
dig canal from slough to railroad tracks." Native people
provided labor at Wilmington and Domínguez Ranch into the
twentieth century.*

world of frontier proprietors was done, swallowed up by big capital. In 1883, the former wagon entrepreneur traveled several times to the railroad's home base, San Francisco, on business. There, he received a prestigious honor. Historian Hubert Howe Bancroft interviewed him for a volume on the state's Anglo founders. At the time, the Southern Pacific—rather than the port—seemed Banning's qualification. That was for better and worse. Banning passed through the city on his way to testify in Ellen Colton's case. The widow had learned that SP executives massively undervalued her inheritance of company stock. To make the point, Ellen released to the court (and newspapers) her dead husband's letters with Collis Huntington, which the latter meant to be burned on receipt. Contents unveiled the SP's extensive frauds and bribery of government officials.[17]

Banning became caught in the scandal. At Huntington's behest, David D. Colton had orchestrated a means of "convincing" ("fixing" was Huntington's preferred term) Arizona legislators to subsidize the trunk line. Lawyers summoned Banning to explain large amounts of money, which the railroad had sent to his expense account, that he paid out to unknown parties. The Colton case proved embarrassing. But, to the most shameless, this was no matter. By enlisting middling proxies, men like Banning and Colton, the SP executives had enough deniability. And that gave the company a monstrous life of its own. In response to a lawsuit over the railroad's unpaid taxes, the US Supreme Court in 1886 would establish the principle of "corporate personhood." This entitled corporations to numerous civil liberties such as equal protection and free speech. The SP gained in its right to evade regulations and to poison politics with dark money. Its interests were protected indeed.[18]

What did the railroad's accumulating power and indignity leave for operatives like Banning? The SP had raised him to greater wealth and prestige. Yet he made trade-offs, reminders of which sometimes rushed back unexpectedly. During one of Banning's visits to San Francisco, he stepped incautiously off a rail trolley and into the path of an oncoming express wagon. Horses and wheels ran him down.

Injuries from the accident compounded his suffering from kidney and liver disease. Banning traveled far and wide—on the railroad—in search for a cure. He never found it. He died in March 1885, in a San Francisco hotel, at age fifty-four. Only days prior, the Los Angeles Supervisors awarded him a franchise for a new rail wharf inside San Pedro's estuary. The Southern Pacific provided muscle behind the scenes and would develop the parcel, which sat at Mormon Island, a high patch of mudflats southwest of Wilmington.[19]

Hansen and the Domínguez family took note. The sisters filed suit in May 1886 to eject all trespassers from the bay. Their target was the Banning claim to Mormon Island and its basis in Ord's meander. The case would contend among the legacies of two recently deceased fathers. Judges would conclude that one—and ultimately *both*—were relics. That would be opportunity for an unexpected claimant to arrive from the distance.

Chapter 25

CONQUEST OVERTURNED, THEN EXPANDED

G EORGE HANSEN SPENT THE LAST DAYS OF JULY 1884 counting chain lengths on a bluff above the estuary. Conflicting documents sent him back and forth, hunting for stakes, posts, and water wells. He hoped they might point to what was now the most valuable property line in the county. Try as Hansen did, the numbers did not add up. As the surveyor recorded in his field book, he found "nothing" to locate the site "with certainty." That cast doubt on the boundary, established by Edward Ord thirty years before, separating Wilmington from the Domínguez family's Estuary Tract. The unknowns set an idea in Hansen's mind as he squinted toward Mormon Island and railroad laborers busy at work.[1]

Hansen's doubts became the lawsuit *Ana J. Domínguez de Guyer, et al. v. William Banning.* The case began as a spat between the Domínguez daughters and Banning sons over real estate. Over the following decade, it evolved into a reckoning of the US conquest of Mexican California, and then again into a prospectus for overseas expansion. Hansen foresaw this vaguely. It would be others who defined US Pacific imperialism and made it a reality. Yet his footsteps on those July days set off a chain of events leading to the nation's capital, farther across the globe, and back again.

WHO OWNED MORMON ISLAND? Litigants disputed this question in narrow terms initially, as they considered the local real
estate market. The Domínguez family petitioned that they had
kept possession of the nineteen-acre parcel as part of their Rancho
San Pedro, that they had paid taxes until the county mysteriously
stopped assessing the land. The Banning and Southern Pacific
companies, they said, trespassed illegally and now owed rent and
damages. William Banning's lawyers countered that he had paid
the most recent taxes on land and improvements, which included
a lumber yard, storage building, and railroad track to Wilmington. They presented Banning's deed and, most important, a patent
from the US Land Office. Banning testified that he had purchased
it from a prior occupant who, in turn, had gained it under federal preemption law. That pointed to a larger controversy. But the

*(Left) Seen here circa 1900, Ana Domínguez de Guyer,
eldest daughter of Manuel Domínguez, was the first
plaintiff in the estuary lawsuit against the Banning sons.
(Right) George Hansen, circa 1860—his work as surveyor
and historian made the suit possible.*

superior court judge ignored this, and in 1888 awarded the parcel to Banning.[2]

Had the lawsuit ended here, it would have remained without much significance. Instead, the Domínguez family appealed based on the testimony of George Hansen. Banning's lawyers had pressed Hansen to answer a simple question: Did Mormon Island lie within the US navigation reserve, as described by Ord's 1855 map? The surveyor replied that, yes, according to the map, the parcel lay within the reserve and not within the rancho. But, he added, Ord's map itself was wrong. Mormon Island had been erroneously removed. And it rightfully belonged to the Domínguez heirs.[3]

Hansen's statement was astounding. It transfigured the case into an appraisal of US actions since annexation. Attorney Sherman Otis Houghton made this the centerpiece of the Domínguez appeal, first for retrial and later to the California Supreme Court. In particular, he questioned the term *navigable* as used in Ord's survey. Since the US Land Office characterized the parcel as an island, Houghton argued, it obviously lay above the tides. It certainly could not be an island *and* navigable water at the same time. The concept of navigability was murky, one whose science was not quite worked out. At the time, it relied on a threshold of two-foot depth (or enough for a small boat to float freely) in areas below the line of "ordinary" high tide. But what was navigable about a place where depths disappeared for much of the day?

By this turn, the court veered into an uncertain question of the nineteenth century: What did conquest among sovereign republics mean? The US had invaded California and imposed its laws. But in doing so, Houghton argued, it assumed preexisting limits and obligations from Mexico. Foremost were provisions of the 1848 Treaty of Guadalupe Hidalgo and the 1851 California Land Act, which stipulated protection of Mexican-era property rights. This overlap of two legal systems kept the US from unilaterally changing property boundaries. And yet, in the case of the Rancho San Pedro, it had. US land commissioners, in 1854, decreed the rancho's Mexican-era lines confirmed and valid, fulfilling the treaty and the will of Congress. Months later, Ord's map emended this validation, based on false claims of navigability. By Houghton's logic, the latter was null and void. More important, if Ord's map could not divest the Domínguez family of Mormon Island, it could not do so of lands elsewhere in the estuary. The argument triumphed in December 1890. In a ruling authored by Justice Charles Fox, the California Supreme Court returned not only Mormon Island, but *the entire estuary* to the Domínguez estate.[4]

But Justice Fox had an agenda, one that had little to do with the Domínguez family's justice. He had already become involved in a similar lawsuit, *United Land Association v. Knight*, which pertained to the Mission Creek waterfront of San Francisco, another land grant and estuary property controlled by the Southern Pacific. It was a far more lucrative parcel than Mormon Island. And Fox knew it. He authored the *United Land Association* opinion in September. Three months later, he used it as precedent for the *De Guyer* case. He then left the bench and hired on as counsel in the appeal of the *United Land Association* case to the US Supreme Court. Here, he attempted to win by citing his own judicial rulings.[5]

It is impossible to know whether the 1890 *De Guyer* decision might have lasted had Fox not schemed so egregiously. With three new members voting, the court set aside his ruling and, in September 1891, unanimously overturned it. Justice John De Haven, who wrote the second *De Guyer* opinion, did not rebut Fox. Instead, he characterized a Domínguez victory as alien to the practice of territorial expansion, the purpose of which was to encourage "settlement by citizens of the United States."[6] Apparently, he did not consider the citizenship promised to Mexican Californians, or the rights of due process and equal protection enshrined in the Constitution. Conquest, De Haven reasoned, gave the US an overriding prerogative over Mexican-era estates, including the power to alter boundaries. The court restored Banning's patent to Mormon Island, and all others derived from the navigation reserve.

Begun decades before by guns and soldiers, conquest culminated with judicial robes. That suggested further implications, as the Domínguez heirs appealed to the US Supreme Court. But the family would not wait. While the case sat on the federal docket, they sold the Estuary Tract, plus any acreage gained by the lawsuit, to a new player: the Los Angeles Terminal Land Company. The sale netted $300,000, an amount twice the appraised value of the *entire* rancho at partition eight years before. The proceeds allowed the sisters and their descendants to avoid selling nearby sections. They would watch as values increased exponentially again. That, too, was result of the deal. Backed by St. Louis capitalists, the Terminal Land Company established a rail line onto Rattlesnake Island, which it renamed Terminal Island. The project was reminiscent of the upstart schemes of William Hyde's era. It aimed to outmaneuver the Southern Pacific, build shipping facilities, and then draw a buyout from competing transcontinental lines. The Terminal Land Company took over as the plaintiff in the *De Guyer* appeal.[7]

WHILE THE *DE GUYER* CASE awaited a federal hearing, sale of the Estuary Tract touched off a related controversy known as the "free harbor fight." At stake were millions of dollars in federal harbor improvements, including an offshore breakwater. The Southern Pacific did not want this work done at San Pedro Bay. Monopoly over access—whether by strategic routes, mountain passes, river crossings, or shoreline—had been a main instrument by which the SP exerted its power. But with the Terminal Company in control of former Domínguez lands, it—not the SP—stood to gain most from a government breakwater at San Pedro.[8]

Collis Huntington could not accept that. Now in his seventies, he had deposed Leland Stanford as company president. But he lost none of his brio for the dark arts of corporate power. From decades as the SP's financier and lobbyist, he had amassed connections on Pennsylvania Avenue and Wall Street. His agents issued stock and travel passes to politicians, judges, and bureaucrats. He counted many as personal friends. In this case, William P. Frye, chair of the Senate's Commerce Committee, proved particularly useful.

Huntington determined to have the breakwater appropriation spent at his favored site, where the company retained monopoly access. In north Santa Monica Bay, at the foot of Potrero Canyon, the SP built a 1,500-foot rail wharf it presumptuously named Port Los Angeles, but more often called the Long Wharf. Two boards of military engineers declined, and recommended San Pedro instead. Huntington sent his hired experts to dispute the findings. They kept Congress from acting on the boards' recommendations. Meanwhile, he shifted SP business to Santa Monica and prepared for victory. The message was clear: he would have his choice, or else no choice would be made.[9]

The railroad's obstructions did not surprise observers like George Davidson, who followed the news from San Francisco. He had long championed science against excessive private greed. But the 1890s seemed to be the conflict's nadir. Davidson had watched the Coast Survey diminish and receive skeptical, even hostile treatment in Washington. It continued to lose ground to the navy. Rail and mining companies built better-funded research units that attracted top talent. The world had changed from mid-century, and Davidson felt on the outs. He wrote that government scientists and engineers—at least those most familiar with the Los Angeles coastline—had been shut out of deliberations about the breakwater. Elected officials these days were more apt to listen to Huntington and his minions. What had the USCS's compromises accomplished?[10]

Davidson's sense of exile increased following his public humiliation. He had hoped to end his career as superintendent of the Coast & *Geodetic* Survey, so renamed to reflect *his* achievements. Having turned sixty in 1885, he did not have many chances left. Obituaries of friends and colleagues gathered among his files. But Congress twice passed over Davidson and awarded the position to non-scientists and political appointees. In 1895, one such appointee, a former coal and railroad executive, forced Davidson to retire. Friend and colleague George Mendell wrote to him in sympathy. "It is simply appalling," the army engineer rued, "that a long and meritorious service ends in this way. It ought not to be possible."[11] Corporations had influenced not merely the votes of Congress and judges. They had reshaped the agencies of federal expertise in their image. Worse still, they made examples of those who disagreed.

One generation of government scientists aged into seeming defeat. Yet beginning in 1892, the Southern Pacific's maneuvers mobilized a new opponent: the Los Angeles business community and its coalition of municipal reformers. Working through the city's recently established Chamber of Commerce, they lobbied Congress to finalize the selection of San Pedro. The group gathered support through community meetings, newspapers, public rallies, and an adjunct association called the Free Harbor League, established in 1895.

For the chamber and league, a "free harbor" was one liberated from monopoly. The SP had become California's largest landowner and employer. It controlled almost ninety percent of the state's rail traffic, plus numerous harbors. With its economic dominance and deep pockets, the SP overran public institutions, becoming California's shadow government. Control over transportation allowed the company to manipulate commerce, defeat challengers, and reduce towns and citizens to dependents. Los Angeles's experience was no different. Business leaders were especially aggrieved by the SP's rates, which suppressed low-cost maritime transportation and forced them to use the railroad for shipping. The twenty miles of track between the city and San Pedro Bay was rumored to be the most profitable railroad per mile in the United States. Merchants found it cheaper to send freight thousands of miles eastward by rail, and then by sea to Europe, than to transport goods from the city to the nearby coast. The business establishment had invited the railroad—subsidized it, even—only to suffer like a colony. They worried that, if the SP got its way and forced a harbor at Santa Monica, the corporation would seal its grip over them for generations to come.[12]

San Pedro's estuary became the basis of the reformers' anti-monopoly program. That made it a basis, too, for efforts to break the SP's political and corporate power nationwide. Most of the estuary was, under law (if not in practice), public lands. Courts so far had affirmed the US reservation for navigable waters. And that space was surrounded by a Mexican rancho that, once it was subdivided, offered numerous approaches to the water.

Estuary anti-monopoly became the message that free harbor champions used to rally Southern Californians and solicit officials in Sacramento and Washington. Most influential of the champions was the lawyer-politician Stephen Mallory White. White's pedigree showed the increasing estrangement of business interests from rail corporations. His father, William F. White, was an Irish immigrant politician. He helped lead the radical Workingmen's Party of California, which threatened to flog SP executives, and was its candidate for governor in 1879. The younger White pursued a more moderate path. By the mid-1880s, he had become counsel to *Los Angeles Times* president Harrison Gray Otis and won election as the city's district attorney. He then formed a prominent law firm with John D. Bicknell, advisor to the Southern Pacific. Bicknell & White managed Phineas Banning's estate, including the SP's tidelands. They represented William Banning in the *De Guyer* suit. The SP's Leland Stanford, while serving as US senator, favored Bicknell for appointment to a federal judgeship. But he hesitated because White was rumored to associate with "anti-monopolists."[13] Bicknell remained in the SP camp. White left their partnership and proclaimed himself an opponent of the railroad. He joined the Chamber of Commerce and the Free Harbor League. Once elevated to state senator, he took the cause to Sacramento. He rose further in 1893, when, at age forty,

Broadway, Los Angeles, 1889—a thoroughfare laid out by Edward Ord four decades earlier. The city's sudden growth raised its interest in San Pedro Bay.

he was appointed to the US Senate, just as Stanford exited office and died. White stayed on as Banning's lawyer for the appeal to the Supreme Court. He saw ahead to the confluence of the free harbor fight and *De Guyer.* He would be ready.

The free harbor movement kept the Southern Pacific from having its way in Washington, yet it could not prevail. In an attempt at compromise, Congress commissioned a third board of army engineers. This time, legislators committed in advance to support the chosen harbor site. The board reported in early 1897 in favor of San Pedro. Engineers found that both Santa Monica and San Pedro were suited to an offshore breakwater, but San Pedro was superior due to its estuary, which afforded further anchorage. Unbowed, the SP delayed again. Secretary of War Russell Alger, another friend of Huntington's, refused to release the breakwater appropriation.[14]

ALGER'S (AND HUNTINGTON'S) recalcitrance caused the free harbor fight to linger into an age of resurgent imperial expansion. The United States modernized and enlarged its navy. It developed harbor concessions at Pearl Harbor and Pago Pago, as well as numerous coaling stations. It claimed dozens of uninhabited guano islands. American settlers in Hawaii usurped power from the Native monarchy. Discovery of gold in the Klondike, meanwhile, raised the nation's interest in the Pacific Far North. Expansionists stepped up calls for a canal across Central America in response to French efforts to build one. With each passing year, the Pacific Ocean increased as a preoccupation of politicians, military strategists, businessmen, farmers, missionaries, and intellectuals.

All of this gave urgency to the Supreme Court's pending decision in *De Guyer*. The justices first heard the case in 1895, and a verdict seemed imminent. Then, unexpectedly, the court reopened arguments at the start of 1897. Senator White reprised his role as Banning's lawyer. In the free harbor fight, the Banning family sided against the SP and Huntington. Lawyer and relative George S. Patton Sr. lobbied in Washington on behalf of the Chamber of Commerce. The split would prove temporary, yet it allowed White to balance his obligations to the Bannings with those to the free harbor crowd.[15]

Upon the suit's reopening, White revealed some new commitments. He previously had refrained from speaking on behalf of Congress or national interests. In fact, the US government had not taken sides in the case. The Army Engineering Board's decision, which became known in early March, changed this. So, too, had President McKinley's inauguration, which energized Washington with Pacific designs. White pivoted. He persuaded newly appointed attorney general Joseph McKenna, a fellow Californian, to bring the government into the *De Guyer* suit. Next, White coauthored with Solicitor General Holmes Conrad the government's amicus curiae brief. The senator's contributions were rumored, but kept off the record. Instead, Conrad and McKenna filed the document in their names.[16]

The amicus brief asserted, for the first time, that the federal government held an interest in San Pedro Bay, "one of the most important points on the Pacific Coast," for a harbor.[17] It argued further that this interest would suffer if the Court deemed the estuary to be part of the Rancho San Pedro. The Supreme Court endorsed the amicus position and even borrowed its language. Its opinion, issued on the last day of the spring session, May 24, 1897, rejected Domínguez claims entirely. Instead, it affirmed as "conclusive" Ord's 1855 survey and the 1858 rancho patent. Justices reasoned that if Manuel Domínguez had found fault with the property boundary, or if the surveyor general had, ample opportunity existed to protest at that time. They learned nothing of the many forces arrayed against the rancher. Nor did they hear of the strange origins and afterlife of Ord's navigation reserve. Those details already were lost. The court buried them for good, averring that "it matters not how the surveyor general arrived at his conclusions."[18]

Justice John Marshall Harlan, best remembered today for his dissents in the *Civil Rights Cases* (1883) and *Plessy v. Ferguson* (1896), authored the unanimous *De Guyer* opinion. He acknowledged the messy history of Mexican-era property in California, but concluded that this should make way for a future American Pacific. Harlan had ties to both. His father, James, a Kentucky Whig politician, was appointed in 1851 to the Land Commission that was to review Mexican grants. However, the elder Harlan turned down the job. Justice Harlan perhaps remembered. He certainly took an interest in US expansion, past and present, and the question of where and when the Constitution extends to annexed territory or subject peoples. Beginning in 1901, the court would negotiate the answer as it applied overseas. It thereby defined territorial expansion for a new age.[19]

Through votes and numerous dissents in the *Insular Cases*, Harlan forged his belief that annexation automatically extends civil rights and limits on congressional power to new lands. In other words, the Constitution should follow the flag. He did not disavow territorial conquest or acquisition. Instead, he warned of its potential harm to equal protection in life, liberty, and property. Harlan never said how he squared the *Insular Cases* with *De Guyer*. But it is worth pondering what might be different had the Domínguez verdict come second and not first. While Harlan's other writings receive more attention, his estuary opinion provided an important step in the court's jurisprudence. It reconceived the Pacific coast as no longer the republic's culmination. Instead, coastal places became the point of transition, where nation-building ended and US global power began.

What did this portend for the Banning patent and all others? *De Guyer*'s outcome upheld the Banning claim by default. The *Los Angeles Times*, relying on an unnamed source, misreported the patent as confirmed. But the high court never said so. Instead, it deemed the question "unnecessary to decide." Harlan and the others seemed inclined to obliterate the Domínguez and Banning claims alike. But Justice Stephen J. Field, a Californian and Southern Pacific man, restrained them. Or so White told his clients. That story comes from his explanation of the court's "very unpleasant bias" against him.[20] Maybe so, in his role as the Bannings' lawyer. But the influence of his amicus brief tells otherwise.

A sheep in wolves' clothing, or a reformer in monopolists' garb? White had climbed a ladder through Los Angeles's class of commerce and industry. Then, by the turn of events, he set up implicitly what his father and the Workingmen's Party had tried, but failed, to achieve. The age of unfettered capital approached an end. And the SP faced coming reversals. As imperial ambitions settled once more upon San Pedro Bay, they would bring back the US Army, too, and complete a circle from the Civil War. But now the stakes would lure a rising metropolis as well.

The past sometimes yields to the present in poetic ways. With a November 1897 obituary entitled "An Old Timer Gone," Los Angeles bid farewell to George Hansen. The surveyor died at age seventy-three. Newspapers recalled him as a pioneer of Southern California. They made no mention of Hansen's four-decade connection to San Pedro Bay, of his 1884 partition survey, or of the Domínguez family's court case, which for a span of months upended US conquest, but pointed ultimately to empires and dispossessions abroad.[21]

Still Teeming with Life

I N THE *DE GUYER* CASE, THE SUPREME COURT TOOK NOTE of San Pedro's estuary as a future harbor for US expansion. Dollars invested, with more on the way, quantified its value. Neither Justice Harlan nor his colleagues viewed the bay with their own eyes. That would have made little difference. After trampling the continent, Americans had just arrived at the idea of environmental conservation. John Muir, with George Davidson and others, founded the Sierra Club in 1892 to save what was left of mountain forests. Avid hunters like Theodore Roosevelt proclaimed the need for nature reserves. Social reformers worried that the citizenry— and the republic itself—were losing their virtue as wild places disappeared, as empires emerged abroad, and as immigrants crowded cities and farmland.

Yet few people thought the estuary's wilderness worth saving, either for its own sake or theirs. What they saw, or didn't, was the crux of the problem. The bay's otters, seals, and bears had been hunted to extinction. Whales offshore swam a gauntlet of harpoons. Migratory birds vanished due to unregulated sport shooting. Collectors targeted herons and egrets for their long plumes, used to embellish women's fashion. In some spots, marsh and mud had dried out entirely. Polluted runoff thinned the kelp forest beyond the entrance bar. In most seasons, the river upstream was no more than a trickle. The estuary seemed to have so little wilderness left. Whether glancing from shore or from Washington, DC, humans deemed it a wasteland.

But they were wrong. The estuary held a great multitude of life. Its smallest creatures excelled at turning sunlight into energy. They fixed carbon and nitrogen into an excess of life-sustaining sugars and proteins—what scientists today call primary production.[1] Algae and diatoms (called "jewels of the sea" for their exquisite single-cell structures) bloomed atop the seemingly vacant mudflats. Phytoplankton flourished in the brine. Bacteria fed voraciously on marsh decay or animal waste.

Billions of microorganisms formed the foundation of a food web that extended next to fish and invertebrates. Tiny goby, killifish, and skeleton shrimp fluttered among blades of eelgrass. Colorful sea slugs and nudibranchs fed off the bottom. Above the waterline, saturated mud kept burrowing animals alive just millimeters deep, even in the estuary's highest reaches. This included fiddler crabs, staphylinid beetles, segmented worms, bubble snails, bent-nosed clams, and a range of protozoa. They dug channels that circulated oxygen, water, and nutrients. Their life and death provided for the higher rung of animal consumers. But these—predators especially—were in short supply.

That was the misunderstanding. Although nature had disappeared to the human eye, much of the estuary's chain of being remained intact. Primary producers, scarcely visible fish, and buried invertebrates waited at the ready to sustain life. Even without thought, without trying, they could conserve the wilderness in ways that Muir and his fellow naturalists could not. Then, the larger birds and mammals might return. But it was not to be. The age of capital had produced legions of new machines. The unseen, microscopic world stood only in their way.

Part Five
EMPIRE

Conquests by Another Name

I N 1898, INVADERS ARRIVED ONCE MORE. THIS TIME, they did so on the opposite side of the Pacific. On May 1, the US Navy's Asiatic Squadron steamed into Manila Bay, at the Philippine island of Luzon. Spain had held the far-off archipelago for three hundred years, ever since the galleon trade in silver and gold. By American reports, the Spanish navy had not changed much. Ships were outdated, with wooden hulls and small-caliber, smoothbore cannons. They were no match for Commodore George Dewey's ironclads. With just nine ships, he destroyed the enemy fleet and left Spain's twenty thousand soldiers stranded inside the harbor's walled city. Congress had declared war one week prior, in response to the mysterious explosion of the USS *Maine* at Havana, Cuba. Both there and in the Philippines, insurrectionists had struggled to throw off Spanish colonial rule. That became an invitation to others.[1]

The Battle of Manila Bay thrust the United States and its world into the twentieth century. By now, Europeans had carried their imperial contests to Africa, the Middle East, and East Asia. The decline of old empires opened doors to new conquest. Recently industrialized nations—Japan, Germany, and the US—entered the competition. Their steam-propelled battleships foretold a shifting international order. Dewey delivered that message. Then it drifted and washed ashore at home.

In the US, victory produced contending notions of empire. Politicians proudly proclaimed they would support colonial independence, for Cuba especially. Republicans had called for as much,
along with protectionist tariffs and the gold standard, in their convention platform of 1896. The party of Lincoln gathered that year
behind its nominee for president, the last Civil War veteran to win
the office. William McKinley's election raised the issue of Cuba to
prominence. The Philippines, by contrast, had little place in public discussion. It is said that most Americans could not locate the
islands on a map. Yet a small group already had. By the mid-1890s,
the Naval Department drew up plans for capturing the Philippines.
McKinley's assistant secretary of the navy, Theodore Roosevelt, quietly (and without the permission of his infirmed superior) instructed
Dewey to proceed, should war with Spain happen. When that contingency came, the United States attacked first at Manila, six weeks
before it moved against targets in the Caribbean. Reaching for what
they saw as low-hanging fruit, US officials revealed their desires,
even when their rhetoric did not.[2]

Inconsistencies grew as war reverberated back toward the American coast. Some initially suggested that the United States should
ransom the Philippines for Cuba's freedom. Instead, thousands of
troops traveled by rail toward San Francisco, where they shipped
out to occupy Manila and invade the surrounding isles. Their
month-long voyage spread imperial designs. Congress revived talk
of annexing the Hawaiian Islands, where the transports stopped for
fuel. In 1893, Anglo settlers had displaced the Native monarchy and
consolidated power under a republican constitution. They asked to
join the US. But royal emissaries protested (as did Japan's government), and doubtful senators refused. The Spanish-American War
overwhelmed their resistance. McKinley asked Congress to annex
the islands out of military necessity. Native Hawaii's freedom was
sacrificed, so the president reasoned, to bring liberty to the Philippines. That contradiction was a sign of more to come. McKinley

declined to recognize a Filipino provisional government. And early in 1899, US forces began to battle against the local independence movement. Voters agreed enough to keep McKinley in office. They reelected him easily over two anti-imperialist candidates: Democrat William Jennings Bryan and socialist Eugene Debs.[3]

With the arrival of its soldiers at Manila, the US assembled a transpacific imperium. That inspired further opportunists beyond the federal government, some of whom would find their way to San Pedro's estuary. Edward H. Harriman foresaw his business empire striking out to the Pacific coast and beyond. Fifty years old in 1898, Harriman had risen from a life of deprivation by his extraordinary shrewdness, hard work, and advantageous marriage, which gave him entry to New York's high society. Wealth gleaned from profitable railroads, or pilfered from non-profitable ones, had elevated this moneyed class. Accountants and bankers siphoned money, hid losses, and pumped up the price of securities, bilking creditors and defrauding government in the process. As result, railroads— and particularly the transcontinentals of the West—suffered from neglected maintenance. Yet that had been the business model. Harriman proved different. He, along with a cohort of rising corporate titans, transformed the rail lines from swindle into proper enterprise. He became a master of reorganization, creating dividends and efficiency in nearly all he touched. That meant profit squeezing, too. While Harriman, Morgan, Carnegie, and Rockefeller pleased shareholders and dampened the boom-bust cycles of yesteryear, creating corporate America as we know it, others paid the price.[4]

The new business order emerged from bold ideas and daring triumph, which Harriman exemplified to a tee. During the 1880s, while still in his thirties, he revitalized the long-troubled Illinois Central Railroad. A decade later, he used the company's good credit to take over the struggling Union Pacific. Harriman entered 1898 as the UP's de facto chairman. He financed an overhaul of the outdated transcontinental line, including track and machinery, and restored profits. It was a stunning achievement, all the more because it came on the heels of severe economic depression.[5]

While the US government eyed Hawaii and the Philippines, Harriman, too, set his mind on the western sea. The UP held controlling interest in the Pacific Mail Steamship Company, whose lines extended to East Asia. Plus, the UP relied on the Central Pacific's tracks to reach San Francisco. Harriman determined to buy out the CP. That required convincing the heirs of the Big Four as well as the line's sole remaining founder. Collis Huntington remained formidable in old age. But his past gambles and trickery had caught up with him. Creditors grew suspicious—withholding, even. That was a problem, because Huntington's finances were a mess. He desperately took a personal loan from Harriman, who in exchange wanted a piece of the CP. Huntington refused. So, Harriman waited. To make known his intentions, he led a well-publicized hunting and scientific expedition to Alaska and Siberia in the summer of 1899. Before long, Harriman would gain what he wanted, along with interests at Los Angeles and its harbor.[6]

The San Pedro faction of reformers also found much to celebrate in news of the Philippines and Hawaii. They refused to suffer the Southern Pacific's corporate empire. Yet, with potential for ocean commerce growing, they set out to build their own city-imperium, a metropolis to rival New York, Chicago, or London. Huntington and his allies obstructed this goal. They continued to stall the San Pedro breakwater project, with the help of War Secretary Russell Alger. But the Spanish-American War changed things. In July 1898, weeks after the annexation of Hawaii, President McKinley ordered breakwater construction to begin. (Alger continued to dither until he lost his job in 1899.) McKinley had been indifferent to the matter before. But the so-called war emergency, with its need for harbors, made up his mind.[7]

So did Los Angeles interests, who smartly linked the free har-

CONQUESTS BY ANOTHER NAME

bor issue to the war abroad. The *Herald*, controlled by San Pedro promoters John T. Gaffey and Thomas E. Gibbon, championed the "War to the West." They were Democrats. The Republican and pro-San Pedro *Los Angeles Times* did the same. Once troops began arriving bound for San Francisco, Pearl Harbor, and Manila, the newspapers stoked public excitement and patriotic rallies. The *Herald* advertised a commemorative set of prints titled *Battleships of the Republic*. (The initial run was forty-five thousand copies.) Hawaiian annexation enthralled the editors, as did talk of a Nicaraguan canal. They hailed an Anglo-Saxon alliance to span the oceans and naively predicted that Japan would assist. Events, they wrote, marked the succession of dying nations by vital ones, of races and civilizations as well. It was familiar talk. The region's US settlers had said as much before.[8]

Not all of the San Pedro faction enthused over Pacific empire, however. Stephen M. White, who led the free harbor crusade in Washington, broke with his allies at the *Herald*. White saw inconsistency in those who cheered the taking of colonies abroad while they battled the Southern Pacific in the name of free government at home. Perhaps his Irish Catholic roots caused him to sympathize with island peoples shackled by foreign rule. In any case, the Philippines and Hawaii were White's bridge too far. He became an anti-imperialist voice in the Senate. California's legislature passed a resolution demanding that he vote in favor of the Spanish-American treaty. He did not, objecting because the accord denied the Philippines its independence. In the end, his protests failed. Newspapers accused him of running "amuck" in disregard for the people's will. White knew he would lose reelection, in what would be the state's first popular vote for federal senators. He retired from office and died in 1901, at age forty-eight. His city and former friends proceeded without him.[9]

As it entered the twentieth century, American imperialism remained a crowd of contending interests. Few were opposed. But even supporters were divided over how a US Pacific would work, and who would benefit most. Just as in the lead-up to the Civil War, the divides impelled expansion, while expansion worsened the divides. In the decade following 1898, the US government, corporate industry, and civic reformers battled over the stakes of a changing world. And they did so at San Pedro Bay. That would bring about an unprecedented finality. The estuary, which was the cause for all so far, would cease to exist.

Chapter 26

ARRIVALS OF
THE SEASON

THE *DE GUYER* VERDICT CREATED A PARADOX. THE
Supreme Court had denied the Domínguez family's claims to
the estuary in favor of government need. Yet private property filled
the estuary regardless. Fifteen patents—three issued by the United
States and twelve by the State of California—spanned the tidal
basin. Most belonged to the Banning family and the Southern Pacific
Railroad. Meanwhile, the Terminal Railway owned the salt marsh
and sandspit that divided the estuary from the sea. Rivals, including
the Southern Pacific, encroached on the barrier now called Terminal
Island. Numerous squatters did, as well. Together, they ignored the
court's ruling. At the same time, they ignored nature and insisted
there was land in spaces half- or fully submerged by water. Only the
deepest tidal channel remained without claim. It was very different
than the harbor imagined by Justice Harlan.[1]

For the moment, private interests prevailed, making San Pedro
Bay a microcosm of the West in industrial-era America. Throughout
the region, government let go of valuable public domain to railroads,
mining companies, ranchers, homesteaders, and reclamation or irrigation schemes. In an age different from our own, citizens viewed
the state as the enabler of property, not its antithesis. They celebrated this—among many things—when, in 1893, they hailed the
closing of the so-called frontier.

But paradox signaled changes yet to come. By giving away natural
resources, government had denied itself power. It accumulated pow-

An 1887 map showing the
estuary as real estate lines.
Public claims and natural
features are absent.

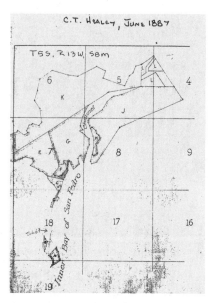

ers nonetheless. Legislatures drafted copious law. Federal agencies flexed their bureaucratic muscle and practiced scientific technique, even in the act of dispensing. By one of many strange twists in American history, the triumph of private property increased government power. When the two were partners, this was of little notice. However, new attitudes had emerged by the twentieth century. Some spoke of an inherent antagonism between public and private. Government, they said, needed to protect the people's interest in order to safeguard liberty. Expansionists would come to a similar conclusion with regard to harbor spaces. That would make the United States' age of globalism an age of reform as well. A government giant slumbered beneath the private claims to San Pedro's mud and marsh. It would stir in the years after the Spanish-American War, awakened by imperial events and the advent of new technology. Still, this was a time of several giants. It remained to be seen which would be the greater force.[2]

James Meyler came to Los Angeles in November 1898, unaware that he brought an answer. Citing the chance of Spanish attack, the War Department sent out officers like him to bolster the nation's

harbor defenses. It also created the Los Angeles District of the
Army Corps of Engineers. The army had directed improvements at
the estuary since 1869, but staff administered the projects from San
Francisco, while contracted civil engineers and laborers did the daily
work. A dedicated district meant a much stronger government presence. It began to construct the offshore breakwater and defensive gun
emplacements. And it set to work on the important task of designat-
ing US harbor lines. These are the master blueprints for channels,
anchorages, and landings that make up a modern port. They include
pierhead lines, which dictate how far wharves or other structures can
intrude, and bulkhead lines, behind which land is reclaimed or reinforced. Together, the lines clarify where federal jurisdiction exists in
a harbor. At San Pedro's estuary they would determine, far beyond
anything before, the balance of private and public.[3]

As the LA District's commanding officer, Meyler took charge
of breakwater construction and the placement of harbor lines. But
he brought something new to the task. After serving in the Engineers' San Francisco bureau in the early 1890s (during which time he
authored reports that were decisive in the free harbor fight and *De
Guyer* case), Meyler transferred to the army projects on Kentucky's
Barren River and Florida's Atlantic coast. There, he gained experience with the cutting-edge practice of hydraulic dredging. His return
to California hinted that the technology soon would follow.[4]

DREDGING, which is the deepening of a waterway to enhance its
human use, developed in the ancient world. Steam engines had mechanized this work by the late nineteenth century. Scoop and scraper
dredges proliferated, just as their motored brethren did in railroads
and factories. They were employed to great effect in construction
of the Suez Canal. But the early generation of dredges was inefficient, expending large amounts of fuel on the repeated motion of
scoop-raise-deposit-return. They were self-defeating, too. By agitat-

ing sand and mud, they merely caused soil to drift and resettle down current. Large jobs required barges to haul away excavated material. That made dredging utterly impossible in the complex shoals and tight, uneven spaces of a coastal wetland. Low-wage labor, with shovel, pickaxe, and wheelbarrow, remained the norm, as used in French-era Panama Canal.[5]

Yet impossibility sparked invention. Tinkerers and woodshed inventors were, alongside moguls like Edward Harriman, the essence of the age of capital. They raced to devise contraptions, flooded the US Patent Office with applications, and hoped to sell their designs to business or government. As of the 1880s, two San Franciscans took credit for inventing the hydraulic (or suction) dredge. Alexis W. Von Schmidt and Alphonzo Benjamin Bowers both designed machines with a centrifugal pump powerful enough to take in and expel sediments out through a disposal pipeline. Both designs featured a cutter with rotary blades to smash and devour bottom materials. Von Schmidt and Bowers fought over patent rights, which climaxed in a seminal case before the US Circuit Court. Bowers won the lawsuit, but Von Schmidt prevailed otherwise. Under contract from the Army Engineers, he put the first machine into use, at Oakland's waterfront. The supervising officer was so impressed that he urged the chief of engineers to showcase the design at the 1884 World's Exposition in New Orleans.[6]

Hydraulic dredging marked a vast leap in capacity, and it portended a sweeping transformation of the world's rivers and coasts. Because the machines worked with a single, perpetual flow of water, they eliminated wasteful movement. Excavation and disposal merged into one seamless task. Without the need for precise visual operation, hydraulic machines might work day and night. And because they eliminated the need for tall, top-heavy mechanisms (like wheels, arms, and booms, which might reach seventy feet in height), hydraulic dredges could be easily mounted atop small, flat-bottomed boats. They could work at sites previously off limits to the

Von Schmidt's hydraulic dredging machine. The Army Corps of Engineers would continue to refine the design.

conventional dredge. Newer piston and turbine engines (fueled by gasoline) had horsepower enough to pierce rock and pump detritus thousands of feet away.[7]

The Army Engineers grasped the implications. In California, the agency had studied siltation damage to the Sacramento and San Joaquin Rivers caused by mining upstream. And for decades, the Engineers had tried to deepen the Mississippi River, with only limited success. Starting in 1894, the army and the Mississippi River Commission adopted and aggressively pursued hydraulic dredge technology. They built a series of prototype machines, each larger and more powerful than the last. Tests showed that hydraulic dredges could remove the fine sediment of rivers, coastal sandbars, and estuary wetlands—exactly where older dredges failed most.[8]

WHEN JAMES MEYLER TOOK command of the Engineers District at Los Angeles, he brought with him innovations that began in California, reached fruition on the Mississippi, and now arrived home again. In 1900, he drew up plans to carve out the estuary's tidal basin entirely, leaving a saltwater lake in its place. The product would be "one of the first-class ports of the world," ready to receive "increasing oriental trade," or that of a future isthmian canal. The work was daunting. "Permanent improvement of a harbor entrance on a sandy coast," Meyler wrote, "has long been recognized as one of the most difficult problems of maritime engineering."[9] William Hyde had suggested such a project to Collis Huntington in 1873, but without support or know-how. By contrast, the army possessed a growing arsenal of machines and the expertise to use them.

But first, the engineers had to clear a path from the sea into the lagoon. And that required the centerpiece of Meyler's phase-one proposal: construction of a government hydraulic dredge. Writing to Chief of Engineers John M. Wilson and Secretary of War Elihu Root, the officer referenced numerous places where army trials had proven the machinery. "At no one of them," he underscored, "are the physical conditions as favorable to dredging as at Wilmington."[10]

In June 1902, Congress approved Meyler's plan. By that time, however, the visionary officer was dead. He had succumbed to pneumonia the previous December, at age thirty-five, during a visit to see his mother in New Jersey. Scientific advances did not come equally. Even an adept of modern technology remained at the mercy of ancient microbes. Los Angeles newspapers reported Meyler's untimely death, as did the *New York Times*. Edgar Jadwin succeeded him at the LA District. He, too, believed that dredging could revolutionize waterways. Five years after taking command at San Pedro, Jadwin would apply his experiences to the Panama Canal's Chagres division as its ranking engineer.[11]

Meyler had anticipated what Jadwin's career would illustrate: simple, unassuming machines could change the world. And the most famed engineering achievements had taken their source in prior, less-known events. So it was that, in early 1903, engineers designed the US dredge *San Pedro*. The 140-foot-long vessel housed tons of equipment, pipes 20 inches in diameter, and a 700-horsepower engine able to remove 500 cubic yards of earth per hour. Observers hailed it as the largest suction dredge ever operated in California. But it marked only the beginning.[12]

All that remained was to power up the machine and set it against the landscape. Between February 1905, when the *San Pedro* began service, and mid-1908, the dredge would remove nearly three million cubic yards from the lower portions of the estuary. Its early accomplishments drew notice. The district received blueprint requests from other army offices, from the Mississippi River Commission, from the Isthmian Canal Commission, and from the government of Mexico. The *San Pedro*'s work prefaced the dredging of Pearl Harbor to house an enlarged Pacific fleet, and similar work in the Philippines, at Manila and Subic Bays. Hydraulic dredge designs, and the officers who knew them, spread throughout the US and to overseas colonies—anywhere that sand or marsh stood in the way.[13]

ESTUARY CLAIMANTS TOOK note of the machines as well. As of 1900, Collis Huntington had not given up on his Santa Monica rail wharf, or on cajoling Congress to pay for it. But he had hedged his bets enough at San Pedro to expect both profit and trouble from the engineers' arrival. The Southern Pacific controlled most estuary lands through its alliance with the Banning Company. Together with the Central Pacific, it held ten thousand miles of track, which serviced the bay and other Pacific harbors. It remained the nation's largest rail line.

Huntington's biggest problem was not that he had lost the free harbor fight—the SP's diverse assets ensured a win of some kind. His woes were financial. He was wildly rich, yet increasingly dependent on bankers and mortgages. Opulence and reputation encumbered him. Even his New York City mansion became collateral for a loan. His company had yet to profit from its freight traffic. The SP was overbuilt and roared through desert without enough customers to squeeze. Meanwhile, the CP owed the government $59 million in interest on construction loans that dated to the 1860s. Huntington had evaded payments so far. But, in 1897, Congress refused to defer any longer. The depression, Wall Street panic, and the Pullman Strike of the 1890s worsened matters. Once standing over six feet tall, Huntington now stooped under age and abuse. Still, one mercy remained. In August 1900, he died at his residence, Camp Pine Knot, in New York's Adirondack Mountains. Unlike other transcontinental developers, he never lost control of his company.[14]

Edward Harriman wanted the Central Pacific alone, but settled for the full Southern Pacific system, which he bought from Huntington's heirs. The acquisition took one year and $40 million. As before, Harriman parlayed his past success to finance the deal. Funds came from newly issued Union Pacific bonds and from the confidence of banks and investors. With the SP and UP, Harriman now controlled two transcontinentals, the majority of mileage within California, and rails into Oregon and Mexico. He owned ocean lines to Havana, Yokohama, and Shanghai. By 1905, he envisioned a round-the-world network that would extend US influence over the Pacific, and across Japan, Manchuria, Siberia, and Russia, to reach the Atlantic Ocean. The industrialist toured the Far East to negotiate concessions. He also took an interest in culture. Harriman sponsored a troupe of Japanese martial artists to follow him home to the US and popularize jiujitsu.[15]

The grappling sport provides a visual for what lay ahead. Harriman was now in the business of Pacific harbors. That brought him to San Pedro just as the federal suction dredge promised to open the estuary to the world. What was paradox before turned into a locking hold between combatants. Harriman said of the SP purchase, "we have bought not only a railroad, but an empire."[16] His statement acknowledged a plain fact: US imperialism could not exist without the global web of ship and rail. Whoever mastered one might master the other. At the moment, he claimed the upper hand. And he had greater plans still.

Chapter 27

LEVIATHANS DIG IN FOR A FIGHT

T HE DREDGE *SAN PEDRO* BEGAN WORK AT ITS NAME-
sake estuary in early 1905. That raised a number of questions
for the Army Engineers to resolve. The Los Angeles District needed
to decide where the vessel could work, and where it could not. And
that depended on the extent of federal harbor lines in the bay's
upper portion. No such lines existed above Smith's Island, where
the entrance widened into an interior lagoon. From here, the bay
spread into two mudflat basins, with Wilmington on high ground in
between. Property claims filled the entirety.[1]

As it considered harbor lines, the Engineers District had to confront issues sidestepped by the government so far. For half a century,
authorities—the US Land Office, Supreme Court, and Congress—
took as an article of faith Edward Ord's 1855 survey, which pronounced the lagoon navigable. They affirmed his findings by logical
leaps and legal technicality, without support of precise maps or tide
data. Further evidence was circumstantial. Shipping demonstrated
that light vessels could pass through the channel to Wilmington.
But what about the thousands of acres on either side? Many people
claimed to know, but government had not concluded scientifically.

Then there was the inconvenient fact: If the lagoon were navigable, why was it cut through with land claims? Land was not water,
and it could not be navigated. Or could it? The Army Engineers
would decide, as they considered how far harbor lines could reach. It
would not be easy; the claims had been consolidated under immense

corporate power. The inquiry coincided, too, with frenzied development at nearby estuaries and a consensus that coastal places were
best surrendered into private hands.

At San Pedro Bay, government already seemed to have given up.
By 1900, numerous firms had built wharves inside the entrance, narrowing its passable zone. The Southern Pacific added to the problem when it stretched its line beyond Mormon Island by way of an
earthen trestle. The extension formed a barrier, cutting the mudflats
off from the tide. Sandbars collected alongside it, which threatened
navigation. Nevertheless, officials hesitated to force the structure's
removal. Both George Davidson and army engineer George Mendell
predicted the SP would win if the US filed suit. Federal jurisdiction
seemed no more than an empty threat.[2]

Harriman's purchase of the Southern Pacific added to the government's woes. It morphed a large transportation line into a far larger
one. Then, he aggrandized it further. With tens of millions in bond
money left over from his SP purchase, Harriman attempted to gain
the Northern Pacific Railroad and its transcontinental line to Puget
Sound. The back-and-forth of failed hostile takeovers sent Wall
Street into a panic. By late 1901, the antagonists reached a détente.
Harriman combined with J. P. Morgan and James J. Hill of the Great
Northern Railway to form the Northern Securities Company. The
accord brought formerly competing roads of the Pacific Northwest
under sway of a single holding company and fifteen directors, making it the largest rail corporation in the world. Because he controlled
the Southern Pacific, Oregon Railroad, and Union Pacific, Harriman
now held lines leading to each of the West Coast's major ports.[3]

At the same time, Harriman strengthened his hand at San Pedro.
In July 1903, he acquired a fifty percent share in the Salt Lake Railroad, which several years before had bought out the Terminal Railway. That gave him control of the Salt Lake Railroad's real estate
(that is, the Estuary Tract once owned by the Dominguez heirs),
plus the SP's properties, the Banning tideland claims, and the only
two rail lines to the bay. With such far-flung business, Harriman

By 1903, Edward H. Harriman dominated the Western railway market and envisioned a business empire to span the Pacific.

The New Colossus of Roads.

himself was not particularly interested in San Pedro or Los Angeles. But he paid others to be interested for him—skillful managers like Julius Kruttschnitt, Robert H. Ingram, and William Sproule. Harriman gave them a massive amount of capital ($242 million over eight years) to refurbish the SP. They shored up the road's political machine and guarded its real estate. The Wilmington–San Pedro branch was so profitable that its earnings still made up for the company's losses elsewhere. If critics thought that the SP's "octopus" had monopolized shipping before, the Harriman system multiplied that problem many times over.[4]

CORPORATE POWER DWARFED the Army Engineers and their *San Pedro* suction dredge. Initially, they deferred. In early 1903, the Los Angeles District launched a study under the direction of Pacific Division commander William H. Heuer. He recently had led the improvement and fortification of ports at Honolulu and Pearl Harbor, and he had scouted canal routes across Nicaragua and Darien decades before. Based on his long experience, Heuer judged the lagoon to be non-navigable. That meant harbor lines were unnecessary and unjustified.[5]

Claimants seized the initiative and began to draft harbor designs of their own. In March 1905, the Southern Pacific, Salt Lake Railroad, and Banning Company unveiled a proposal to develop their massive bayside holdings. The Peninsula Plan, as it became known, asked the US to conform dredging work to private property. In exchange, the corporations pledged to donate small amounts of real estate to public navigation channels, paid for at government cost. Of course, the channels would serve a privately held waterfront. Top SP executives visited Los Angeles to approve the plan. Proponents of a free harbor, particularly the Chamber of Commerce, complained in Washington. The town council of Wilmington protested, too, in fear of losing access to the sea.[6]

Under pressure, the War Department scrapped Heuer's decision and ordered a second study of the estuary. Ostensibly, the Engineers District would broker a compromise to balance public and private interests. That task fell to Amos Fries, the district's new commander, and David E. Hughes, its ranking civil engineer. Both men represented a shift underway inside the district. Hughes was an autodidact—an example of an old American tradition that was disappearing in his time. He taught himself mathematics, when not laboring on farms and quarries, then began teaching at a local technical college. In 1892, while in his mid-thirties, Hughes gained international renown for his research on the "Sickle Curve" and the performance of railroad brakes.[7]

Notoriety led to a job with the Army Engineers. But Hughes retained his independent streak. He promoted practicality and detested pointless bureaucracy. As his work on railroad braking suggests, he believed that numbers and slide rules might create a safer, more humane world. That idea was shared by Harriman's corporate managers. Yet Hughes's democratic quality sent the Engineers District in a different direction. He and Fries assumed the lagoon was navigable until proven otherwise. That was a complete reversal from the Engineers' prior approach. It also meant that they saw private claims as a threat to presumed navigability. Land claimants would need to dissuade them. With new information from the district, the War Department in 1906 declared the upper estuary open to federal harbor lines.[8]

Now, the Engineers District would begin the slow, exacting work of placing the lines. Depending on how they did, the SP and its allies might either win or lose big. If the engineers located a navigation channel or bulkhead through a parcel, dredging could erase that portion into a public watercourse. If they deposited sediments to fill in land, this could landlock claimants behind. If harbor lines bisected a structure, such as the SP's trestle, the owner bore the cost of removing the obstruction. Alternatively, the same actions could benefit a claimant. Small differences in a harbor line or degree of angle might mean acres of waterfront property preserved, gained, or lost.

For this reason, the corporations tried to preempt the outcome. In December 1906, they offered a second Peninsula Plan. This accepted federal jurisdiction over the bay, at least in part, but proposed that government and claimants each finance and develop separate harbor facilities. The railroads and the Banning Company would combine their parcels and build a large waterfront, based around Mormon Island, protruding into the lagoon. The US would dredge a channel to the site, from which private terminals would link to transcontinental rail lines. At the same time, the government would build a public harbor on the lagoon's west shore that SP tracks would also service. The second Peninsula Plan, like the first, aimed to keep the bay under private control and even wagered to increase this. Claimants had reason for confidence; San Pedro Bay's history so far was one in which private gain won out repeatedly. Government officers had been no match.[9]

EVENTS NEARBY SUGGESTED the corporations might succeed. Los Angeles suddenly went mad for coastal real estate at all of the region's smaller estuaries. Schemes circulated through banks and society clubs. They swept a wave of investment seaward, like an outgoing tide of dollars and concrete, over the shore and into the tidal margin. Most important, the Army Engineers seemed to yield before it.

Some projects had started decades ago, when transcontinental railroads first brought long-distance travel to the Los Angeles coast. Developers built resort hotels, sanatoriums, housing subdivisions, and amusement parks overlooking the sandy beaches at Santa Monica and Redondo. Similar locales sprung up north and south. Tens of thousands of tourists journeyed from the Midwest and Eastern cities to seek health and pleasure by the seaside. Many stayed to build new lives. Advertisement of the region's citrus crops, Hispanic and mission heritage, and Mediterranean climate increased the appeal.

Estuaries passed unscathed through this early period because of the difficulty and cost of developing them. This changed when

The US Dredge San Pedro *(left) and its rotary cutter (below). The vessel would demolish the estuary in areas of federal jurisdiction.*

the dredge *San Pedro* exhibited how machines and money might conquer the shoreline. Entrepreneurs raced to cash in. In mid-1905, projects began simultaneously at Ballona and Alamitos Bays. Backed by significant investment, the "Venice of America" project broke ground within Ballona. The "Naples" project did the same at Alamitos, financed by Collis Huntington's heir and nephew Henry E. Huntington. Both syndicates would employ dredging to sculpt nature into man-made islands and canals. These soon featured arched bridges, pleasure boat docks, gondola rides, and other entertainments. Meanwhile, improvements started at Cerritos Slough, a distinct marsh area east of Wilmington, at the mouth of the Los Angeles River.[10]

The Engineers District approved the private schemes at the same time that it mulled over San Pedro Bay's fate. In each case, federal officers complied with the needs of speculators. They drew up harbor lines for Cerritos Slough, enabling investors to gain a congressional subsidy. Army engineers refused harbor lines for Ballona Bay; nonetheless, they permitted the Venice syndicate to erect a breakwater to protect against wave damage. Likewise, the district encouraged the work of real estate companies inside Newport Bay, while declining to set harbor lines. (One decade later, the office reconsidered and did so.) Only in the case of the Bolsa Chica wetlands, at the southeast end of Anaheim Bay, did the engineers object to a development. Here, the district protested the Bolsa Chica Gun Club, which had dammed the tidal inlet in order to attract ducks for shooting. Courts rejected the complaint, and the district backed down. The dam lasted, in various forms, until 2006.[11]

Los Angeles in 1905 was a mix of currents. Yet they tended to drift one way. Private harbor plans hatched in many places and did so because of the Army Engineers' work at San Pedro Bay. The district approved development at the smaller lagoons, all of which had navigable portions enough to merit a public interest. But, in each case, big money won out, and private waterfront would be the result. That was one of several ironies. The coastal land rush involved many of the city and commercial leaders who pushed for a free harbor at San Pedro. As members of exclusive groups like the Bolsa Chica Gun Club, the free harbor contingent shot shoulder to shoulder with the Bannings and Southern Pacific men. They smoked cigars and locked arms in speculation. And all stood to gain as estuaries disappeared under the onslaught. Events would show that, in a region hell-bent on expansion, sides—like good and bad—were never pure or mutually exclusive.

David E. Hughes later marveled that 1905 seemed to usher in a "psychological moment" when Southern California's marshes and strand beaches were reconceived, parceled, and brought to market as "water-front lots."[12] Hughes did not object, as he watched from his drafting desk. By demonstrating the possibilities, he and other federal technicians endorsed the land rush and counseled as psychologist to its latent desires. So far, the engineers had done little to prove that government might do otherwise. It was an inherited predicament. Why should San Pedro's estuary be any different?

Yet it would be. Hughes would see to it.

Chapter 28

THE PEOPLE'S PORT

A FRENZY FOR ESTUARY PROFITS GRIPPED SOUTHERN California. Yet San Pedro's lagoon remained the greatest prize. And its fate was undecided, as the Army Engineers considered where to set US harbor lines and whether big capital would dictate the design of the port-to-be. The district did so while caught between two uncertain developments. Officials in Washington, DC, pursued, then halted their anti-monopoly policies. At the same time, Los Angeles began an aggressive territorial expansion that made it the estuary's preeminent speculator. Given these variables, army engineers had to decide just how independent their scientific work could be. Their actions would weave together local and national events.

The district's *political* significance was unexpected. The office originated in 1898, the byproduct of a Supreme Court case and foreign war. As the endpoint in a command chain extending from Washington, DC, it received and carried out orders. Both the army and the War Department had become more centralized and involved in distant affairs. Still, engineering matters rarely mixed with political controversy. This changed, however, as the administration of President Theodore Roosevelt attempted to police the excesses of industry. For several years, Roosevelt had talked up anti-monopoly reforms to reverse the concentration of power in business. Tough talk made him popular with voters. Otherwise, it made little difference. Corporate mergers continued, creating combinations like

US Steel and Standard Oil. Courts interpreted interstate commerce and antitrust laws narrowly to prohibit only the most blatant price-fixing. And corporations were too smart to make that mistake.[1]

Then, in March 1904, the administration won a court order to dissolve the Northern Securities Company, the combination of Harriman, Morgan, Hill, and Rockefeller lines that controlled half of the country's rail transportation. The verdict marked only the second time the government had successfully used the Sherman Antitrust Act of 1890. It had never before broken up a large holding company. Wall Street was stunned. When financier J. P. Morgan learned of the suit, he rushed to Washington to counsel the young, misguided president. He held forth about how the good of business equaled the good of the nation. Roosevelt endured the lecture, then replied. American liberty—the people's interest—he insisted, required him to rein in the wayward captains of industry.[2]

Roosevelt believed he was right. But the Supreme Court made it so. Justice Harlan wrote the *Northern Securities* decision, concluding that the combination posed an illegal restraint of trade. His opinion empowered the federal government to dismember corporations whose size or actions adversely affected markets, even when they did not fix prices. Trusts could no longer shield themselves behind concepts of contract or personhood. They objected, of course. Harriman sued to block the dissolution of Northern Securities and worried about coming attacks on his transportation empire. In November 1906, the Interstate Commerce Commission announced it would investigate his control of the Union Pacific and Southern Pacific systems. Not that the industrialist shied away from a fight. With his short stature (perhaps five feet, four inches), pugnaciousness, and thick moustache, Harriman resembled a lightweight boxer of the period. He had bested nearly all so far.

Northern Securities did not directly threaten Harriman's holdings at San Pedro. But the case offered a model for how the army engineers might view the lagoon and its land claims. Would they dare to do so? The Banning brothers thought they might, and feared that SP executives would abandon the claims in order to bargain with federal regulators. Even so, the Bannings saw a more urgent problem. Anti-monopoly fervor had gripped Los Angeles and turned the city racing toward the bay.[3]

TERRITORIAL GROWTH WAS nothing new to the people who settled Los Angeles. Most were from somewhere else. Nearly all had participated in the great taking of American lands. That story stretched back a hundred years. Still, a new dynamic began in the 1890s, that decade of closing frontiers. Los Angeles pursued conquest at its borders. The effort was without violence, but not without force or imperative. A campaign to cleanse the city of corruption and monopoly impelled its growth. Reformers, taking their cue from cities of the Northeast, conceived of an oppositional corporation—the municipality. This enlarged, bureaucratic city would assume delivery of services—utilities and transportation especially—divested from the private sector. That made civic anti-monopoly different than the federal or state variety. Finances did as well. The municipality's power would depend on its taxpayers and credit to issue bonds. Bonds, in turn, necessitated building infrastructure that would generate revenue to pay off debt. Because Los Angeles lacked the urban density and mass immigration of other cities, it chased power by incorporating territory. It remained true to its frontier beginnings.

Los Angeles's mix of reform and spatial growth intensified into the twentieth century. By 1910, the city would triple its area and sprawl over one hundred square miles. Within five more years, it enlarged to 290 square miles. Expansion had an added dimension: the advent of city agencies. Following initial, minor annexations, Los Angeles made its first move against a monopoly utility. Starting in 1898, it regained rights and delivery systems (including *zanjas* that dated to the Spanish era) from a private water company. To take over these duties, the city set up a Department of Water and Power (DWP). DWP's superintendent, William Mulholland, had once labored at San Pedro's estuary and the Domínguez Ranch. He would become one of Los Angeles's great builders. While DWP drafted its projects, the city in 1906 established a board of public works to administer construction. More commissions and offices would follow.[4]

An ethos of reform and bureaucracy distinguished municipality-building from earlier town developments like Wilmington. Yet they shared a familiar, speculative aspect. Each annexation raised the number of residents and ratepayers to shoulder city debt. That allowed for more bonds and more ambitious proposals, making it possible to service larger areas, which incentivized further growth. Yet people—not numbers—were behind this. It was city makers who promoted causes, made decisions, and often led the agencies. A commercial establishment pushed for expansion and profited as a result. By the twentieth century, its reach spanned Los Angeles's adjacent valleys and as far as the eastern Sierra, Colorado River, and mines and oil fields in Mexico.

US Pacific expansion promised even greater opportunity. That directed Los Angeles toward San Pedro Bay. Amid the coastal land rush, city makers determined to expand south to the lagoon. In November 1906, voters annexed the "shoestring," a sixteen-mile-long (and one-mile-wide) strip of unincorporated land leading to the town boundaries of San Pedro and Wilmington. For the first time, Los Angeles gained a minor portion—several thousand feet worth—of waterfront marsh. But the estuary beyond was crowded with private parcels. The city waited to see if federal officials might open the space to its ongoing growth.[5]

THE ARMY ENGINEERS' US harbor lines would determine whether federal and municipal efforts might accord. Meanwhile, the district's new, untested leadership made this question even more uncertain. Its officers had come and gone without much difference

so far. Then, for three years starting in 1906, Amos Fries commanded the post. Unlike the older generation of engineers whose
formative experience came from the Civil War or frontier service,
Fries's career began in a time of US global power. He graduated
from West Point in the class of 1898, just as the Spanish-American
War began. In 1901, the army sent him to the Philippines. Here,
victory against Spanish forces had descended into bloody suppression of Filipino rebels. Fries served two years with John J. Pershing's command, which worked to pacify the island of Mindanao
and its Muslim Moro tribes. Pershing, in turn, answered to General Adna R. Chaffee, supreme commander in the US-Philippine
War. After the defeat of Filipino insurgents, the army assigned
Fries to navigation projects on the Columbia River, then to Southern California.[6]

The Los Angeles District offered Fries a semi-independent command. In time, this would reveal his didactic, even contrarian streak.
Before entering the US Military Academy, he had taught school in
rural Oregon. He and wife Elizabeth Wait held a lifelong interest in
education. They later would build ties to groups that policed school-
ing, such as the Anglo-Saxon Christian Association, American
Legion, Daughters of the American Revolution, and Friends of the
Public Schools of America. Fries relished the chance to impart young-
sters with national pride and to fight "everything un-American," as
one newspaper put it.[7] His political leanings were clear. Yet his zeal
and inclination to instruct others are the more important trait. It
set Fries apart from the typical army engineer and equipped him to
leave a lasting mark at San Pedro.

Land claimants saw things differently. Confident that Fries was
weak and amenable, the Southern Pacific, Salt Lake Railroad, and
Banning Company pressured him to conform US harbor lines to
their estuary parcels. They also lobbied in Washington, DC, where
superiors might compel the young commander. Most of all, the companies pushed for a seemingly innocuous document known as the
quit-claim deed. By signing the documents, claimants would relin-

quish portions of their land to public navigation. The government, to its benefit, would avoid delays and the hassle of settling legal disputes. All the same, claimants would gain the greater reward. Quitclaim deeds would validate their property as well as their contention that the lagoon was not truly navigable (if it were, deeds would not be necessary). That would confront money-minded US officials with the threat of punitive damages. If claimants could obtain the documents, they knew they could force the government to negotiate.[8]

The pressure proved effective. Fries did not object when the corporations began to reclaim and build above navigable waters. In several cases, he allowed them to dredge parcels below the tide line. Whenever Fries denied permits, he did so cordially. The district, he explained, was sympathetic to property owners so long as their plans did not impinge upon harbor design. Of course, in their race to make improvements, claimants hoped to present him with a fait accompli. That seemed to be happening.[9]

AT THE SAME TIME, the railroads and Banning Company attempted to win over Los Angeles municipal interests, the Chamber of Commerce most of all. The chamber had led the free harbor fight a decade earlier. But, as of January 1907, it had yet to take a position on the corporations' Peninsula Plan.

That was a sign of subtle, though important, changes. Since the harbor fight, reformers had undercut the SP's power with a series of amendments to the city charter. The measures established competitive bidding for contracts, merit-based civil service, recall elections (of corrupt officials), and boards of commissioners in place of graft-ridden appointments. Reform victories had an adverse effect, however. Without the old SP to unite against, Los Angeles grew fractious and preoccupied by labor conflict, municipal debt, and public hygiene (to regulate race and eugenics). City leaders also realized their need for railroad services. Agriculture and real estate benefited from SP advertisements and investment. The SP provided Los Angeles with its commercial hinterlands. The sides shared an array of interests and seemed to agree that private property could exist in the tidal margin. Claimants to Wilmington's lagoon held land patents, just as development syndicates did at other coastal wetlands throughout the region.[10]

Compromise, of some sort, was a real possibility. The SP knew this, and tried to better its odds by appealing to the free harbor crowd. Harriman's operation was not the piratical, short-sighted company of Collis Huntington's heyday. Business managers now were savvy, forward-thinking, and interested in profitable partnerships. Their Peninsula Plan offered a number of concessions to create public waterfront amid an otherwise privately owned harbor. It allowed for municipal wharves. It promised to place the reclaimed peninsula under third-party ownership. And a respected civic reformer set about selling the plan to city leaders. Thomas Gibbon was a Chamber of Commerce lobbyist and former chair of the Free Harbor League. In the 1890s, he had been vice president of the Terminal Railway Company, Huntington's rival at San Pedro Bay. By 1906, Harriman's consolidation of the industry moved Gibbon into the role of chief counsel for the Salt Lake Railroad and consultant to the SP. His job was to convince Angelenos that Harriman's was a reformed, enlightened octopus.[11]

However, the Southern Pacific retained just enough of its old ways to go awry. During the 1870s and 1880s, Huntington's SP had built itself into the de facto government of California and numerous cities and counties, including Los Angeles. By the time of the Peninsula Plan, the SP's political machine was not what it used to be. Waning influence tempted it to take risks and to fail massively. In 1906, railroad agents were caught bribing delegates at the state party conventions. The SP suffered a wave of negative press (interrupted briefly by its relief efforts following San Francisco's earthquake and fire and the Colorado River floods). Statewide, the company's great disaster would come with the 1909 direct primary law, which empowered voters (not bosses or lobbyists) to choose candidates for the general election. Within two more years, the SP's political machine lay broken entirely. That finale had its preview in Los Angeles, where anti-SP candidates triumphed in the 1906 city elections.[12]

For Gibbon, selling the Peninsula Plan would not be easy.

WASHINGTON, DC, WAS a different world. Here, things looked brighter for the Southern Pacific's harbor proposal because of an unexpected event. Estuary claimants arrived to lobby for the Peninsula Plan in late 1906, just as Wall Street fell into a severe panic. The financial crisis would last until the following November. Banks verged on default, and the world on economic depression. Corporate leaders pointed the finger at Roosevelt's trust-busting. The president countered with corresponding blame. But, although government had grown considerably, it remained too small to prevent a collapse. The US Treasury pledged a paltry $25 million to stabilize markets and restore liquidity.

Roosevelt was forced to ask the help of big business, and specifically of J. P. Morgan, architect of the nation's largest, most notorious trusts. Morgan offered to contain the crisis using his personal fortune and good name with stock traders and bankers. In exchange, he asked Roosevelt to call off the anti-monopoly program. The president had little choice. Yet a further loss of face awaited: the monopolist won assurance that he could enlarge US Steel while the president, for time being, stood down. Corporations basked in their victory. Morgan, who had sat out the Civil War to enrich himself while a hired substitute fought in his place, earned praise for his patriotism. Harriman also counted his luck.[13]

The pause in federal regulation showed in the War Department's friendly hearing of the Peninsula Plan. The SP sent both Thomas Gibbon and company superintendent Robert H. Ingram (Harriman's former assistant) to meet with the army's chief of engineers and War Secretary William Howard Taft. The chief of engineers expressed doubts about the lagoon's navigability and whether harbor lines were necessary at all. He and Taft were particularly wary that the Los Angeles District would provoke costly litigation. They seemed to favor using quit-claim deeds. This was exactly what the railroads and Bannings wanted to hear: that government had no business in the question of estuary ownership.[14]

FEDERAL OFFICIALS hesitated to take on the nation's largest corporations. And this put the Los Angeles Engineers District in a difficult position as to how to proceed. David E. Hughes bristled at the "erroneous" ideas of his Washington superiors. But Fries appeared to comply. He agreed in conversation, so SP lobbyists reported, to let private parcels alone. What else could he do? Without stronger evidence to the contrary, the War Department would approve the Peninsula Plan—and overrule the district, if necessary.[15]

Inside the office, matters were more decided. Hughes's papers reveal that he and Fries never viewed the Peninsula Plan as viable. It offered too little public access. That suggested the district's forthcoming plan would make a political statement, one committed to anti-monopoly ideas and government authority. Fries and Hughes took to it with surprising conviction, as they criticized the Peninsula Plan and private claimants in sweeping, non-technical terms. "Is the public who pays for these improvements to have no benefit?" Fries wrote in Socratic style, "except such as a great money-gathering corporation shall see fit to grant them? Is that republican government? We think not."[16] The engineer honed his beliefs around a perceived struggle for liberty, one evocative of the Civil War and the Jacksonian battle against the Bank of the United States. The public technician carried on this fight. "As government officials we are the representatives of the people," he explained. "We are false to our trust if we do not guard the public's interest as we would our own."

Hughes thought likewise. "We must remember," he implored a correspondent, "that the Government is to endure into time when the corporations will be more amenable to the people who created

them, and after the real estate men of to-day have presented their
last claims to Saint Peter."[17] With lofty words, the spirit of disinter-
ested, republican science—which had suffered on the frontier and
through the Gilded Age—returned in force. It proclaimed San Pedro
Bay the people's port.

Times were not so simple, however. What did republican science mean in the era of big capital and bigger government that now
included the municipality? Modern society was made up of large
institutions, whose capacity and functions rendered them more
abstract and distant from the "people." Fries and Hughes attempted
to sort through this. Yet the quandary was unavoidable. Public interests were a form of interest. And power vested in public authority
was power nonetheless. It could and would be wielded against others. This put the Engineers District in league with global expansionists and Los Angeles boosters, who chased colonies abroad and at
home with patriotic slogans and promises of prosperity.[18]

To CONVINCE OFFICIALS in Washington, the Engineers District
still needed to show the weakness of private claims. Fries assigned
Hughes to investigate. The brilliant, self-taught engineer faced a formidable task. He had to out-reason the SP's well-heeled lawyers and
the drift of real estate development over five decades. Also, government records were scattered and required extensive sleuthing. No
one knew the estuary's full history. Hughes perhaps came closest.
He versed himself in Spanish and Mexican grants, state and federal
land policy, Ord's reservation for navigable waters, the De Guyer
case, and tidelands law. Hughes became so expert that he launched a
side career as a consultant in coastal property cases. His conclusions
about San Pedro's lagoon earned that reputation.[19]

Hughes determined that the private claims within the estuary had
no lawful basis. He reasoned that, when the United States and California assumed title to the bay in 1848 (and 1850), governments were
bound to keep it in trust—and clear of property—for the purpose of

public navigation. Surveyors and land offices broke with this fiduciary task. But the Army Engineers could correct them. Engineers could wield harbor lines to obliterate false claims, even those outside of federal waters. The goal was to restore the bay to the moment *before* any parcels existed—the moment of Ord's 1855 survey—when pure government authority overspread it like a Pacific king tide. By his varied talents and independent mind, Hughes cut a path forward, an exit from the trap of property and the self-doubt of anti-monopolists.[20]

The Engineers District released its proposed harbor lines on February 21, 1907. Two war secretaries, William Howard Taft and Luke E. Wright (both former governors-general of the Philippines), would approve the lines in the summer of 1908. Taft did shortly before he resigned to run for president. Approval meant defeat for the Peninsula Plan. But the Southern Pacific and allied corporations were not beaten yet. Instead, conflict shifted from navigable waters to a new venue, with a new antagonist. Harbor lines were the opportunity the City of Los Angeles had waited for.[21]

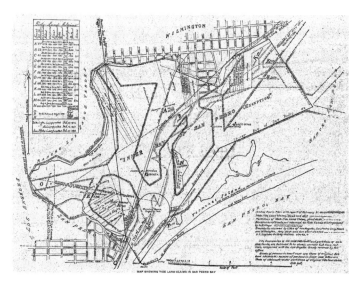

US harbor lines, proposed by the Los Angeles Engineers District and adopted by the War Department. Lines at top and right follow Ord's 1855 inner bay exception.

Chapter 29

TIDELANDS IMPERIALISM

RAMATIC LAND GRABS PUNCTUATE THE HISTORY OF US expansion. Yet they culminate in a less-known acquisition, of a place that few Americans think about. This extended from the estuary's high tide line to its navigable waters—that is, to US harbor lines. People called it the "tidelands." While small in size, it had gained global importance. Ambitions that previously reached across the continent and over the Pacific now waded beyond the shoreline onto mudflats partly submerged by sea. But the ambitions came in new form, carried by a new arrival. Starting in 1908, Los Angeles would take control of the tidelands and assert itself as rightful heir to a century of imperialism.

The tidelands' allure was all about connection. Los Angeles coveted them as a bridge between ocean and earth. It did even more so because tidelands combined an expansionist past with opportunities of the future. US conquest, scientific exploration, Civil War, and railroad capital had transformed the coast and inland region. Government had enlarged and elaborated, a key development, as tidelands were state—not federal—domain. By linking the various stories, the tidelands would enable Los Angeles to step beyond its roots as a borderlands society to become a world metropolis. Tidelands held power. That is exactly why Los Angeles pursued them and their destruction.

When did observers first see a city and nation's destiny in the estuary shoreline? For decades, some suspected, even professed to know, that private claims obstructed some larger purpose. Army engineer George Mendell first questioned the claims in the 1870s. Lawyer J. D. Bicknell did, too, on behalf of the Southern Pacific and Bannings. He reported that his clients likely did not have title. The problem, Bicknell observed, was that nearly all of the properties lay within two miles of the incorporated town of Wilmington. Under the 1879 California constitution, this left parcels null and void and subject to government use. It was claimants' secret to keep. In 1891, however, Thomas Gibbon of the Terminal Railway concluded the same. Gibbon notified the Chamber of Commerce just as the free harbor fight erupted. Then nothing happened.[1]

The railroads and Banning family held the parcels in spite of what critics said. Defeat in the harbor lines contest surprised them. But claimants prepared for such contingency and planned their next move. In May 1908, after last-ditch lobbying in the federal capital, company representatives met at New York's Belmont Hotel. The newly finished building stood at Park Avenue and Forty-Second Street, opposite the old Grand Central Station and not far from today's United Nations headquarters. In its time, the hotel was a comparable seat of power. Here, the Southern Pacific, Banning Company, and Salt Lake Railroad huddled in war council. William Banning attended on his family's behalf. SP executives were there, too: vice presidents Julius Kruttschnitt, Robert Ingram, and others. Edward Harriman presided over final talks.[2]

Matters at the estuary had at last become pressing enough to draw Harriman's involvement. The parties negotiated an agreement, signed June 16, to sort out titles, ownership, and improvement of tidelands. Claimants would fight on. But by distributing assets, the accord signaled the end of close coordination. The SP faced big problems elsewhere. At the end of January, US Attorney General Charles J. Bonaparte—grand-nephew of the French emperor—filed suit to break up Harriman's control of the Union Pacific, SP, and Salt Lake Railroad. President Roosevelt took a keen interest in the case, even as he prepared to leave office for an African hunting safari. A series of affronts left him with a grudge against the rail magnate. Face-to-face meetings did not help. Newspapers likened them to jiujitsu combat. The president began to distinguish between trusts that were harmful to the public interest and those that were not. He put Harriman's squarely among the former. The government named the industrialist as defendant—an unusual step in anti-monopoly prosecution—in a conspiracy to stifle rail competition west of the Mississippi River. Harriman suffered under the stress, which worsened his pulmonary disease and a number of mysterious ailments.[3]

On paper, the Banning family held nearly seven-eighths of the lagoon, masking the Southern Pacific's stake. The 1908 agreement rearranged things. The Banning Company sold the SP 266 acres, mostly in the lagoon's western half. The Bannings kept the bulk of the eastern lagoon, 170 acres including Mormon Island. The remainder went to the Salt Lake Railroad. The parties agreed to share access to future waterfront. All hurried to complete their work before the government dredge approached and before the Engineers District could require permits.[4]

Cartoon of a meeting between Harriman and President Theodore Roosevelt, based on their common interest in martial arts. The president prevailed in ink and in real life.

Yet the agreement could not resolve the bigger legal problem: California's ban on tideland patents near incorporated cities. The companies worried that regulators might start to enforce it. Then they caught a break. California's supreme court voided Wilmington's incorporation of 1872 (thus, also the town's disincorporation in 1887). By no coincidence, the latter prefaced a final rush of estuary patents before Wilmington incorporated once again in December 1905. The court's verdict did not quite legitimize the patents, but it strengthened them and gave claimants a window in which to improve the parcels. It also added to their case for common law possession.[5]

WHILE SP EXECUTIVES GRAPPLED with the tidelands situation in wide frame, the Bannings grew attuned to local events. What seemed a remote province to the Harriman Lines was the Bannings' everything. Estuary parcels were a family investment that dated to the 1860s. Lighterage and wharfing remained the brothers' main business. Surviving telegrams from the New York summit indicate their worry that a new threat had emerged. To fend off betrayal or surveillance, the brothers sent messages in code: "Expertous on ost galbandom castment." (Decoded: "Negotiations on most satisfactory basis.") The Southern Pacific Company was ciphered as "Tangent"; individual Banning brothers as "Mingle" and "Millstone." William Banning saw this as a necessary step. So far, claimants had "resisted every assault" made upon their titles. But he warned that "assertions by the public bodies" soon would become "vexatious" and "burdensome."[6]

Banning's suspicion proved correct. The New York summit was indeed monitored by the City of Los Angeles, whose desire for tidelands had received a massive boost from US harbor lines. By coincidence, the former city attorney was present at the Belmont Hotel. His official mission was to sell $2 million in bonds to Wall Street investors. These were for the Los Angeles–Owens River aqueduct. The task disguised his unofficial mission. On June 5, the agent

received a message from James A. Anderson, outgoing head of the
Board of Public Works, the bureau created to build the aqueduct.
But Anderson detoured into estuary matters. Regardless of harbor
lines, private claimants might yet "dictate harbor development" by
controlling the tidelands.[7] This, of course, was the Southern Pacific,
Salt Lake Railroad, and Banning Company's purpose for gathering at the same hotel at that exact moment. Anderson informed the
agent that the municipality, in alliance with the Chamber of Commerce, was ready to launch a legal foray. "These tideland patents
are undoubtedly illegal and void," he wrote. "A number of influential men are contemplating the formation of an organization" to
prosecute the claimants "with all diligence to a finality." Anderson
wanted the message relayed to federal officials.

The organization formed to defeat the tideland patents was the Los
Angeles Harbor Commission. It convened at the start of 1908, after
gaining an inside man: the SP's former lobbyist and lawyer Thomas
Gibbon. Gibbon returned to civic reform months after leaving the
employ of estuary claimants. He reemerged as a leading voice on
harbor issues, in particular his long-standing belief that the patents
were unlawful. He and associates took a controlling share in the *Los
Angeles Herald* and stepped up its attacks on the Southern Pacific.
Did claimants know about Gibbon's views when they enlisted him
to promote their Peninsula Plan? The idea that opponents could be
salaried and bought as friends pervaded business of the era. But it
had exceptions, too. And this one proved fatal to the corporations'
plans. Gibbon's résumé remains one of the story's strangest facets.[8]

When the inaugural Harbor Commission began work, Los
Angeles had only a minor toehold at the lagoon. Elsewhere, the
city remained separated from the sea by the incorporated towns of
Wilmington and San Pedro. The commission's first task, therefore,
would be a new round of territorial conquest. Gibbon would lead a
campaign to consolidate Los Angeles with the bayside towns. San
Pedro and Wilmington were not the ultimate object, however. The
real prize lay just beyond, in the adjacent tidal zone.

Reformers like Gibbon had come to see tidelands as the solution to rail monopoly. As the Coast Survey had noted fifty years before, the Los Angeles region lacked a fully protected natural harbor. Instead, it featured several coastal bays, too shallow for navigation, spread among lengths of exposed strand beach. The latter are the pride of Southern Californians today. But at the start of the twentieth century, commercial interests blamed the shoreline for their exploitation by the Southern Pacific and San Francisco-based shipping lines. Development of an artificial harbor—protected by breakwater and deepened by dredge machines—offered liberation. But it did so only if that harbor fell under control of a single, expanded municipality, "Greater Los Angeles," which would encompass the lagoon and consolidate the waterfront towns.[9]

The Army Engineers threw their support behind the city's territorial aims. District Chief Fries had become a hero to bayside residents and Angelenos alike because of his harbor plan and deft handling of the SP. He was ideally suited to bring the three communities together. Beginning in the fall of 1908, the officer lent his name to the consolidation movement. He became a fixture at events, frequently as a keynote speaker or guest of honor. Fries, the teacher, took the issue to the people—women's clubs and schoolchildren among them. He propounded about patriotic duty, restraint of corporate power, and the need for municipal expansion. "I am strongly for consolidation," he explained, "because I believe that is the only thing that will make San Pedro a great public harbor."[10] His colleague, D. E. Hughes, agreed. Both stretched farther beyond their roles as federal technocrats. They shared insights—archival research especially—with city lawyers. They ventured deep into civic affairs and local property disputes.

Fries's effort on behalf of consolidation earned him praise in Los Angeles. But the army considered it improper. In the summer of 1909, the War Department announced it would redeploy him. Reformers cried foul and blamed estuary claimants. The city protested and even offered to put Fries on its payroll. But he refused to leave the army. Before departing, he made an appearance at the final pro-consolidation rally. Fries then served as engineer-in-charge at Yellowstone National Park, where he laid out roads and bridges still used by visitors today. During World War I, he resumed his partnership with John Pershing. Fries commanded the army's Gas Service Section (renamed the Chemical Warfare Service in 1918) under Pershing's American Expeditionary Force. Hughes would keep him informed of harbor affairs for decades after.[11]

Voters approved bayside consolidation in August 1909, creating a larger municipal Los Angeles. Edward Harriman's death weeks later made a remarkable counterpoint. Business competition, anti-monopoly lawsuits, and relentless criticism from newspapers took a toll on the sixty-one-year-old railman. Due to the extent of his holdings, he faced battle on a thousand fronts. In response, Harriman overworked himself. Doctors advised him to relinquish day-to-day management to save his life. He could not do it. "You were right," he confessed to a confidant in June 1909, "I ought to have quit and laid-back. But it is too late now. I am in deeper than ever, and it must go on." Journalists pried about his sickly appearance. However, Harriman was too inured to his investors' confidence. He lied and said he had never felt fitter. The corporate imperialist and master of all succumbed to gastric ulceration. One giant's demise enabled the rise of another. Los Angeles officials knew their moment had come. They proclaimed that consolidation finally had established a city "large enough and strong enough" to develop San Pedro Bay "into one of the great municipally owned and controlled harbors of the world."

"By the efforts of the patriotic citizens," they effused, "her commercial future is assured."[12]

TIDELANDS IMPERIALISM had only begun. With the acquisition of the waterfront towns completed, Los Angeles leaders began a campaign to place the city at the center of US global power. Harbor Commissioner Gibbon publicized a set of interrelated ideas—geographic concepts—for a maritime city-state of the future. First, Los Angeles would take charge of the North American West, or, rather, the portion he termed the "Pacific Slope." The region extended from Southern California, through the Southwest to New Mexico; north to Wyoming, Montana, and parts of the Dakotas; and back again through the intermountain basin of Idaho, Nevada, Utah, and western Colorado. It was the same region taken from Native Americans since the Civil War. It now produced the majority of the nation's precious metals, citrus, and winter vegetables. It contained a large share of US range-fed beef, timber resources, coal, and petroleum. Gibbon foresaw this expanse of eight hundred thousand square miles rivaling the hinterlands of Chicago or St. Louis, whose reach he knew from his boyhood in Arkansas.[13]

Second, Los Angeles would connect this inland empire to the markets of East Asia via the "Great Circle Route," the major ocean current (or gyre) that flows clockwise along the north Pacific littoral. Today's readers may know it for gathering a massive patch of plastic garbage. But for Gibbon, it presented Los Angeles with the opportunity to command transpacific and colonial trade. The current passed just seventy miles offshore. From there, it flowed south, down the coast of Latin America toward the equator, and then west to the "great Oriental ports" of Yokohama, Shanghai, and Hong Kong. En route were the US harbors at Pearl Harbor (by detour), Samoa, Manila, and Guam. From Asia, the circle returned north, then east to North America before repeating. With this seaborne highway, Los Angeles had access to eight hundred million customers. Raw materials would arrive from Asia and US island possessions. Industry would flock to the city and export finished products. The idea that manufacturing might flee by the same currents seemed unthinkable then. Positioned ideally between the Pacific Slope and Great Circle, Los Angeles would rise astride continent and ocean.[14]

Or two oceans? The Panama Canal, a construction project administered by US Army Engineers and a federal commission, began in 1903 to connect the Atlantic and Pacific via dredges, human labor, and a system of mechanical locks. The project took hold over Los Angeles's calculations. It obsessed city globalists like Gibbon. San Francisco had announced its intent to host the Panama-Pacific World's Fair. But tidelands would determine which city received the canal's business, and that gave Los Angeles the advantage. Unlike San Francisco's outdated and inefficient waterfront, Los Angeles could shape the tidelands to the canal's specifications. That included room for the largest ocean-class ships (Gibbon cited the thirty-five-foot-draft RMS *Lusitania*, launched in 1906), sufficient fuel stocks, and facilities to move high volumes of cargo at the lowest market rates.[15]

Gibbon roused his audiences to see the opportunity before them. A modern harbor, completed in time for the Panama Canal's opening (set for 1913), would position Los Angeles as the link between Asia and Europe, between Pacific and Atlantic worlds, among the whole of humanity. Combined, the Pacific Slope, Circle Route, and Panama Canal augured a new world of international friendship and shared prosperity. It was oddly similar to Harriman's ambitions—to late twentieth-century neoliberal thinking, as well—and yet different in expression.[16]

With grandiosity, Gibbon seated American expansion—past, present, and future—squarely on the Los Angeles coast. Others had tried before. From the nation's birth, merchants and city builders of New York, Boston, New Orleans, St. Louis, Chicago, and San Francisco dreamed similarly. They opened the Erie Canal, blazed fur and cattle trails, built mines and railroads, and sailed ships to Hawaii and beyond. Yet Gibbon's vision for Los Angeles spoke to the nation's status as a world power as much as to shifts in regional power. His ideas relied on industrial-era transportation, fossil fuels, and US naval reach. Most of all, Gibbon premised his arguments on the artificial harbor at San Pedro Bay, a conceit impossible a decade earlier. They proved truer than most. Los Angeles would achieve the globalism he predicted. It would carry the United States and world with it.

Federal officials responded. They began to include San Pedro Bay tidelands within the nation's imperial plans. Bayside consolidation took place as the Taft administration pushed for reduced tariffs to open transpacific and Latin American trade. President Taft visited San Pedro Bay in October 1909, two months after the consolidation vote. Gibbon's *Herald* celebrated the visit as inaugurating an "epoch" of "far western opportunities and responsibilities. . . . Westward the course of American empire has taken its way, and westward the nation gazes."[17]

Taft's visit followed those by the navy's Pacific Squadron and Atlantic Battleship Fleet. The latter group, sixteen dreadnoughts famously known as the Great White Fleet (for their non-combat paint), visited San Pedro Bay before crossing the ocean on a round-the-world voyage. Fleet Day, in April 1908, drew tens of thousands of Southern Californians. Military force, or friendly displays of it, were part and parcel of the imperial age. Officials like Taft and Angelenos like Gibbon underestimated the twelve-inch guns that sailed, too. They foresaw neither the carnage of the Great War nor the delays to commerce and canal building that would result. In that sense, they were much like the prior generation of Americans who

The first battleship to pass from the Atlantic to the Pacific via the Panama Canal docked at Wilmington, Los Angeles Harbor, in August 1915.

talked their way through conquest, only to find themselves bloodied by Civil War.[18]

The city's strongest claim to US imperialism came with its hiring of retired general Adna R. Chaffee. Chaffee had long experience with expansion. He enlisted in the Union cavalry and was twice wounded during the Civil War. He rose to command during the Indian wars, and briefly administered San Carlos Agency in Arizona. When Apache escaped the reservation (provoking "Geronimo's War"), Chaffee subdued one group at the Battle of Big Dry Wash in July 1882. The officer remained in the army for the Spanish-American War, when he served in Cuba as brigadier and chief of staff to the occupation governor. Chaffee then commanded the US contingent that helped suppress the Boxer Rebellion in China. Promoted further, he commanded all forces in the Philippines during the war against nationalist insurgents. His military life culminated with a two-year term as army lieutenant general and chief of staff. That forged a working relationship with Taft while the latter served as Philippine governor-general and secretary of war. Chaffee retired in 1906. Two years later, Los Angeles's mayor appointed him to lead the Public Works Board.[19]

Chaffee's hiring raised the city to global stature. He provided a direct channel to the president and War Department, and he would remain the go-to emissary to receive imperial visitors. The *Herald* remarked on the two "old friend[s]" during Taft's 1909 visit to the waterfront. "The moment he saw General Chaffee," the paper said of the president, "he was on his feet, his face wreathed in smiles and his hand extended in glad surprise. It was the first meeting in many days of the former secretary of war and his right-hand man."[20] The mayor, Army Engineers, and Harbor Commissioner Gibbon looked on with delight at Los Angeles's good fortune.

Yet America's newest metropolis faced challenges still. Its expansion to the shoreline remained moot as long as private claimants, with the State of California's indifference, sat atop marsh and mud. To give its ambitions life, Los Angeles needed to destroy the tideland patents, and then the tidelands itself. Litigation began. It would not be quick, however. In addition to facing powerful corporations, the city, in its hubris, had engendered a self-destructive force. Strange divisions and overlapping duties gathered over the harbor-to-be. Construction remained within the purview of the Public Works Board, led by Chaffee. Administration belonged to the Harbor Board, led by Gibbon. Then there was the city engineer, whose chief sat on both boards. His office had two hundred employees, but none had ever worked on a harbor project.[21]

Each center of power had its say. Each was defensive of turf. It was a recipe for bureaucratic disaster, staved off only so long as a common foe existed. For now, tideland claimants were the enemy. But by Los Angeles's hand, that would come to an end.

Chapter 30

A CITY RETURNS
TO CIVIL WAR

THE YEAR 1911 MARKED THE CIVIL WAR'S GOLDEN
anniversary. Many veterans, military and political leaders especially, were already gone. The rest edged closer to life's end. These
disappearing survivors would meet for reunions and remembrance.
As the season approached, Los Angeles outbid rival cities to host
the Grand Army of the Republic's forty-sixth national encampment,
scheduled for September 1912. The Union group would commemo-
rate events of a half century past: Lee's Maryland invasion and the
bloody, suspenseful weeks between Second Bull Run and Antietam.[1]

The aged soldiers would do so in Southern California, of all
places. It seemed an odd choice. Los Angeles was far from the war's
main events. Californians made up a small contingent of troops in
the East. Most of those were army regulars who happened to be stationed on the West Coast in 1861. It was not quite their home. But
since then, the United States had changed. The nation now leaned
to the West and beyond. Los Angeles grew by an influx of Northerners *and* Southerners, arriving by railroad and in search of Pacific
opportunities. One was *Los Angeles Times* publisher Harrison Gray
Otis, a Union veteran from Ohio who ran his paper with martial dis-
cipline. All became Angelenos. Fascinated by romantic, sentimen-
tal histories of the region, they appropriated Native and Hispanic
pasts as their own. Yet their coalition joined most wholeheartedly
around future profits. The city's immigrant elite determined to build
a metropolis atop the former frontier town, Mexican pueblo, and

Gabrieleño-Tongva village. Estuary tidelands were the key. They allowed for sectional reunion unlike anything in the New South or industrial Northeast. In that regard, Los Angeles was a perfect place to memorialize the Civil War and the difference it made.

At the same time, Los Angeles exemplified the nation's growing amnesia about the war, its effects, and causes. A large part of this pertained to race and the legacy of emancipation. Southern California's experiments with segregation (in schools, marriage, and housing) were just beginning. Likewise, Los Angeles would show the ease with which Americans could forget, then reenact, the hazards of ambition. Driven by its haste to build a harbor in the coastal shallows, it pushed to an analogous place of risk. By 1912, frustrations touched off a municipal conflict that threatened to tear the city apart.

Los Angeles's descent into disunion began by its rapid success. Bayside consolidation fired up imperial designs. At the same time, consolidation intensified the city's great problem: it still did not own the tidelands needed for harbor development. Annexation of Wilm-

As it remade itself into a Pacific imperial city, Los Angeles would host a national encampment for Civil War veterans. Unfortunately, the city was mired in conflict over its port.

ington provided a mere thirty-seven acres. Elsewhere, 1,200 acres remained under private claim; several hundred more at the former town of San Pedro. Los Angeles needed to confront claimants in court. That was the only way to win title and rush ahead in time for the Panama Canal. In October 1908, Wilmington filed the first lawsuit. It argued that the state constitution forbade tideland patents within two miles of incorporated city limits. Los Angeles endorsed the complaint. It provided financial support and its best lawyers. The Chamber of Commerce did the same.[2]

Like the tidelands itself, however, the lawsuits became a mire of uncertainty. Following consolidation in late 1909, Los Angeles took over as plaintiff and filed fourteen more suits. By 1911, the total had grown to twenty-seven. But delays accumulated. The county court declined to merge the complaints. Each would have its own fact-finding, arguments, trial, evidence, verdicts, and appeals. The Banning Company and Southern Pacific's lawyers offered a steady stream of demurrers and objections. And they announced they would appeal any adverse decision. Their strategy was to hit Los Angeles at its vulnerable point. Claimants read Gibbon's speeches and about Chaffee's backslapping with visiting dignitaries. They knew to withhold what the city wanted most. They would bottle-neck port development, frustrating municipal expansionists and forcing them to some kind of deal. Despite the rush of enthusiasm that accompanied the suits, Los Angeles's growth would turn slow and painful.[3]

The first tidelands trial opened before Judge Walter Bordwell of the Superior Court on October 5, 1909. Cases did not last long by today's standard. The first closed after just one month. The *Times* noted the trial was "devoid of sensations" and "dull from the lay-man's point of view."[4] There were noteworthy moments, however. Opening day featured reminiscences of the estuary by old-timers like G. M. Lopez of Wilmington, persons described by the paper as "Mexican boys who skulled across the inlets and lazy channels" and "men who hunted and fished along the shores." William Mulholland, of the Department of Water and Power, gave featured testimony, as he recalled his days laboring on the entrance jetty in 1878.

In anticipation of a verdict, Los Angeles added further lawsuits, including an action to stop the SP's and Banning Company's improvements at Mormon Island. Yet the entirety of 1910 passed without a ruling. Judge Bordwell, descendant of the Puritan theologian Jonathan Edwards, worked studiously on his opinion. His devotion left Los Angeles in a holding pattern. The city could not proceed with construction. Nor could it sell the $3 million in harbor bonds that voters had approved in January. California's attorney general advised cities not to file any more tideland complaints until Bordwell gave a decision.

As the estuary cases sat in limbo, Los Angeles's coalition of reformers and businesses began to bicker. Trouble first appeared during the lead-up to consolidation. Reformers alleged that Mayor Arthur Harper, previously thought their ally, had condoned graft and vice among saloon interests. San Pedro's estuary became entangled in the scandal. Critics charged that the corruption reached to the Public Works Board, jeopardizing harbor development. A recall campaign began. Harper's allies countered with their own allegations, a libel suit, and threats to recall *their* opponents. The mayor resigned in March 1909 to stand trial. Reform candidates swept the municipal elections, and voters passed numerous charter amendments. Ill will and suspicion lingered, however.[5]

The mayoral scandal spiraled Los Angeles into conflict for the first time since the defeat of the Southern Pacific machine three years prior. It shifted the focus from corruption to general discontent with the city's establishment. Then, violence struck. After midnight on October 1, 1910, an explosion tore through the *Los Angeles Times* building downtown. The collapsing structure and heavy machinery

broke open vats of flammable ink. Damage severed the gas main, setting off fire and more explosions. Twenty persons died, and others
were injured, some after jumping from burning, smoke-filled win-
dows. The bombing took place against the backdrop of an ironworkers' strike and countercampaign led by the *Times*. Six months later,
police arrested brothers John (J. J.) and James (J. B.) McNamara.
They were said to be part of a nationwide "dynamite conspiracy,"
for which three dozen labor leaders would receive convictions. The
McNamara trial entered the docket of Judge Bordwell just as he concluded the first tideland cases.[6]

The bombing split Los Angeles into angry, alarmed camps. Business groups attacked labor for inciting murder and mayhem. Unions
and radicals accused police of framing the McNamaras. Moderate
voices were caught in between, yet they blamed the trouble on city
leaders. Harsh open-shop policies had kept out organized labor at
nearly all cost instead of resolving industrial conflict, as progressives
elsewhere tried to do. Violence, critics argued, was the inevitable
price. Samuel Gompers of the respectable American Federation of
Labor hired famed attorney Clarence Darrow to defend the brothers. Many liberals and socialists (some Marxist, some not) agreed.
The latter, in this moment prior to the Bolshevik Revolution, were
a significant part of the reform movement. The nation's preeminent
socialist, Eugene Debs, had sparred with the *Los Angeles Times*
(a "leprous sheet," in his words) ever since the Pullman Strike of
1894. He saw the 1910 bombing as the city's comeuppance. "It is
my deliberate opinion," Debs wrote in the radical press, "that the
Times and its crowd of union-haters are themselves the instigators,
if not the actual perpetrators of the crime." Rather than serving the
people's interest, their intransigence had brought "class war" to "its
acutest stage."[7]

The McNamara trial put Los Angeles under the scrutiny of world
opinion. No matter that the brothers were convicted. The city's pre-
tense of good government became the bombing's added victim. In
the 1911 mayoral primary, socialists challenged the pro-business

establishment and earned the highest tally of votes. The McNamaras' confessions of guilt undercut the socialists just before the general election, handing the victory to incumbent George Alexander.
But a wider number of residents soon would share the radicals' dis-
satisfaction. Among their complaints was the city's broken promise
to build a harbor for the benefit of all. The former town of Wilmington thought so. Its leaders published an open letter rebuking city
hall for its failure to uphold consolidation vows. Lofty rhetoric,
whether from Fries or Gibbon or others, returned to Los Angeles
as indictment.[8]

At the start of 1911, just as city politics approached a crisis, the
superior court issued its first tidelands verdict. The case, *People
v. Southern Pacific Railroad Company, et al.*, dealt with a small
area, 18.49 acres located on the estuary channel. With his twenty-
thousand-word opinion, Judge Bordwell invalidated the claimants'
patents and declared that tidelands useful for navigation could never
be alienated into private ownership. Such coastal lands formed a
public trust, which the state was bound to keep in perpetuity for the
good of its citizens. "The state," he wrote, "is not to be deprived of its
right to assert title of which it has never been divested."[9] Bordwell's
words would have a two-sided legacy. They prepared the estuary for
total destruction. At the same time, they strengthened California's
public trust doctrine, used since the 1960s to preserve coastal lands,
ecology, and public access.

Los Angeles hailed the verdict as clearing the major obstacle to
port construction. In May 1911, the state legislature relinquished its
title and granted tidelands to the city. In June, the Board of Public Works and the city engineer unveiled a dredging plan. The first
round of harbor bonds, approved in early 1910, would pay for the
work. The city, meanwhile, broke ground on its first municipal
wharf, a 330-foot structure located in Wilmington. At the same
time, the Army Engineers continued to dredge navigation channels
to open the lagoon. The US dredge *San Pedro* suctioned lands near
Terminal Island, displacing numerous squatters and a multiethnic

community of fishermen and harvesters. That was progress enough for Commissioner Gibbon, who announced the city's "great artificial harbor" to the national press.[10]

But each step forward was undone by another. Despite the verdict, Los Angeles had only the slightest capacity to develop a harbor. Up until 1913, the city possessed only two waterfront sites. And officials could not agree on which of the two to reclaim first. Furthermore, Los Angeles could not find enough buyers for its harbor bonds. The city earmarked $2 million of the bond issue for the outer harbor, $1 million for the estuary, or inner harbor. Neither amount was sufficient to do much of anything. Investors and the big financial houses balked. The city seemed over-leveraged. It had spread its credit too thin among several massive projects at once. Plus, tideland claimants had appealed the county verdict to the state supreme court. Eventually, they turned to the US Supreme Court. Bonds were a tough sell while the lawsuits remained unresolved. That created a cyclical problem: Without cash, the city could not pay its contractors. Without payment, the contractors refused to work. And that spooked bond purchasers even more.[11]

Municipal agencies involved in harbor development grew frustrated and at odds. Each dollar spent came at the cost of another spent elsewhere. Divergent priorities made it impossible to compromise. Public Works rushed to build the aqueduct. The city engineer prioritized streets and a new sewer outfall. The harbor commissioners promoted the harbor, of course. Estuary tidelands, they insisted, were the most cost-effective place to create accessible waterfront. Both Public Works and the city engineer disagreed. Meanwhile, Gibbon demanded structures of concrete and steel, hallmarks of modern ports like New York, Liverpool, and Hamburg as well as the Panama Canal. But until more bond money arrived, Public Works would allow only wooden, temporary construction.[12]

Opportunity was slipping away. President Taft visited Los Angeles again in October 1911. But he avoided going to the harbor. Taft spoke instead about tariff reduction, disarmament, and an international court to "relieve our future" of "wars which we have had in the past."[13] He praised Booker T. Washington in a speech to the city's African American community. Journalists seemed more interested in Taft's plans for the Supreme Court. Justice John Marshall Harlan—author of the Domínguez estuary opinion and dissents on imperialism, corporate monopoly, and racial segregation—had died two days before. The press soon found a more sensational story. On the night the president's train left Los Angeles, a Southern Pacific watchman discovered twenty-one sticks of dynamite beneath a bridge west of Santa Barbara, along Taft's scheduled route. Newspapers pointed to labor radicals as the likely suspects.

It was an inauspicious sign. Gibbon, who followed the steady progress of the Isthmian Canal, lost patience. He filed a formal letter of complaint with the city clerk, reprimanding the Public Works Board and Adna R. Chaffee. This was the opening salvo of a civil war. Public Works and the city engineer—even the mayor—were incensed. They retaliated by diverting funds away from the estuary. Gibbon and the harbor commissioners shot back with a letter demanding that Public Works account for its "obstinate refusal" to meet the canal deadline. The harbor commissioners volunteered to lead construction. However, under the city's division of authority, they held no control over spending or improvements.[14]

Puzzled and sensitive to a feud that might paralyze the city, reform and business groups stepped in to keep the harbor on schedule. Their answer was to add layers of decision. Under a model of joint governance enacted in January 1912, the harbor commissioners gained the power to draw up plans and select locations for improvement. Public Works would agree in deference. Then, an additional Harbor Advisory Board would give final approval. If that were not complicated enough, the city added several ad hoc bodies—a Bureau of Harbor Improvement (within the Board of Public Works), the city council's harbor committee, and a municipal Art Commission (to approve aesthetics). Reformers set up between *six and ten* centers of power, depending on who counted. Their idea was to harmonize development and guard it from corruption. But that meant dozens of ambitious, strongheaded individuals took a hand in the task.[15]

Two individuals, Gibbon and Chaffee, dominated the field. As a newcomer to the region, Chaffee had tenuous ties to Los Angeles's progressive and commercial groups. He was accustomed to a chain of command and information shared on a need-to-know basis. He was not a trained engineer, yet he was proficient from five decades as an army officer. He dismissed as novices the white-shoe lawyers, bankers, and businessmen who insisted on directing harbor construction. That made him unpopular with most civic groups, which tried to block his reappointment. Both the mayor and city engineer stepped in to stop this. The failed attempt to get rid of Chaffee put Gibbon and his allies more on the outs.[16]

Worse still, the Harbor Commission's enhanced powers meant it would take the blame once trouble reemerged. In January, a special

Broadway, 1915—heir to colonial conquests and western expansion, Los Angeles had gained territory, water, inland commerce, and a harbor.

investigatory committee reported on the causes of delay and their remedy. It found that the Harbor Commission, Board of Public Works, and city engineer had "reached a point of serious conflict."[17] The agencies disagreed fundamentally, vetoed each other's proposals, and selectively used evidence to justify their tactics. Investigators recommended that the city hire an independent consultant whose harbor plan would be binding on all involved. However, the mayor and city council rejected the report outright. With support from the *Los Angeles Times*, they portrayed the Harbor Commission and Gibbon as the problem. "No one man, nor any two men," the mayor pronounced, "can deadlock the harbor improvements. . . . If they try to, they will have to get out of the way." The council was more direct: "Failure to formulate harbor plans is a failure of the Harbor Commission." Critics suspected the mayor and Chaffee of orchestrating a setup. In June, the council stripped the Harbor Commission of authority and returned this power to the Board of Public Works. Gibbon and fellow commissioners abruptly resigned in a show of no confidence.

THE MUNICIPAL MACHINERY, set up by reformers to pursue their anti-monopoly and imperialist aims, no longer worked. While the mayor, Chaffee, and the Board of Public Works tried to consolidate authority over the harbor, former commissioner Gibbon fought on. He promised to tell the whole truth of the city's problems. And he levied charges of incompetence and profiteering against Public Works, the Harbor Advisory Board, the city engineer, and the mayor. Gibbon traded glares and insults with the mayor during public meetings. Chaffee reportedly shook his fist in Gibbon's face. Citizens grew bellicose and lined up on respective sides. A coalition of progressives and socialists campaigned to scrap the municipal charter and replace it with a commission system of government. This proposed to put the harbor under a single commissioner, or czar. Gibbon was the silent candidate.[18]

Each time, reformers lost. Voters rejected most of the charter refer-
enda in an election plagued by low turnout. Further attempts, in 1914,
1915, and 1916, failed. Los Angeles's elite—real estate men, lawyers,
manufacturers, businessmen, Northerners, and Southerners—who
had made the harbor possible and hitched it optimistically to imperial expansion, had lost their monopoly on the public interest. They
would be too angry at one another to celebrate when the Supreme
Court finally ordered separation of the Union Pacific and Southern
Pacific railroads.[19]

The city's breakup climaxed in the fall of 1912. Mayor Alexander, once hailed as an exemplar of efficient government, became the
second executive in four years to face impeachment. He survived the
recall, but had only months left in his term. Before that, in December,
Chaffee stepped down from the Board of Public Works and ended
his half century in government service. There was no fanfare for
the old soldier as he retired once more. Alexander denied that Chaffee's resignation had anything to do with the state's investigation
into aqueduct finances or with any coming investigation of harbor
affairs. Yet the next mayor put the costs of grand ambition squarely
at issue. "The question," he asked, "is whether we shall have an economic administration of the city's finances or whether we shall continue in office the same powers who have misused and squandered
the people's money." The harbor and its promises seemed not worth
the price.[20]

VETERANS OF THE Grand Army of the Republic and their rebel
counterparts, "grizzled" and "bent with the weight of years and
exposure," entered Los Angeles in early September 1912.[21] An estimated fifty thousand persons attended. Many arrived via San Pedro
Bay, though one with particular ties to the bay did not. George
Davidson, who spent the war guarding maps and harbors for the
Union, died the previous December at his home in San Francisco.

George Davidson, 1903

He was the last of the four young coast surveyors who had arrived there in 1850. Never quite wealthy, the scientist left a modest estate valued at $10,000 (several hundred thousand dollars today). Time had proven his thoughts on San Pedro right, but also wrong. It was no place for a harbor. And that was precisely the point.

Whether they landed at the bay or stepped off railcars, the war's living veterans marched on to a program of parades, speeches, and lemonade picnics. They celebrated forgiveness and rekindled nationality. But they did so in a city collapsing under the weight of unrealized pretensions, placed foremost upon its shoreline. "City Stricken by Incompetency Blight," roared the *Los Angeles Times*. "Administration's Continuous Failures Cost Taxpayers Millions."[22] Stragglers may have noticed the headline and squinted curiously at the small print. As the "world event" of the Panama Canal approached, the *Times* lamented, the city had fallen short. Its experiment with self-government verged on failure, its civic bonds strained to the point of rupture. To those decrepit veterans, it seemed strange, yet also strangely familiar. Only this time, the weary army shouldered its knapsacks and rode comfortably for home. Together, they passed into history.

So, too, did a nearby estuary. While Los Angeles fell into disunion that summer, federal contractors attended to their work. They dredged another million cubic yards from the lagoon's navigable zone, opening a channel from the lower end of Mormon Island to Wilmington, and eastward into the marshes. Los Angeles would gain title to much of this waterfront the next year, when California's supreme court decided in its favor. Meanwhile, the Army Engineers' *San Pedro* dug away at adjacent turning basins. The technicians' job was to follow plans. If they had disputes, they resolved them with numbers, coffee, and interchangeable parts.

Even as the city's politicians fought, machines of the modern state and corporate capital tore apart the lagoon. A harbor took shape, heedless of events on land. Similar developments occurred across an imperial world, and especially at the Panama Canal, whose opening—luckily for Los Angeles, but tragically for so many others—became delayed by world war. The promise of inevitable commerce gave the city an easier path to peace. It was profoundly different than America's experience in 1861. And the estuary and its life were disappearing for it.

People of the Earth

B Y THE TIME THE DREDGES BEGAN THEIR WORK, THE estuary had been many things. But before, and ever since, it remains a Gabrieleño-Tongva place.

The word *Tongva* translates as "People of the Earth." Yet they wove together an indigenous world (Tovaangar), sprawled over ocean and continent. Persons, goods, and beliefs circulated among a population of five thousand. Villages, with domed thatch and tule structures called *kich*, stood near springs, streams, marsh, or back bays. Trade connected villages—a dozen large communities, plus dozens of smaller ones—to deserts and mountains encircling the Los Angeles Basin. Using a *ti'at*, a sewn-plank canoe, they moved among the mainland and islands of Santa Catalina (Pimu), San Clemente (Kinkipar), and San Nicolas (Haraashngna). They navigated long distances up and down the coast. Sea canoe technology distinguished the Gabrieleño-Tongva, along with their neighbors the Chumash, among North American indigenous groups. Polynesians are the only other people of the eastern Pacific known to use sewn-plank vessels rather than the dugout variety.[1]

Within the Gabrieleño-Tongva world, San Pedro provided a center for transit and gathering. Surplus food, harvested from grasslands and oaks, moved across the offshore channel. Steatite, a workable soapstone, arrived from Catalina, where it was especially plentiful. High-nutrition fish and shellfish traded inland to the Mojave. The estuary and adjacent beaches provided a source for shell-bead currency, timber logs (including redwood preferred for ti'at), and asphaltum tar that washed ashore. San Pedro Bay served as a manufacturing site, as suggested by tule rushes grown in the marsh and deposits of tools made from stone and mammal bones. The people occupied at least five villages near the estuary at the time of Spanish settlement. Suanga (or Suangna) was located on the lagoon's northwest bluffs. Chaawvenga stood west, on San Pedro's open shore, at the base of the Palos Verdes Hills. A number of settlements lay upstream. The largest was Yaagna, situated along the river near a uniquely large sycamore (called El Aliso after the Spanish founded Los Angeles pueblo). The tree was known as far away as the Colorado River. Middens unearthed near San Pedro contain Hohokam ceramic pottery, believed to have come from the Gila Valley.

The bay was a place of creative and destructive forces. The people consider Catalina the "breast" or "belly button," a ceremonial center and source for power, medicine, and sorcery. East from the estuary by ten miles, over the hill from Alamitos lagoon, stood the settlement of Puvungna, or the "place of emergence." A Franciscan missionary recorded this as the birthplace of Chingichngish (Chinigchinich), the creator god who gifted the religious code shared by the Gabrieleño-Tongva and neighboring Acjachemen (Juaneño). A variety of watcher animals—raven, rattlesnake, bear, puma, tarantula, and stingray—enforced the moral order. The coast brought together human, natural, and spiritual existence.

European invaders arrived by the same interconnected space. Their powers—a blend of gunpowder, crucifix, microbes, and invasive livestock—unraveled the Gabrieleño-Tongva world and formed a new one. By the 1880s, colonization had stripped signs of indigenous life from the landscape. Construction and farming unearthed innumerable objects and the remains of ancestors. As Los Angeles grew taller, machines pressed the removal process deep into the earth. Evidence of the Gabrieleño-Tongva claim disappeared into downtown fills or simply was lost. Collectors picked over "Indian relics" wherever they could. At the region's estuaries, dredging

*Mission San Gabriel, 1828, showing Gabrieleño Mission
Indians and traditional* kich*, at bottom right*

obliterated the historical record held within. Objects mixed with
sand and sediment to form the concrete foundations of a city and
its harbor. Still, the evidence was incontrovertible. Southern Californians knew they occupied—and still do occupy—Gabrieleño-
Tongva land. The people, they insisted, had disappeared. Some said
this optimistically. Others spoke of tragedy. They erected museums,
rebuilt missions, formed historical and anthropological societies,
and advocated reform for the Indian survivor. All were boosters,
convinced of Southern California's progress. Through their promotion of real estate and a narrative of regional growth, they took the
story for themselves.[2]

But, beneath a metropolis, earth and water remain. The
Gabrieleño-Tongva claim proved too pervasive to disappear, sacred
objects too prevalent to cease showing. When construction spread
during the late twentieth century, signs of the indigenous past were
revealed all the more. This time, state and federal law mandated
recording and acknowledgment. Excavations for the Harbor Free-
way, completed in 1970 to speed truck access to the port, exposed

middens and archaeological sites. In the face of progress, the Earth threw back fishhooks of shell and bone, stone weights once used for nets, partially worked scallop and abalone, and the remains of hunted birds, deer, rabbit, whales, and otter. The bones of ancestors emerged from Southern California's last undeveloped parcels. They interrupted their rest to tell the opposite of disappearance. They halted construction and postponed permits. Descendants took a leading role in conservation of the twenty-two-acre Puvungna site and remnant coastal bays of Ballona, Bolsa, and Newport. They won tribal recognition from the State of California and revived cultural practices. Ti'at launched into the sea once more. Gabrieleño-Tongva claims to ocean and earth continue.

The story is unfinished despite the catastrophe of Southern California's past. During the 1910s, while construction of the Port of Los Angeles accelerated, a San Gabriel Mission Acjachemen told his remembrances to a linguist sent by the Smithsonian Institution. José de los Santos Juncos (aka Kuhn), nearly one hundred years old and having lived among Gabrieleño-Tongva for most of it, spoke of powerful shamans who once occupied Catalina Island. Two of the most potent sent a contagious fever (likely the 1801 diphtheria outbreak) to kill the mainland population. Their sorcery took the form of a sand painting, a powerful artifice of shoreline sediment, onto which they sketched an interconnected world. Upon this, the malignant ones rendered destruction and plague. Ravaged but aware of the source—for he had commissioned the work, yet now feared it—a Gabrieleño-Tongva *tomyaar* (chief or captain) navigated the people to the conjurers' stronghold. They destroyed the sand painting, along with those who had made it. With that, Juncos explained, the plague ended.[3]

The meaning remains for us to consider today.

Conclusion

S AN PEDRO'S ESTUARY NO LONGER EXISTS. IT DISAP-
peared beneath machines a century ago, well beyond living
memory. By the 1930s, US Army Engineers had completed their initial construction of the inner harbor. City departments had dredged
an additional twenty million cubic yards, and private interests
five million more. Shipping lanes reached thirty feet of depth and
accessed forty miles of waterfront based on the Fries-Hughes plan.
Los Angeles had reclaimed half of the 1,600 acres of tidelands it had
gained title to so far. Lease revenue helped the city pay off bonds,
float new debt, and pursue more development. Already, the port had
surpassed San Francisco as the Pacific coast's leader in cargo ton-
nage. Petroleum dominated its exports. Lumber, canned fish, tex-
tiles, farm crops, and industry provided other important traffic.[1]

As the United States grew more involved in the Pacific, the bay
continued to change. Despite the role of army engineers in design-
ing the port, the military never found a large or lasting presence in
Los Angeles Harbor. Instead, commerce and consumerism became
its defining force. That was prescient, since the nation's global power
would rely on softer forms of influence, on markets especially. Southern California prospered by association. The region experienced an
astounding population boom between the 1920s and 1970s, which
raised demand for inbound goods. The Second World War created shipbuilding and aviation jobs—later in aerospace, too—that
attracted a large influx of residents. Japan, the port's main trading

partner before 1940, emerged as the United States' rival for hegemony over the Pacific. After its defeat in war, Japan again became essential to Los Angeles shipping, lifting exports of oil, agriculture, and machinery.[2]

Over the last decades of the twentieth century, containerization and the expanding volume of transpacific commerce recast the bay. Much of this has occurred since the 1990s, with freer trade policies and the entry of China into global markets. San Pedro Bay became the receiving point for a web of logistics and offshore industry now centered on East and Southeast Asia. Multibillion-dollar international firms like Yusen, Maersk, China Ocean Shipping Company, and Yang Ming own and operate terminals on land leased from the city. Municipal ownership—the "people's port" hoped for by progressive reformers of the early twentieth century—was only ever partially realized. Instead, today's port is public and private at the same time. That comes with a catch. The public once venerated by reformers—the "people" of the "people's port"—is now a mass of impulse spenders. Imported products move through mechanized terminals before appearing in front of Americans, on shelves or in stock, as if by magic. The unseen workings of the world economy, like the closed-door deliberations of its overseers, mystify the process as well as its problems. The harbor looks much like the one drafted on old blueprints, and yet nothing like it—if we see it at all. As with past iterations, it one day will be gone and reinvented, too.

Some things are unchanged. In spite of the estuary's destruction and the industrial complex in its place, high water still occurs twice a day. Pacific tides push under docks and against concrete and steel. It is a reminder that, even in the most unnatural of settings, we are not free of nature and its basic cycles. The tides can also help us understand history and the turns brought by human action. Events rise, wash over us, and then fade into memory. Yet they leave their mark. Very often, that is a gap between what might have been and what is. Our capacity to build by design and with best intentions has limits.

That is one lesson of the bay's history, and of the United States' engagement with the Pacific since the mid-nineteenth century. From that founding moment onward, the western coast became a site onto which Americans projected their ideas of national and individual fulfillment. Expansion to the ocean would resolve the country's divides. It would forge unity from fragments, stability from impermanence, and establish the republic among the community of nations. Or so Americans hoped. Some searched for wealth on a frontier of abundant, dispossessed land. For them, expansion promised property, profits, and material security. Settlers sought to escape social inequalities, but they also gained by the misfortune of others. For railroad corporations, land and property were the means of business dominance. Federal and city reformers challenged them by capturing the shoreline and imposing their own dominance.

The republic's territorial growth was rapid, its creation of material prosperity unprecedented. Nonetheless, few characters in this story got exactly what they wanted. Their judgments proved right, but also wrong at the same time. And expansion repeatedly pushed the nation and its people into great volatility or divisions. That became a risk they accepted. In the worst instances, it brought personal ruin, political fracture, and civil war. All the while, it drew the nation and people further westward. Despite success, many individuals' hopes were left unfulfilled. In the Port of Los Angeles, their predicament became a foundation for the United States and the way Americans live today.

The people who made a harbor of San Pedro Bay and first grappled with its consequences are gone. But the machine they created remains. As a marvel of human invention, it obliterates distance to connect the pieces of our world. It defies scarcity to provide for our material needs. Yet it entangles us in a system that obscures the capacity to see, think, or choose otherwise. The port, with all it represents, is a scarcely visible thing. That may be because it works so well. It may be also because it seems beyond our reach to steer.

Hence, this book is one of origins, but also possible endings. For two generations, US historians have concluded their books with questions about fulfillment. They ponder when and how the nation's founding ideals—the promise of liberty and popular government—will be realized. That is a hallmark of some of the best historical writing of the past half century, work without which this one could not exist. Those writings came out of a moment of American pre-eminence in the world and great confidence in democracy, economic integration, and abundance. It is worth revisiting such questions in a moment of less confidence and to disentangle ideas that have become intertwined, even convoluted. As this book ends on the shores of Southern California, looking out to the horizon, we should ask: What is the relationship today between democratic promise and material promise? Because one is often mistaken for the other. This was certainly true in the bay's history. Yet the same history reveals how frequently these promises are at odds.

So it is for us, too. We can forswear territorial expansions of the past. However, expansionist legacies continue to shape our existence. And our yearning drives a similar search for fulfillment, enabled this time by economic growth and material objects. How that affects our nation's civic health and long-term stability remains unclear. It should not remain unexamined. US democracy and material secu-rity show signs of strain in our moment, the nation and world trou-bled again by fragmentation, yet Americans look all the more to the Pacific for some kind of deliverance. Fulfillment and its pursuit con-tinue to tilt us west, and beyond the horizon.

That is San Pedro Bay's enduring paradox in American history. It suggests that a rising tide is the moment to be sure of steady ground and solid footing: to sustain democratic ideals, to direct the course of our longings, and to square intentions with consequence as best we may. If not, we may find the high-water mark of our time—and of much more—in the ruins of an estuary that once was.

Acknowledgments

I owe a great debt to the archives that care for historical sources. The Bancroft Library and Huntington Library house many of the collections used to write this book. The Huntington's Reader Services department deserves special thanks along with Stephanie Arias, Linde Lehtinen, Li Wei Yang, Catherine Wehrey-Miller, and Juan Gomez. I am grateful to John Cahoon and Betty Uyeda at the Seaver Center for Western History Research, Los Angeles County Museum of Natural History; the Water Resources Collections and Archives at University of California, Riverside (at Berkeley during my research); Michael Miller at the American Philosophical Society; Scott Daniels at the Oregon Historical Society; Paul Spitzzeri at the Workman and Temple Family Homestead Museum; and Jennifer Hill, Greg Williams, and Thomas Philo of Gerth Archives and Special Collections, California State University Dominguez Hills.

Thank you to staff at the Library of Congress including Bruce Kirby of the Manuscript Division; the National Archives campuses at Riverside, College Park, San Bruno, and Washington, DC; Kay Peterson and Erin Beasley of the Smithsonian Institution museums; Benjamin Stone at Stanford University Special Collections; Erik Johnson at the Theodore Roosevelt Center, Dickinson State University; Jeff Kato of the California State Lands Commission; UCLA Special Collections; University of Oregon Special Collections; Lynda Crist at the Jefferson Davis Papers, Rice University; University of Southern California's Digital Library and Special Collections; Kimball

Garrett of Los Angeles County Museum of Natural History; Terri Garst of the Los Angeles Public Library; Michael Holland at Los Angeles City Archives; Jane Newell of the Anaheim Public Library; and Debra Kaufman of the California Historical Society.

Several archivists went above and beyond to supply me with digital scans. I wish to acknowledge Ted Jackson, of Georgetown University's Booth Family Center for Special Collections, and Tim Noakes, of Stanford University Special Collections. Both helped me access the Ord family's letters. Likewise, Albert "Skip" Theberge, Jr., captain (retired) and historian of the National Oceanic and Atmospheric Association, graciously offered his writings and transcriptions of primary sources.

I am fortunate to have outstanding mentors and teachers: Eric Foner, Steven Hahn, Elizabeth Blackmar, Michael Parrish, David Gutiérrez, Herbert Sloan, Alice Kessler-Harris, and Alan Brinkley. Each of them is present in the book in some way. I also am grateful to the western and Los Angeles historians who supported this project: Bill Deverell, Merry Ovnick, and Becky Nicolaides most of all; participants in the National Endowment for the Humanities "View from the East" seminar; Greg Robinson; Justin D. Spence; John Mack Faragher; Geraldine Knatz; Juan de Lara; and prize committees of the Western History Association, Historical Society of Southern California, and American Society for Environmental History. I long ago had coffee with Elliott West, who told me to "follow the people." That research advice proved more fruitful than either of us imagined.

Brilliant friends sustained my work. Those at Columbia and Princeton broadened my interests by sharing their own: Molly Loberg, Eric Yellin, Eduardo Canedo and Yael Schacher, Matt and Cindy Wisnioski, Jason Governale, Chris Capozzola, David Greenberg, Gwen Shufro, Philipp Stetzel, Nick Turse, Patrick Whittle, and Ted Wilkinson among others. This book may not have reached completion at all without Christopher Thrash, Richard Belkin, and Jeffrey Riker. I also thank Rob and Amanda Cattivera and family,

Jeanie and Andy Greensfelder, Frank and Mary Harris, the Finley family, Scott Oh, Matt Smith, Luis Álvarez-Mayo, Sarah Silkey, the Kraemer family, and Adrienne Miller. My own family—Charlene and Bill Dale, the Lobergs, Alekai, Mercedes, Nasir and Sabira, Riaz and Kari, and Zarin and Talal—kept a well of patience.

The bay's story, as I have told it, would be incomplete if not for the generosity of Chief Red Blood Anthony Morales, Tribal Chairman and Chief of the Gabrieleño San Gabriel Band of Mission Indians. Chief Anthony gave his time to enhance my knowledge and discuss my draft writings. Thank you to Kim Kolpin of the Bolsa Chica Land Trust for the introduction to Chief Anthony.

Above all, this book would have been impossible without several persons to whom I am deeply obliged. Eric Foner encouraged the project and advised me over its many years. Elise Capron, my literary agent, was an invaluable guide in finding a publisher. Thank you as well to Sandra Dijkstra. Numerous people, whom I never met, worked behind the scenes at W. W. Norton and deserve my gratitude. Zeba Arora helped through the entire production process. Dan Gerstle, Norton's vice president and editor-in-chief, brought his experience and keen eye to the manuscript, improving it by leaps and bounds. Dan gave this book its opportunity and set its high standard. Any remaining shortfall is mine.

I benefitted from an unofficial reader and critic who saw this project from start to finish. Molly Loberg read draft chapters as well as the full manuscript too many times to count. Her insights became a foundation for the book. Pascale heard all about it and gifted me her drawing of the mustached Union general she had come to know. Among life's many finds, I am most thankful for those we find together.

Cast of Characters

Alden, James—US Navy and Coast Survey officer, in command of hydrographic survey ship USS *Active*, 1850s. He and Davidson often clashed over their work.

Alexander, Barton Stone—Army engineer, in command of Pacific Engineers Board at San Francisco, 1870s. Secured US rights to Pearl Harbor, Hawaii.

Bache, Alexander Dallas—Coast Survey superintendent, 1843–67. Scientist and political insider. Great-grandson of Benjamin Franklin.

Banning, Phineas—San Pedro Bay freight entrepreneur and developer of New Town/Wilmington.

Banning, William—Eldest son of Phineas. Defendant in the suit brought by the Domínguez daughters over estuary property.

Benton, Thomas Hart—Missouri senator. Patron and father-in-law of explorer John Frémont.

Brent, Joseph Lancaster—Los Angeles Democratic Party leader, 1850s. Attorney for the Domínguez family. New Town investor.

Carleton, James H.—US Army. Expert on desert cavalry. Commanded Southern California and New Mexico during the Civil War.

Carson, Christopher "Kit"—Frontier guide and trapper, popular hero, New Mexico settler. Worked for Frémont during the 1840s and for Carleton during the Civil War.

Chaffee, Adna R.—Veteran of the Civil War, Apache Wars, and Philippine-American War. Hired by Los Angeles City to oversee harbor construction.

Cota de Domínguez, Engracia—Wife of Manuel Domínguez. Landowner. Received Edward and Molly Ord during visits.

Davidson, Ellinor Fauntleroy—Wife of George Davidson. Granddaughter of reformer Robert Owen.

Davidson, George—Coast Survey scientist. Protégé of Bache. One of the US's top geodetic surveyors.

Davis, Jefferson—War secretary and senator from Mississippi. Friend of Superintendent Bache. President of the Confederate States during the Civil War.

Davis, Varina Howell—Wife of Jefferson Davis. Admirer of Superintendent Bache. Popular Washington, DC, hostess, 1850s.

Domínguez, Manuel—Patriarch of the Domínguez family. Heir to the Rancho San Pedro land grant.

Domínguez de Guyer, Ana J.—Eldest daughter of Manuel Domínguez. Plaintiff in the suit against the Banning sons over estuary property.

Frémont, Jessie Benton—Wife and biographer of John Frémont. Daughter of Thomas Hart Benton.

Frémont, John C.—The "Pathfinder." Controversial explorer of the West, army officer, and California senator.

Fries, Amos A.—US Army engineer, in command of Los Angeles District. Coauthor of the 1907 federal harbor lines plan.

Gibbon, Thomas E.—Attorney and railroad executive. Los Angeles Harbor Commission. Visionary of LA's global commerce.

Grant, Ulysses S.—Army officer on the Pacific frontier. Civil War general, US president, and friend of Edward Ord.

Gwin, William M.—Landowner and California senator, 1850s. Momentary patron of the Ord family.

Hansen, George—Southern California land surveyor and engineer. Hired by the Domínguez family to map the Rancho San Pedro.

Harlan, John Marshall—US Supreme Court justice. Wrote the opinion in the estuary case *De Guyer v. Banning* (1897), in addition to numerous well-known cases.

Harriman, Edward H.—Railroad magnate. Briefly (1901–09) controlled lines to major Pacific harbors, including San Pedro Bay.

Hassler, Ferdinand—Founding superintendent of the US Coast Survey (to 1843).

Hughes, David E.—Civil engineer in the Army Engineers' Los Angeles District. Coauthor of the 1907 federal harbor lines plan.

Huntington, Collis P.—Founder of the Central Pacific Railroad and president of the Southern Pacific Company. A pioneer of corporate influence in politics.

Hyde, William B.—Southern Pacific Railroad agent at Los Angeles and Wilmington, 1869–73. Drafted a forward-looking plan to develop the estuary.

Meigs, Montgomery C.—US Army engineer. Quartermaster general during and after the Civil War.

Ord, Edward O. C.—US Army and Coast Survey. Surveyor for hire and aspiring land speculator. Civil War general.

Ord, Molly Thompson—Wife and confidant of Edward Ord.

Ord, Pacificus—Older brother of Edward Ord. US attorney for California's Southern District, 1850s.

Roosevelt, Theodore—US president and opponent of industrial monopoly.

Sherman, William T.—Army officer. San Francisco banker. Civil War general and friend of Edward Ord.

Sibley, Henry H.—Former US Army officer. Led the Confederate invasion of New Mexico, 1862.

Stanford, Leland—Founder of the Central Pacific Railroad, and president of the Southern Pacific Company. California's first Republican governor, 1862–63.

Stevens, Isaac I.—US Army Engineer and Coast Survey. First governor of the Washington Territory. Classmate of Edward Ord.

Stockton, Robert—US Navy. Led the invasion of Mexican California and Los Angeles, 1846.

Sumner, Edwin V.—US Army. Commanded the Department of the Pacific at San Francisco, 1861.

Taft, William Howard—Governor-general of the US-occupied Philippines. Secretary of war. US president.

Wheeler, George M.—US Army. Explorer of the Southwest. Led logistical study of desert forts, 1869.

Whipple, Amiel W.—Army engineer. Commanded the 35th parallel transcontinental railroad survey to Los Angeles, 1853–54.

White, Stephen M.—California senator. Led the fight in Washington, DC, to make San Pedro Bay a free harbor, 1890s.

Williamson, Robert Stockton—Army engineer. Commanded the War Department railroad survey to find passes into Southern California, 1853–54, and the study of San Pedro's harbor, 1869.

Wilson, Benjamin D.—Los Angeles Democratic Party leader, 1850s. Early US settler in Mexican California. New Town investor.

Notes

ABBREVIATIONS

Archives

BANC—Bancroft Library, University of California, Berkeley.

BOOTH—Booth Family Center for Special Collections, Georgetown University.

GERTH—Gerth Archives & Special Collections, California State University, Dominguez Hills.

HUNT—Huntington Library, San Marino, California.

NARA—National Archives & Records Administration.

SEAV—Seaver Center for Western History Research, Los Angeles County Museum of Natural History.

STANF—Department of Special Collections, Stanford University.

WRCA—Water Resources Collections & Archives, University of California, Riverside.

Manuscript Collections & Published Primary Sources

ADB—Correspondence of Alexander Dallas Bache, Superintendent of the Coast and Geodetic Survey, 1843–65, National Archives Microfilm Publications, Microcopy No. 642, RG 23, M642.

RBC—Records of the Banning Company, 1859–1948, Huntington Library.

GD—George Davidson Correspondence and Papers, 1845–1911, C-B 490, Bancroft Library, University of California, Berkeley.

JD—*The Papers of Jefferson Davis*, 13 vols., Haskell M. Monroe, Jr., James T. McIntosh, and Lynda Lasswell Crist, eds. (Baton Rouge: Louisiana State University Press, 1971–2008).

TEG—Thomas E. Gibbon Papers, Huntington Library.

DEH—David E. Hughes Papers, WRCA 083, Water Resources Collections and Archives, University of California, Riverside.

WBH—William Birelie Hyde Papers, 1861–1896, M0267, Department of Special Collections, Stanford University.

LAACR—Los Angeles Area Court Records, Huntington Library.

ORDC—Edward Otho Cresap Ord Correspondence and Papers, 1850–1883, C-B 479, Bancroft Library, University of California, Berkeley.

ORDF1—Ord Family Papers 1, Booth Family Center for Special Collections, Georgetown University.

ORDF2—Ord Family Papers 2, Booth Family Center for Special Collections, Georgetown University.

ORDF3—Ord Family Papers 3, Booth Family Center for Special Collections, Georgetown University.

ORDL—Edward Otho Cresap Ord Letters, 1854–1885, M210, Department of Special Collections, Stanford University.

ORDP—Edward Otho Cresap Ord Papers, 1840–1887, 2002/93, Bancroft Library, University of California, Berkeley.

RSP—Rancho San Pedro Collection, Gerth Archives and Special Collections, California State University, Dominguez Hills.

SR—Solano-Reeve Papers, 1849–1910, Huntington Library.

WoR1—War of the Rebellion, Part 1 (Washington, DC: Government Printing Office, 1897).

WoR2—War of the Rebellion, Part 2 (Washington, DC: Government Printing Office, 1897).

INTRODUCTION: EXCAVATING THE LOST COAST

1. San Francisco Call, December 3, 1911. Historians have yet to write a full-length biography of Davidson. Select accounts include: Henry R. Wagner, "George Davidson, Geographer of the Northwest Coast of America," Quarterly of the California Historical Society 11, no. 4 (December 1932), 299–320; Charles B. Davenport, "George Davidson," Biographical Memoirs 18, Ninth Memoir (Washington, DC: National Academy of Sciences, 1937), 189–217; and Oscar Lewis's brief George Davidson: Pioneer West Coast Scientist (Berkeley: University of California Press, 1954).

2. 2022 statistics. Annual Facts and Figures Card (Port of Los Angeles, 2023-0084). The "port complex" groups the adjacent, separately administered ports of Los Angeles and Long Beach.

3. For an overview of borderlands scholarship and its influence on US historical writing, see Herbert E. Bolton, The Spanish Borderlands: A Chronicle of Old Florida and the Southwest (New Haven: Yale University Press, 1921); Jeremy Adelman and Stephen Aron, "From Borderlands to Borders: Empires, Nation-States, and the Peoples in between in North American History," American Historical Review 104 (June 1999), 814–41; and Samuel Truett and Pekka Hämäläinen, "On Borderlands," Journal of American History 98 (September 2011), 338–61.

4. Walt Whitman, "Passage to India," in Leaves of Grass (Philadelphia: David McKay, 1891–92), 316, 323.

1. *Daily Alta*, June 20, 1850. James S. Lawson, *Autobiography* (1879), USCGS .091 1879-2, provided by NOAA's Central Library and Albert Theberge Jr.

CHAPTER 1: EUROPEAN MEASURES, WESTWARD AMBITION

1. "Ferdinand Hassler," in *Appleton's Cyclopaedia of American Biography*, ed. James T. White (New York: Appleton, 1889). Florian Cajori, *The Chequered Career of Ferdinand Rudolph Hassler, First Superintendent of the United States Coast Survey* (Boston: Christopher Publishing House, 1929), 35–45.
2. Committee of Twenty, *Report on the History in Progress of the American Coast Survey* (Washington, DC: Robert Armstrong, 1858), 11–14.
3. Richard D. Cutts, US Coast Survey, *Fieldwork of the Triangulation: Methods, Discussions, and Results* (Washington, DC: Government Printing Office [GPO], 1877), 3–4, 39–41. George Davidson, "The Pacific Coast and Geodetic Surveys," *Californian* 1 (January 1880), 61–62.
4. National Ocean Service, "What Is the National Spatial Reference System?," National Oceanic and Atmospheric Administration, oceanservice.noaa.gov. "The Foundations of the National Spatial Reference System" and "Geometry of Making Surveys," *NOAA Celebrates 200 Years of Science, Service, and Stewardship*, noaa.gov. Davidson, "The Pacific Coast and Geodetic Surveys," 63–66.
5. Twenty, *Report on the History . . . Coast Survey*, 21.
6. Cajori, *Chequered Career of Ferdinand Rudolph Hassler*, 207.
7. Report of the Superintendent . . . 1850, H. Exec. Doc. No. 12, 31st Cong., 2nd Sess. (1850), 5–7. Twenty, *Report on the History . . . Coast Survey*, 18–20.
8. "Obituary: Alexander Dallas Bache," Carton 19, Memorial Resolutions, GD, BANC. Merle M. Odgers, *Alexander Dallas Bache* (Philadelphia: University of Pennsylvania Press, 1947), 12, 17.
9. Hazard Stevens, *Life of General Isaac I. Stevens* (Boston: Houghton Mifflin, 1901), 253.
10. Twenty, *Report on the History . . . Coast Survey*, 21–22.
11. Davidson, *History of the Coast Survey on the Pacific* (1863), Carton 19, GD, BANC. Report of the Superintendent of the Coast Survey . . . 1849, S. Exec. Doc. No. 5, 31st Cong., 1st Sess. (1849), 8. Report of the Superintendent . . . 1850, 5–12.
12. Lawson, *Autobiography* (1879), NOAA. Susan Lee Johnson, *Roaring Camp: The Social World of the California Gold Rush* (New York: W. W. Norton, 2000).
13. Lewis A. McArthur, "Pacific Coast Survey of 1849 and 1850" (1915), provided by NOAA/Theberge. Twenty, *Report on the History . . . Coast Survey*, 35–37.
14. C. P. Huntington in "Biographical Material for Bancroft's Chronicle of the Builders," C-D 773 and ov Box 21, B-C 7, BANC.
15. Provided by NOAA/Theberge: William Gibson to W. P. McArthur, September 21, 1849. Order of Thomas ap Catesby Jones, October 19, 1849, in Erwin G. Gudde, "Mutiny on the Ewing."
16. *Daily Alta*, March 7, 1850. Report of the Superintendent . . . 1850, 11–12, 49–50. Davidson, *History of the Coast Survey on the Pacific* (1863), Carton 19, GD, BANC.
17. *Daily Alta*, April 17, 1850, and August 23, 1850.
18. Franklin Spencer Edmonds, *History of the Central High School of Philadelphia* (Philadelphia: J. B. Lippincott, 1902), 79.
19. In GD, BANC: Bache to Davidson, letters, 1844, Box 28; Davidson to Belle, July 31,

1853, Box 59. Oscar Lewis, *George Davidson: Pioneer West Coast Scientist* (Berkeley: University of California Press, 1954), 9.
20. Lawson, *Autobiography* (1879), Oregon Manuscripts, P-A 44-54, BANC. "George Davidson," *Proceedings of the California Academy of Sciences* (April 9, 1914), 10. In Box 28, GD, BANC: Bache to Davidson, March 14, 1849; May 1, 1850; May 10, 1850.
21. *Daily Alta*, June 20, 1850.
22. Davidson, *History of the Coast Survey on the Pacific* (1863), Carton 19, GD, BANC.
23. Bache to Davidson, September 24, 1850, Box 28, GD, BANC. *Report of the Superintendent, 1853* (Washington, DC: Robert Armstrong, 1854), 78. On the Pacific coast, one surveyor noted, there is "no such thing as a lee shore." W. P. Trowbridge to Bache, July 10, 1854, Roll 107, ADB, NARA.

CHAPTER 2: AN HEIR OF INVASION WELCOMES THE NEXT

1. Father Antonio de la Ascensión, "Account of the Voyage of Sebastian Vizcaino," in *Spanish Voyages to the Northwest Coast of America*, ed. Henry R. Wagner (San Francisco: California Historical Society, 1929), 237. Lois Roberts, *Los Angeles–Long Beach Harbor Area Cultural Resources Survey* (Los Angeles: US Army Corps of Engineers, 1978), 16. Homer Aschmann, "Evolution of a Wild Landscape and Its Persistence in Southern California," *Annals of the Association of American Geographers* 49, no. 3 (September 1959), 47–55.
2. Hugo Reid, whose wife was Gabrieleño-Tongva, refers to Suanga in 1852, during his term as US Indian Agent. Reid, *Indians of Los Angeles County*, ed. Robert F. Heizer (Los Angeles: Southwest Museum, 1968), 14, 22–23. William McCawley, *The First Angelenos: The Gabrieliño Indians of Los Angeles* (Banning, CA: Malki Museum Press, 1996). Kelly Lytle-Hernández, *City of Inmates: Conquest, Rebellion, and the Rise of Human Caging in Los Angeles, 1771–1965* (Chapel Hill: University of North Carolina Press, 2017), 16–26.
3. Box 1, Folder 11, Robert C. Gillingham Papers, GERTH. Gillingham's *Rancho San Pedro* (Los Angeles: Cole-Holmquist Press, 1961) and papers remain the indispensable history of the Domínguez family.
4. Stephen G. Hyslop, *Contest for California: From Spanish Colonization to the American Conquest* (Norman, OK: Arthur H. Clark, 2012), 243–56. John Mack Faragher, *Eternity Street: Violence and Justice in Frontier Los Angeles* (New York: W. W. Norton, 2016), 68–74.
5. Benjamin D. Wilson, "Observations on Early Days in California and New Mexico, 1877," RB 249995, HUNT. Robert Glass Cleland, *The Cattle on A Thousand Hills: Southern California, 1850–1880* (San Marino: Huntington Library, 1975), 184–207.
6. Box 2, Folders 9, 23, 25, Gillingham Papers, GERTH. In Rancho San Pedro: Documents 170/041, UCLA Library Special Collections, UCLA: "Papers Related to a Suit . . . against . . . Avilla and Manuel Gutierrez" and "Transcript . . . Relating to the Transfer of 1000 Varas from Rancho San Pedro to Rancho Los Palos Verdes."
7. Gillingham, *Rancho San Pedro*, 134–42, 177–78.
8. Amy S. Greenberg, *A Wicked War: Polk, Clay, Lincoln and the 1846 US Invasion of Mexico* (New York: Knopf, 2012). Brian Delay, *War of a Thousand Deserts: Indian Raids and the US-Mexican War* (New Haven: Yale University Press, 2008).
9. "Narrative of Benjamin D. Wilson," in Robert Glass Cleland, *Pathfinders* (Los Angeles: Powell, 1929), 396. "Benjamin David Wilson's Observations on Early Days in California and New Mexico," ed. Arthur Woodward, *Annual Publication of the Historical Society of Southern California* 16 (1934), 104.

10. J. M. Guinn, "Battle of Domínguez Ranch," *Annual Publication of the Historical Society of Southern California* 4 (1899), 261–66. Faragher, *Eternity Street*, 122–30.

CHAPTER 3: TO FIX A CONTINENT ADRIFT

1. Lawson, *Autobiography* (1879), BANC. Report of the Superintendent . . . 1850, 52.
2. Lawson, *Autobiography* (1879), BANC. While an army officer, Anastacio Carrillo defeated the Chumash Revolt of 1824. Hyslop, *Contest for California*, 196–97.
3. Report of the Superintendent . . . 1850, 52–53. *Report of the Superintendent, 1851,* 83. *Alta* clipping in J. A. Alden to Bache, February 1, 1852, Roll 69, ADB, NARA.
4. *Report of the Superintendent, 1852,* 52–53. Davidson to Bache, January 2, 1852, and October 15, 1852, Roll 69, ADB, NARA. Twenty, *Report on the History . . . Coast Survey,* 30–33.
5. Jesse Applegate to Joseph Lane, October 5, 1851, and J. L. Parrish to Joseph Lane, January 14, 1852, Joseph Lane Correspondence, Oregon Historical Society. Lawson, *Autobiography* (1879), NOAA. Davidson to Bache, January 20, 1852, Roll 69, ADB, NARA.
6. Davidson to Bache, December 7, 1851, and December 15, 1851, Roll 69, ADB, NARA. *Report of the Superintendent, 1852,* 50–51. *Report of the Superintendent, 1853,* 78.
7. Davidson to Bache, February 19, 1852, Roll 69, ADB, NARA.
8. Davidson to Bache, February 28, 1852, and March 27, 1852, Roll 69, ADB, NARA.
9. Cutts, *Fieldwork of the Triangulation,* 4, 39–41.
10. Bache to Alden, Cutts, and Davidson, September 24, 1852, Roll 69, ADB, NARA.
11. History of the Coast Survey's *Active* provided by NOAA. The original *Active* was built by order of the Continental Congress. Nathaniel Philbrick, *Sea of Glory: America's Voyage of Discovery, the US Exploring Expedition, 1838–1842* (New York: Viking, 2003), 100–103, 171–74, 311–12.
12. Roll 69, ADB, NARA: Bache to Alden, May 19 and June 7, 1852; Alden to Bache, February 14 and June 30, 1852. Letter of F. K. Murray, June 3, 1851, provided by NOAA/Theberge. Appendix No. 53, *Report of the Superintendent, 1851,* 533–40.
13. Latitudes, Reconnaissance, 1853, Carton 19, GD, BANC. Davidson to Bache, March 27, 1852, May 13, 1852, and June 30, 1852. Alden to Bache, May 15, 1852, and June 30, 1852. Bache to Davidson, July 16, 1852, Roll 69, ADB, NARA.
14. Davidson to Bache, May 13, 1852, ADB, NARA. Twenty, *Report on the History . . . Coast Survey,* 26.
15. In ADB, NARA: Davidson to Bache, January 20, 1852, Roll 69; Alden to Bache, August 7, 1852, Roll 69; Davidson to Bache, July 8, 1854, Roll 107. Lawson, *Autobiography* (1879), BANC. Philbrick, *Sea of Glory,* 223–31.
16. Lawson, *Autobiography* (1879), BANC. Joshua Reid, *The Sea Is My Country: The Maritime World of the Makahs* (New Haven: Yale University Press, 2015), 117–23.

CHAPTER 4: PROSPECTING IN PROPERTY

1. Treaty of Cahuenga, January 16, 1847. Treaty of Guadalupe Hidalgo, February 2, 1848.
2. See Leonard Pitt, *Decline of the Californios: A Social History of the Spanish-Speaking Californias* (Berkeley: University of California Press, 1966). Lisbeth Haas, *Conquests and Historical Identities in California* (Berkeley: University of California Press, 1995).
3. On the "great coincidence," see Elliott West, *Continental Reckoning: The American West in the Age of Expansion* (Lincoln: University of Nebraska Press, 2023).

4. *Daily Alta*, December 23, 1853. Studies of local borderlands society include William F. Deverell, *Whitewashed Adobe: The Rise of Los Angeles and the Remaking of Its Mexican Past* (Berkeley: University of California Press, 2004). Anne F. Hyde, *Empires, Nations, and Families: A New History of the North American West, 1800–1860* (Lincoln: University of Nebraska Press, 2011). David Torres-Rouff, *Before LA: Race, Space, and Municipal Power* (New Haven: Yale University Press, 2013).

5. Benjamin Madley, *An American Genocide: The United States and the California Indian Catastrophe* (New Haven: Yale University Press, 2016), 155–58. Box 2, Folder 24, Gillingham Papers, GERTH.

6. J. Ross Browne, *Report of the Debates in the Convention of California* (Washington, DC: John T. Towers, 1850).

7. Faragher, *Eternity Street*, 214, 229.

8. J. L. Brent to Edward Brent, April 16, 1851, Box 1, Folder 1, Joseph Lancaster Brent Papers, HUNT.

9. At GERTH: Box 1, Folder 10, Gillingham Papers; Domínguez, Mortgage . . . 1855, Box 1, Folder 40, RSP. In Rancho San Pedro: Documents 170/041, UCLA: "Deed, July 1, 1850, from María Victoria Domínguez to Manuel Domínguez"; "Deed, June 12, 1852, Manuel Rocha to Manuel Domínguez." Gillingham, *Rancho San Pedro*, 178–80.

10. *Daily Alta*, October 23, 1852.

CHAPTER 5: DR. GWIN BRINGS GOOD NEWS

1. Leonard L. Richards, *The California Gold Rush and the Coming of the Civil War* (New York: Alfred A. Knopf, 2007), 97–110. Robert Elder, *Calhoun: American Heretic* (New York: Basic Books, 2021), 514–21.

2. William J. Cooper, Jr., *Jefferson Davis, American* (New York: Vintage, 2000), 217. Davis, Speeches, May 20, 1850, October 23–24, 1850, December 7, 1850, and July 11, 1851, in *JD*, vol. 4.

3. Galley Proof of Biographical Sketch, William McKendree Gwin Papers, C-B 378, BANC.

4. Joseph Davis Howell to Mother, November 21, 1845, in *Jefferson Davis: Private Letters, 1823–1889*, ed. Hudson Strode (New York: Harcourt, Brace & World, 1966). Jefferson Davis to James A. Pearce, August 22, 1852, in *JD*, vol. 4.

5. Browne, *Report of the Debates in the Convention of California*, 445. Appendix, Constitution, iv. "Memorial of US Senators and Representatives Elect," March 12, 1850, xix.

6. William H. Ellison, ed., "Memoirs of Hon. William M. Gwin," *California Historical Society Quarterly* (1940), 277.

7. Gillingham, *Rancho San Pedro*, 58. Karen Clay and Werner Troesken, "Ranchos and the Politics of Land Claims," in *Land of Sunshine: An Environmental History of Metropolitan Los Angeles*, ed. William F. Deverell and Greg Hise (Pittsburgh: University of Pittsburgh Press, 2005), 55.

8. "Memoirs of . . . Gwin," 157, 162. *Act to Ascertain and Settle the Private Land Claims, March 3, 1851* (9 Stat. 631). *Congressional Globe*, February 6 and March 3, 1851, 31st Cong., 2nd Sess., 826, 829, and 840.

9. Browne, *Report of the Debates in the Convention of California*, 11. "Memoirs of . . . Gwin," 172.

10. *Daily Alta*, December 15, 1851, January 1, 1852, and January 3, 1852.

11. *Daily Alta*, March 8, 1852, and October 23, 1852. In *US Commission for Ascertaining and Settling Private Land Claims in California*, Letterbook, 70/66 c FILM,

BANC: Hiland Hall, Harry I. Thornton, and James Wilson to Alexander H. H.
Stuart, May 31, 1852; Hall and Thornton to Elisha Whittlesey, December 15, 1852.
12. *Los Angeles Star,* July 24, 1852. *Daily Alta,* September 19, 1852, and October 30,
1852. Deposition of José Antonio Carrillo, October 21, 1852, in "Petition, October
19, 1852" and "Transcript of the Proceedings in Case No. 398," Land Case Files
273 SD, BANC.
13. *Los Angeles Star,* October 16, 1852.

CHAPTER 6: A PROFESSOR'S GAMBLE

1. *Report of the Superintendent, 1853,* 80–81 and Appendix 1.
2. Twenty, *Report on the History . . . Coast Survey,* 82–85. This 1858 review of Bache's
tenure included praise from the Geographical Society of London, Royal Geographical Society, and Alexander von Humboldt.
3. "Memoirs of . . . Gwin," 165. Bache to Alden, August 16, 1852, Roll 69, ADB, NARA.
4. Bache to Davidson, August 6, 1852, Roll 69, ADB, NARA.
5. Bache to Davidson, September 24, 1852, and November 15, 1852, Roll 69,
ADB, NARA.
6. Davidson to Bache, October 15, 1852, and November 15, 1852, Roll 69, ADB,
NARA. *Daily Alta,* November 3, 1852. In GD, BANC: Joseph Ruth to Davidson,
July 31 and October 8, 1852, Box 52; Davidson to Ed Hall, October 31, 1852, Box
59; Bache to Davidson, September 22, 1853, Box 28.
7. Davidson to Bache, October 15, 1854, Roll 107, ADB, NARA.
8. Davidson to Tom Davidson, March 8, 1854, Box 59, GD, BANC. In ADB, NARA:
Davidson to Bache, February 19, 1852, Roll 69; Davidson to Bache, October 15,
1854, Roll 107; Alden to Bache, November 18, 1855, Roll 125. *Report of the Superintendent, 1854,* Appendix 60. Lawson, *Autobiography* (1879), BANC.
9. In Roll 69, ADB, NARA: Davidson to Bache, October 25, 1851; December 7, 1851;
January 20, 1852; May 2, 1852; and November 15, 1852. "George Davidson," *Proceedings of the California Academy of Sciences* (April 19, 1914), 11. *Report of the
Superintendent, 1854,* 80. Davidson to Belle, August 2, 1853, Box 59, GD, BANC.
10. In ADB, NARA: Davidson to Bache, November 15, 1853, Roll 88; Davidson to
Bache, March 23, 1854, Roll 107. Ulysses S. Grant to Julia Dent Grant, February
3, February 6, and March 6, 1854, *Memoirs and Selected Letters* (1886; New York:
Library of America, 1990), 949–55. Davidson returned to Humboldt between late
March and June 1854, coinciding with Grant's stay.
11. Davidson to Bache, March 23, 1854, Roll 107, ADB, NARA. One former assistant
addressed Mr. and Mrs. Bache as "dear grandpa" and "grandma." John Fries Frazer
to Bache, March 8, 1861, Box 26, William Jones Rhees Papers, HUNT.
12. In ADB, NARA: Bache to Davidson, April 16, 1852, Roll 69; Davidson to Bache,
July 19, 1853, and Bache to Davidson, October 17, 1853, Roll 88; Alden to Bache,
December 30, 1853, Davidson to Bache, April 13, 1854, Roll 107; Davidson to Bache,
May 13, 1852, Roll 69.
13. Davidson to Bache, December 31, 1852, and Alden to Bache, December 31, 1852,
Roll 88, ADB, NARA.
14. *Report of the Superintendent, 1853,* 77. For a general description of baseline measurement, see *Report of the Superintendent, 1854,* 78. Twenty, *Report on the History . . . Coast Survey,* 13.
15. Davidson, "Memorandum for the Use of the Committee on Hypsometry," in the
appendix to *Report of the Superintendent, 1893.* Davidson, April 6, 1853, Letters to

Jane Dale Owen Fauntleroy, 74/176c, BANC. In Roll 88, ADB, NARA: Davidson to
Bache, April 22, 1853, and Lawson to Bache, May 3, 1853.
16. Bache to Alden, June 4, 1853, Box 28, GD, BANC. Lawson, *Autobiography*
(1879), NOAA.
17. Bache to Davidson, May 19, 1853, and June 1, 1853, Roll 88, ADB, NARA.
18. Bache to Davidson, July 14, 1853, Roll 88, ADB, NARA.

BEWARE OF SWINDLERS

1. E. O. C. Ord to Manuel Domínguez, [n.d.], Box 9, Folder 244, RSP, GERTH. *Daily
Alta*, September 6, 1854.

CHAPTER 7: PERIL IN THE SIERRAS, INTRIGUE IN WASHINGTON

1. Hampton Sides, *Blood and Thunder: The Epic Story of Kit Carson and the Con-
quest of the American West* (New York: Anchor Books, 2007), 63–77. On Frémont's
life and expeditions, see Tom Chaffin, *Pathfinder: John Charles Frémont and the
Course of American Empire* (New York: Hill & Wang, 2002).
2. Jessie Benton Frémont, "Great Events During the Life of Major General John C.
Frémont," 110–18, Box 6, Frémont Family Papers, C-B 397, BANC. Winston Groom,
Kearny's March: The Epic Creation of the American West, 1846–1847 (New York:
Alfred A. Knopf, 2011), 259–64. Regarding the Pathfinder's popularity, see Steve
Inskeep, *Imperfect Union: How Jessie and John Frémont Mapped the West,
Invented Celebrity, and Helped Cause the Civil War* (New York: Penguin Press,
2020). Yosemite remained unknown to settlers before 1851.
3. Edward Ord to James Lycurgus Ord, April 13, 1848, Box 1, Folder 3; and November
10, 1850, Box 1, Folder 12, ORDF2, BOOTH. "Ord's Contract with the City, July
22, 1849," Box 3, ORDP, BANC. William Tecumseh Sherman, *Memoirs of General
W. T. Sherman* (1885; New York: Library of America, 1990), 96–105.
4. Williamson to "Sisters," September 28 and October 28, 1849, R. S. Williamson Correspondence, 2014/125, BANC.
5. In Box 1, Folder 8, ORDP, BANC: E. R. S. Canby, Special Orders No. 50, September
21, 1849, and Canby to Ord, October 1, 1849.
6. Ord Report, S. Exec. Doc. No. 47, 31st Cong., 1st Sess. (1850), 118–21. In Box 3,
ORDP, BANC: "Memorandum, Reconnaissance to San Bernardino, ca. 1849" and
"Notes on Report of Examination . . . to San Bernardino, 1849."
7. Ord to Canby, December 30, 1849, in *Congressional Globe*, January 27, 1853.
Report, December 30, 1849, Box 1, Folder 8, ORDP, BANC.
8. *Daily Alta*, December 31, 1849, March 7, 1850, and May 21, 1850.
9. Edward Ord to James Placidus Ord, February 27, 1845, Box 1, Folder 23, ORDF1,
BOOTH. In ORDF2, BOOTH: Edward Ord to James Lycurgus Ord, April 13,
1848, Box 1, Folder 3; and June 14, 1848, Box 1, Folder 5.
10. "E. O. C. Ord Statement, ca. 1846–47," Box 3, ORDP, BANC.
11. Davis to [Secretary of War], June 27, 1850, and Davis to Joseph G. Totten, February 14, 1851, *JD*, vol. 4. Ord to C. M. Conrad, November 4, 1851, Box 1, Folder 10,
ORDP, BANC.
12. *Congressional Globe*, January 27, 1853. "Memoirs of . . . Gwin," 173–80. David M.
Potter, *The Impending Crisis, 1848–1861* (New York: Harper Perennial, 1976), 149–54.
13. Edward Ord to James Lycurgus Ord, February 24, 1850, Box 1, Folder 10,
ORDF2, BOOTH.

14. Bache to Ord, January 12, 1853, Box 1, Folder 12, ORDP, BANC. *Report of the Superintendent, 1854*, Appendix 1, Report No. 2, 8.

CHAPTER 8: ORD FINDS AN AUSPICIOUS APPOINTMENT

1. Henry Washington to Samuel D. King, December 10, 1852, US General Land Office Reports by California Deputy Surveyors, C-I 54-55, BANC. *Star*, October 23, 1852.
2. *Manual of Instructions for the Survey of Public Lands* (Washington, DC: A. O. P. Nicholson, 1855), 2.
3. "An Act to Ascertain and Settle the Private Land Claims in the State of California," 9 Stat. 631, 31st Cong., 2nd Sess. (March 3, 1851).
4. George Hansen, Field Books, vol. 4, SR, HUNT. Depositions, 1856, Land Case Files 273 SD, BANC.
5. Twenty, *Report on the History . . . Coast Survey*, 11–12, 77–79.
6. In ADB, NARA: Ord to Bache, November 12, 1853, and December 21, 1853, Roll 107; Davidson to Bache, April 5, 1853, Roll 88.
7. In Roll 88, ADB, NARA: Ord to Bache, August 31, 1853; Bache to Ord, November 5, 1853. John Wilson to Jefferson Davis, November 11, 1853, E.171, Box 6, RG 77, NARA.
8. Ord to George Stephens, August 31, 1853, and Ord to Bache, August 31, 1853, Roll 88, ADB, NARA. James Lee McDonough, *William Tecumseh Sherman: In the Service of My Country: A Life* (New York: W. W. Norton, 2016), 168–72.
9. In ADB, NARA: Bache to Ord, September 16, 1853, October 10, 1853, and November 5, 1853, Roll 88. Bache to Ord, February 1, 1854; Alden to Bache, December 30, 1853, Roll 107.
10. Ord to Bache, December 21, 1853; Ord to Bache, July 1, 1854, Roll 107, ADB, NARA.
11. Twenty, *Report on the History . . . Coast Survey*, 27.
12. *Sacramento Daily Union*, November 1, 1853, and November 15, 1853. Scott O'Dell fictionalized Juana María's life in his Newbery Award–winning *Island of the Blue Dolphins* (1960).
13. Ord to Bache, February 20, 1854, April 29, 1854, and July 1, 1854, Roll 107, ADB, NARA.
14. *Daily Alta*, March 9, 1853, April 10, 1853, October 10, 1854, and August 7, 1862.
15. Davis, May 3, 1846 (Appendix III, Addenda, 1846); Davis to Bache, March 21, 1849, May 20, 1850, and February 26, 1851, JD, vol. 4.
16. Varina Davis, *Jefferson Davis: A Memoir, by His Wife*, vol. 1 (New York: Belford, 1890), 261–63. Jefferson Davis to Varina Davis, August 28, 1853, and Bache to Davis, July 2, 1858, in *Jefferson Davis: Private Letters, 1823–1889*. Davis to A. A. Humphreys and G. K. Warren, June 14, 1858, JD, vol. 6. Cooper, *Jefferson Davis*, 282, 311.
17. M. C. Meigs to Bache, September 15, 1856, Box 17, Rhees Papers, HUNT. Nearly all of the Lazzaroni were proponents of scientific racism and viewed slavery as a tool of uplift.

CHAPTER 9: A FAMILY ON THE MOVE AND ON THE MAKE

1. *Journal of the Senate*, 32nd Congress, 1st session, 1851–52 (Washington, DC: A. Boyd Hamilton, 1852), 546, 598, and 602. *Statutes at Large*, 32nd Congress, 1st session, 91. *Statutes at Large*, December 1, 1851–March 3, 1855, 32nd Congress, 2nd session, ed. George Minot, vol. 10 (Boston: Little, Brown, 1855), 205.

2. Ord to Bache, April 29, 1854, and July 1, 1854, and Bache to Ord, July 29, 1854, Roll 107, ADB, NARA.

3. On James Ord's birth and parentage, see Saul David, *Prince of Pleasure: The Prince of Wales and the Making of the Regency* (New York: Atlantic Monthly Press, 1998), 75–78.

4. In Box 1, ORDF1, BOOTH: Edward Ord to James Placidus Ord, November 20, 1845, Folder 24; April 7, 1852, Folder 33; and November 20, 1856, Folder 35. In Box 1, ORDF3, BOOTH: James Ord to James Lycurgus Ord, March 2, 1850, Folder 4. Los Angeles, San Francisco, and Washington, DC, each have an Ord Street named in Edward's memory.

5. In Box 1, ORDF1, BOOTH: Edward Ord to James Placidus Ord, January 4, 1859, Folder 38, and January 27, 1859, Folder 39. In ORDF2, BOOTH: Edward Ord to James Placidus Ord, March 5, 1854, Box 1, Folder 17. Ord to Bache, November 12, 1853, Roll 107, ADB, NARA.

6. James Ord to Edward Ord, September 16, 1854, Box 2, Folder 34, ORDP, BANC.

7. At BOOTH: Edward Ord to James Lycurgus Ord, November 10, 1850, Box 1, Folder 12, ORDF2; Edward Ord to James Placidus Ord, October 14, 1851, Box 1, Folder 32, ORDF1. At ORDP, BANC: Ord to "Dear Dr." [Jas. L. Ord], November 23, 1851, Box 2, Folder 32; Ord to D. C. De Leon, August 1852, Box 1, Folder 11; and James Ord to Edward Ord, June 28, 1853, Box 2, Folder 33.

8. Ord to Bache, September 1, 1854, Roll 107, ADB, NARA. Lawson, *Autobiography* (1879), BANC. Davidson took pride that Survey officers were not "carried away" by the region's "universal fever of speculation." *History of the Coast Survey on the Pacific* (1863), Carton 19, GD, BANC.

9. Ord to James Placidus Ord, January 27, 1859, Folder 39, Box 1, ORDF1, BOOTH.

10. Ord to Bache, March 5, 1854, and Bache to Ord, April 18, 1854, Roll 107, ADB, NARA. Ord had begun to complain to his Coast Survey peers months earlier. W. P. Trowbridge to Davidson, December 23, 1853, Box 55, GD, BANC.

11. Edward Ord to James Ord, December 18, 1853, Box 1, Folder 14, ORDF1, BOOTH.

12. Bache to Davis, May 22, 1854, *JD*, vol. 5.

CHAPTER 10: SECRETARY DAVIS'S MASTERSTROKE

1. William H. Goetzmann, *Army Exploration in the American West, 1803–1863* (New Haven: Yale University Press, 1959). West, *Continental Reckoning*, 83–89. Cooper, *Jefferson Davis*, 275–77.

2. "Memoirs of . . . Gwin," 172–75, 259–62.

3. Potter, *Impending Crisis*, 145–76. Eric Foner, *Free Soil, Free Labor, Free Men: The Ideology of the Republican Party before the Civil War* (1970; New York: Oxford University Press, 1995), 124–33, 159–66, 190–99.

4. Instructions, May 6, 1853, in *Reports of Explorations and Surveys*, vol. 5 (Washington, DC: Beverly Tucker, 1856).

5. Report of the Secretary of War, H. Exec. Doc. No. 1, 33rd Cong., 2nd Sess. (1855).

6. James Gadsden to Davis, May 9, 1853, and Davis to J. J. Abert, June 13, 1853, *JD*, vol. 5.

7. Benton Frémont, "Great Events During the Life of . . . John C. Frémont," 178–80, Box 6, Frémont Family Papers, BANC. Allan Nevins, *Frémont: Pathmarker of the West* (New York: D. Appleton-Century Co., 1939), 405–7.

8. Abert to Davis, April 28, 1853 and April 29, 1853, *JD*, vol. 5.

9. *Daily Alta*, July 16, 1853.

CHAPTER 11: "SAN PEDRO HAS NO HARBOR"

1. The Tulare is a 120-mile-long section of the San Joaquin Valley. In the nineteenth century, it lay between the massive Tulare Lake (or Tache Lake) and the interconnected Kern and Buena Vista Lakes. George H. Derby, Report of the Tulare Valley, 1850, S. Exec. Doc. No. 110, 32nd Cong., 1st Sess. (1852).

2. G. H. Stoneman to O. Cross, August 23, 1853, in Gwin, Speeches, BANC. Los Angeles Star, August 3, 1853.

3. Reports of Explorations and Surveys, vol. 5, 13. Daily Alta, August 10, 1853. Davis to Gwin, March 25, 1854, JD, vol. 5.

4. Galley Proof of a Biographical Sketch, Gwin Papers, BANC.

5. Williamson to Gwin, August 30, 1853, in Gwin, Speeches, December 12, 1853, BANC.

6. San Francisco Herald, August 8, 1853. Benjamin Madley, An American Genocide, 161–71. San Joaquin Republican, December 22, 1853. Statement of Joel H. Brooks to Superintendent Beale, September 21, 1852, in S. Exec. Doc. No. 57, 32nd Congr., 2nd Sess. (1853).

7. T. H. Benton to Davis, March 16, 1853, Jefferson Davis, Constitutionalist: His Letters, Papers and Speeches, vol. 2 (Jackson: Mississippi Department of Archives and History, 1923). Gwinn Harris Heap, Central Route to the Pacific . . . Journal of the Expedition of E. F. Beale (Philadelphia: Lippincott, Grambo, 1854). Sacramento Daily Union, February 22, 1854. Los Angeles Star, June 18, 1853, and September 3, 1853.

8. Reports of Explorations and Surveys, vol. 5 (1856), 16.

9. Daily Alta, July 21, 1853.

10. Benton wrote to Beale on November 2, 1853, that he and Frémont hoped to find a road across the Sierra just north of Walker Pass. Reprinted in Stephen Bonsal, Edward Fitzgerald Beale (New York: G. P. Putnam's Sons, 1912), 171. Benton Frémont, "Great Events During the Life of . . . John C. Frémont," 181, Box 6, Frémont Family Papers, BANC. Alta, April 21, 1854. James F. Milligan, Journal of Frémont's Fifth Expedition, 1853–1854, ed. Mark J. Stegmaier and David H. Miller (Glendale, CA: Arthur H. Clark, 1988), 136.

11. Williamson to Davis, August 31, 1853, and January 12, 1854, JD, vol. 5.

12. Gwin, Reply to Benton, National Intelligencer, n.d. [December 15, 1853], Box 1, Gwin Papers, BANC. Washington Union, June 16, 1857, in E.726, Box 2, RG 48, NARA. In JD, vol. 5: Williamson to Davis, January 30, 1854; Gwin to Davis, March 17, 1854; Davis to Gwin, March 17, 1854.

13. Maps and Profiles, E.31 and E.34, RG 48, NARA. Reports of Explorations and Surveys, vol. 5, 29–36. "New Pass" and "San Fernando Pass" today are called Soledad Pass and Newhall Pass.

14. Madley, An American Genocide, 245–46. Bonsal, Edward Fitzgerald Beale 191–94. Yokut Rancheria suffered harassment and murder until its forced removal in 1861. The mining town of Keyesville was set up near Walker Pass. There, troops massacred Indians in April 1863, during the Owens River War.

15. Reports of Explorations and Surveys, vol. 5, 42.

16. Report of the Secretary of War, March 17, 1854, S. Exec. Doc. No. 52, 33rd Cong., 1st Sess. A. A. Humphreys to Davis, September 22, 1854, E.726, Box 1, RG 48, NARA.

17. Reports of Explorations and Surveys, vol. 5, 42.

18. Reports of Explorations and Surveys, vol. 3, 135–36.

CHAPTER 12: GUESTS OF THE DOMÍNGUEZ FAMILY

1. Letter of Secretary of War to the President, December 4, 1854, in H. Exec. Doc. No. 1, 33rd Cong., 2nd Sess. Davis, January 9, 1855, *JD*, vol. 5.

2. *Reports of Explorations and Surveys*, vol. 1, 25, 29–30.

3. US Coast Survey, "Received, Benicia, September 28, 1854," Box 1, Folder 1, ORDC, BANC. Ord to Bache, March 5, 1854, and April 29, 1854; and Bache to Ord, July 29, 1854, and November 21, 1854, Roll 107, ADB, NARA.

4. Ord to Domínguez, [n.d.], Box 9, Folder 244, RSP, GERTH. *Daily Alta*, September 6, 1854. Edward Ord to Molly, January 11, 1855, and February 10, 1855, Box 1, Folder 1, ORDL, STANF.

5. Ord to Bache, October 19, 1854, Roll 107, ADB, NARA. Inset of "San Pedro and Vicinity," in "Sketch J, Showing the Progress of the Survey on the Western Coast" (1854), HM, 2007.158.1.1, Workman and Temple Family Homestead Museum Library. Cutts, *Fieldwork of the Triangulation*, 4.

6. Ord Diary, 1854, October 21, ORDC, BANC. In Box 1, Folder 1, ORDL, STANF: [1854, October 5], NP. Frag. Ord to [Mary] and "Goliah," [n.d.].

7. Edward Ord to Molly, January 11, January 24, and February 19, 1855, Box 1, Folder 1, ORDL, STANF. Ord Diary, 1854, October 21 and November 16, ORDC, BANC.

8. Ord to Domínguez, [n.d.], Box 9, Folder 244, RSP, GERTH.

9. Report of Referees, June 27, 1884, Box 24, Folder 7, SR, HUNT. *Los Angeles Star*, February 11, 1854.

10. In Box 23, Folder 15, SR, HUNT: Indenture [of] "Manuel Domingues and Gracia, his wife," to Benjamin D. Wilson, William T. B. Sanford, John G. Downey, Henry R. Myles, Joseph L. Brent, December 22, 1854; Indenture [of] Myles to Banning, July 20, 1858; Indenture [of] Wilson, Brent, McFahrland, Downey, Myles, and Domínguez to Banning, May 5, 1859; Map of New San Pedro by E. O. C. Ord.

11. See Box 2, Folder 20, "Professional, Accounts—US Coast Survey, 1854," ORDP, BANC.

12. Report, December 30, 1849, Box 1, Folder 8, ORDP, BANC.

13. Torres-Rouff, *Before LA*, 65–74, 91–93. Joshua Simon, *The Ideology of Creole Revolution: Imperialism and Independence in American and Latin American Political Thought* (New York: Cambridge University Press, 2017).

14. In Box 1, Folder 1, ORDL, STANF: Edward Ord to Molly, January 11 and 24, 1855, February 7, 1855, and April 2, 1855. In Report of Referees, June 27, 1884, Box 24, Folder 7, SR, HUNT: Deeds, December 22, 1854, December 27, 1854, and April 19, 1855.

15. Ord to E. R. S. Canby, November 6, 1849, in S. Exec. Doc. No. 47, 31st Cong., 1st Sess., 127.

16. January 27, 28, 29, 1851, Ord Diary, 1850-51-52, ORDC, BANC.

17. Edward Ord to Molly, January 11 and 28, 1855, Box 1, Folder 1, ORDL, STANF. Mid-September was the time for reports. Coast Survey, Circular, [n.d. 1850s], Box 28, GD, BANC. San Francisco had experienced numerous great fires, including seven between 1849 and 1851.

18. "Map to Navigable Water of Inner Harbor, January 1855, by . . . Ord, US Army," Box 24, Folder 7, SR, HUNT. This document, which shows only a fragment of the estuary, is a reproduction submitted by Banning to the County Recorder in 1860.

CHAPTER 13: TRIANGLES OF A DIFFERENT SORT

1. Edward Ord to Molly, February 10, 14, and 16, 1855, Box 1, Folder 1, ORDL, STANF.

2. San Gabriel Mission records show three possible women baptized "Catarina." Steven W. Hackel, Anne M. Reid, et al., *Early California Population Project: Version 1* (Huntington Library, San Marino, CA, 2006). Ord acknowledged his employment of Mission Indians as "both private servants and on public service" in *City of the Angels and the City of the Saints* (1856; San Marino: Huntington Library Press, 1978), 32.

3. Edward Ord to Molly, January 24 and February 10, 1855, Box 1, Folder 1, ORDL, STANF.

4. Edward Ord to Molly, February 16 and 19, 1855, Box 1, Folder 1, ORDL, STANF. Edward Ord to James Lycurgus Ord, February 7, 1855, Box 1, Folder 20, ORDF2, BOOTH.

5. Edward Ord to Molly, January 11, February 7, and February 19, 1855, Box 1, Folder 1, ORDL, STANF.

6. Edward Ord to Molly, January 11, 1855, Box 1, Folder 1, ORDL, STANF. Ord to "Mary," n.d. [1855], Box 2, Folder 35, ORDP, BANC.

7. Edward Ord to Molly, February 7, 1855, Box 1, Folder 1, ORDL, STANF. Ord to Adjutant General, February 5, 1855, Registers of Letters Received, M22, Roll 81, RG 107, NARA.

8. Edward Ord to Molly, February 7, 1855, Box 1, Folder 1, ORDL, STANF.

9. *Daily Alta*, October 29, 1855. Edward Ord to Molly, February 16, 1855, Box 1, Folder 1, ORDL, STANF.

10. New Town's estuary lands were variably measured at 34.40 acres, 36.32 acres, 43.40 acres, etc. By the twentieth century, 37 acres became the accepted number. See Box 24, Folder 7, SR, HUNT. Resolution No. 171 (January 30, 1909), Box 658881, Wilmington Municipal Records, Los Angeles City Archives.

11. "*Probablemente el squatter ha hecho trato con Brent—a sacar su cosecha.* [The squatter likely has made a deal with Brent—to take your bounty.]" The sentence shows the challenge of translating Ord's imperfect and imprecise Spanish.

12. Trowbridge to Bache, March 13, 1854, Roll 107, ADB, NARA.

13. Gwin to Davis, February 27, 1855, and Davis to Gwin, August 29, 1855, JD, vol. 5. Ord to Pacific Railroad Office, April 12, 1855, Registers, M22, Roll 81, RG 107, NARA. Davis to J. B. Weller, et al., September 14, 1855, Roll 125, ADB, NARA.

14. P. Ord to Gwin, January 3, 1855, E.27, Box 8, RG 60, NARA.

15. P. Ord to Cushing, March 8, 1855, and "Report . . . Cases on Appeal, 1856–1857," E.27, Box 8, RG 60, NARA. Notices of Appeal, June 13 and July 21, 1855, Land Case Files 273 SD, BANC. "Order Vacating Order of Appeal, June 4, 1857," in *De Guyer v. Banning*, No. 4991, LAACR2, HUNT.

16. Edward Ord to Molly, February 7, 1855, Box 1, Folder 1, ORDL, STANF. List of Conveyances, Box 24, Folder 4, SR, HUNT.

17. Mystery persisted decades later. A transcription of the rancho's 1857 final survey papers (for harbor engineer David E. Hughes) contains the handwritten query "Outer boundary by Hansen in March 1855, Wilmington frontage by ____?" Field Notes of the Final Survey . . . December 11, 1857, Box 2, Folder 54, DEH, WRCA. Other documents credited the rancho's outer boundary to Hancock. See "Survey of Boundary Lines of the Rancho San Pedro," Box 23, Folder 15, SR, HUNT.

18. Ord to Bache, March 29, 1855, Roll 125, ADB, NARA.

19. Ord to Bache, September 13, 1855, Roll 125, ADB, NARA. Appendix, Report No. 26, *Report of the Superintendent, 1855*, 183.

CHAPTER 14: DAMAGES DONE

1. *Daily Alta*, April 16, 1855.
2. In James Mandeville Papers, HUNT: Gwin to Mandeville, February 18, 1855, Box 5; Gwin to Mandeville, December 29, 1856, Box 8; Gwin to Mandeville, March 19, 1857, Box 9. W. R. Isaacs MacKay to Davis, January 13, 1857, *JD*, vol. 6.
3. Davis to A. A. Humphreys or G. K. Warren, July 14, 1858, *JD*, vol. 6.
4. *Washington Sentinel*, September 3, 1854.
5. H. L. Abbot to G. K. Warren, March 2, 1855, E.724, Box 4, RG 48, NARA. The lithographers were Napoléon Sarony and Bien & Sterner.
6. At STANF: *Map of Routes for A Pacific Railroad* ([Bien & Sterner], 1855), Barry Lawrence Ruderman Map Collection, and *Map of the Territory . . . to Accompany the Reports of the Explorations for a Railroad Route*, 1855, David Rumsey Map Collection. Various letters, February 20, 1855, to November 26, 1856, and "Map of Routes for a Pacific Railroad, 1855," E.724, Box 4, RG 48, NARA.
7. Davis, March 3, 1857, *JD*, vol. 6. Davis cited Ord's gallantry in Oregon during 1856.
8. Decree, *Domínguez, et al., vs. US*, February 10, 1857, Land Case Files 273 SD, BANC. In RG 60, NARA: P. Ord to J. S. Black, August 19, 1858, E.9, Container No. 74; J. Ross Browne to James Guthrie, April 5, 1856, E.27, Box 1. Throughout the Civil War, Edward Ord believed that Stanton disliked him and obstructed his promotion, a feeling that stemmed perhaps from Stanton's scrutiny of Pacificus Ord. In Box 1, Folder 3, ORDL, STANF: Ord to Molly, November 7 and November 15–18, 1862.
9. Davis to Ord, May 25, 1855, *JD*, vol. 5. Edward Ord to James Placidus Ord, January 27, 1859, Box 1, Folder 39, ORDF1, BOOTH. D. R. Jones, Special Orders No. 86, October 17, 1855, Box 1, Folder 14, ORDP, BANC.
10. Ord to W. W. Mackall, August 22, 1856, and Ord, "Statement, Presidio," September 30, 1856, Box 1, Folder 17, ORDP, BANC.
11. Ord, *City of the Angels*, 5, 7, 23, and 32. Ord to Mackall, August 22, 1856, Box 1, Folder 17, ORDP, BANC.

SECESSION AND LIVES INTERRUPTED

1. Bache to Davidson, August 10, 1859, and September 16, 1859, Box 28, GD, BANC.
2. *Alta*, November 22, 1860. In Box 29, GD, BANC: Bache to Davidson, March 2, 1860, July 7, 1860, and September 5, 1860. In Rhees Papers, HUNT: Stephen R. Mallory to Bache, January 22, 1861, Box 25; Benjamin Apthorp Gould to Bache, March 6, 1861, Box 26.
3. Bache to Davidson, May 5, 1861, and May 15, 1861, Box 29, GD, BANC.
4. *Star*, February 12, 1859.
5. Davidson to Thomas Davidson, February 10, 1861, Box 59, GD, BANC.
6. Frémont ordered killings of Wintu natives on the Sacramento River near Reading's Ranch (Redding), Klamath people at Klamath Lake, and three Californio prisoners at San Quentin. Sides, *Blood and Thunder*, 106–10, 119–21; Madley, *American Genocide*, 45–50.
7. Almira Russell Hancock, *Reminiscences of Winfield Scott Hancock, by His Wife* (New York: Charles L. Webster, 1887), 46. Varina Davis, *Jefferson Davis: A Memoir*, vol. 1, 582.

8. Varina Davis, *Jefferson Davis*, vol. 2 (New York: Belford, 1890), 19.
9. "E. O. C. Ord Statement, ca 1846–47," Box 3, ORDP, BANC. Edward Ord to Molly, March 19, [1862], Box 1, Folder 3, ORDL, STANF.
10. Molly Ord to Edward, June 22, [1862?], and Edward Ord to Molly, December 2, 1859, Box 1, Folder 3, ORDL, STANF. Edward Ord to Molly, December 10, 1861, Box 1, Folder 21, ORDP, BANC. Edward to James Placidus Ord, December 11, 1859, Box 1, Folder 43; ORDF1, BOOTH.
11. Edward Ord to Molly, March 9, 1859, Box 1, Folder 3, ORDL, STANF. Edward Ord to James Placidus Ord, January 4, 1859, Box 1, Folder 38; Pacificus Ord to Edward Ord, n.d., 1859, Box 3, Folder 48, ORDF1, BOOTH.
12. Edward Ord to James Placidus Ord, May 23, 1864, Box 1, Folder 48, ORDF1, BOOTH.

CHAPTER 15: A REGION OF REBELS

1. James F. Brooks, *Captives and Cousins: Slavery, Kinship, and Community in the Southwest Borderlands* (Chapel Hill: University of North Carolina Press, 2002). Andrés Reséndez, *Changing National Identities at the Frontier: Texas and New Mexico, 1800–1850* (New York: Cambridge University Press, 2005). Juliana Barr, *Peace Came in the Form of a Woman: Indians and Spaniards in the Texas Borderlands* (Chapel Hill: University of North Carolina Press, 2007). Pekka Hämäläinen, *The Comanche Empire* (New Haven: Yale University Press, 2008).
2. Steven Hahn, *A Nation without Borders: The United States and Its World in the Age of Civil Wars* (New York: Viking, 2016), 4–5, 153–67. West, *Continental Reckoning*, 36–50.
3. Edward Ord to Molly, May 9, 1856, Box 1, Folder 2, ORDL, STANF.
4. The destruction of Mackanootney by Ord's Company B occurred on March 26, 1856. Ord letter, March 28, 1856, Box 1, Folder 2, ORDL, STANF. William E. Birkhimer, "Third Regiment of Artillery," *Army of the United States*, ed. Theodore F. Rodenbough and William L. Haskin (New York: Maynard, Merrill, 1896), 343. Edward Ord to James Placidus Ord, January 27, 1859, Box 1, Folder 39, ORDF1, BOOTH.
5. Madley, *An American Genocide*, 206–30.
6. *Star*, December 26, 1857. *Alta*, November 12, 1857.
7. Richards, *The California Gold Rush and the Coming of the Civil War*, 66–81; Stacey L. Smith, *Freedom's Frontier: California and the Struggle over Unfree Labor, Emancipation, and Reconstruction* (Chapel Hill: University of North Carolina Press, 2013), 237–45.
8. *Star*, October 23, 1858, December 10, 1853, April 22, 1854. *Daily Alta*, October 21, 1854. Albert L. Hurtado, "Empires, Frontiers, Filibusters, and Pioneers: The Transnational World of John Sutter," *Pacific Historical Review* 77, no. 1 (February 2008), 19–47. Rachel St. John, "The Unpredictable America of William Gwin: Expansion, Secession, and the Unstable Borders of Nineteenth-Century North America," *Journal of the Civil War Era* 6, no. 1 (March 2016), 56–84.

CHAPTER 16: BANNING SEES AN OPPORTUNITY

1. *Southern Vineyard*, October 2, 1858. *Star*, October 2, 1858.
2. *Star*, October 9, 1858. Of the known "California hurricanes," only the 1858 storm reached Los Angeles at hurricane strength.
3. Phineas Banning, Settlement of Wilmington, 1883, C-E 139, BANC. Tom Sitton,

Grand Ventures: The Banning Family and the Shaping of Southern California (Berkeley: University of California Press, 2010), 39–46.

4. *Alta*, January 31, 1855, February 25, 1855, April 26, 1855, and March 27, 1858.
5. Wilson, Brent, McFahrland, Downey, Myles, Domínguez to Banning, May 5, 1858, and Myles to Banning, July 20, 1858, Box 24, Folder 4, SR, HUNT.
6. Appendix, Report No. 44, *Report of the Superintendent, 1858*, 303–4.
7. *Star*, January 30, 1858, November 6, 1858, and February 12, 1859.
8. Davidson to Bache, January 17, 1859, Letterbook, vol. 3, Carton 1, Davidson Correspondence, BANC. Davidson to Bache, April 4, 1859, Roll 208, ADB, NARA.
9. *Star*, February 12, 1859.
10. Alden to Bache, February 4, 1859, Roll 208, ADB, NARA.
11. Robert O. Tyler to Thomas Swords, January 31, 1858, and May 31, 1858, E.225, Box 99, RG 92, NARA. *Star*, December 25, 1858. *Alta*, March 27, 1858, and May 29, 1858.
12. Rigg to Carleton, March 6, 1862, WoR1. Banning charged fifty-three cents per pound versus the lowest bid of fifteen cents. "This fact speaks for itself," Rigg commented.
13. *Star*, February 12, 1859. *Alta*, January 5, 1859, and January 11, 1859.
14. *Star*, June 11, 1859. Logbook of USS *Active*, vol. 6, E.102, RG 23, NARA. Alden to Bache, August 26, 1859, Roll 208, ADB, NARA.
15. *Star*, February 5, 1859, and September 24, 1859. *Alta*, January 16, 1861, October 8, 1861, and October 13, 1861.

CHAPTER 17: BECOMING THE DESERT SCOURGE

1. Aurora Hunt, *James Henry Carleton: Western Frontier Dragoon* (Glendale, CA: Arthur H. Clark, 1958).
2. Cooper, *Jefferson Davis*, 163–66. Greenberg, *A Wicked War*, 157–61.
3. Carleton, *The Battle of Buena Vista* (New York: Harper & Bros., 1848), 74.
4. Davis to E. O. C. Ord, May 25, 1855, *JD*, vol. 5. Hunt, *James Henry Carleton*, 165–70. Sophie Wolfe Carleton was a granddaughter of Benjamin Wolfe, Continental Army officer during the American Revolution, and grandniece of the British general James Wolfe, killed in the Seven Years' War and commemorated by Benjamin West's painting.
5. Report of the Secretary of War Communicating the Report of Captain George B. McClellan, S. Exec. Doc. No. 1, Special Sess. (1857), 77–83; 124–30. Richard Delafield, *Report on the Art of War in Europe in 1854, 1855, and 1856* (Washington, DC: G. W. Bowman, 1860).
6. Report of the Secretary of War . . . Respecting the Purchase of Camels for the Purposes of Military Transportation, S. Exec. Doc. No. 62, 34th Cong., 3rd Sess. (1857).
7. Hunt, *James Henry Carleton*, 171–74.
8. *Star*, April 30, 1859, May 28, 1859, April 28, 1860, and July 14, 1860. *Alta*, August 18, 1861.
9. Alvin M. Josephy Jr., *The Civil War in the American West* (New York: Alfred A. Knopf, 1991), 236–38.
10. T. T. Teel, "Sibley's New Mexican Campaign," *Battles and Leaders of the Civil War*, vol. 2, ed. R. U. Johnson and C. C. Buel (New York: Century Co., 1887), 700. Josephy, *Civil War in the American West*, 36–52.
11. Hyde, *Empires, Nations, and Families*, 354, 443–45; Sides, *Blood and Thunder*, 126, 341–57.
12. *Alta*, June 22, 1861, September 30, 1861, and October 2, 1861. John W. Robinson, *Los Angeles in Civil War Days* (1976; University of Oklahoma Press, 2013), 24–31, 50–55. Faragher, *Eternity Street*, 296–300, 376, 381–84.
13. Hancock, *Reminiscences of Winfield Scott Hancock*, 60–61, 67. *Alta*, May 18, 1861.

CHAPTER *18*: UNION AND DISUNION IN LOS ANGELES

1. The event's advance program appeared in *Star*, May 18, 1861. Robinson, *Los Angeles in Civil War Days*, 55–61.

2. *Alta*, October 2, 1861.

3. *Alta*, March 27, 1860. Ord, *City of the Angels*, 7, 32.

4. *Alta*, March 15, 1862. Faragher, *Eternity Street*, 344–72. Torres-Rouff, *Before LA*, 120–29.

5. *Star*, November 8, 1856, August 28, 1858, and August 11, 1860.

6. *Alta*, June 6, 1861, and September 30, 1861.

7. *Alta*, December 3, 1863. In WoR1: J. W. Davidson to D. C. Buell, August 13, 1861; J. W. Davidson to R. C. Drum, September 9, 1861; E. A. Rigg to Carleton, October 28, 1861; Carleton to Rigg, November 4, 1861.

8. In WoR1: C. E. Bennett to E. V. Sumner, August 6, 1861; Edwin A. Sherman, et al., to J. W. Davidson, September 14, 1861. *Alta*, November 12, 1857, and April 29, 1858. *Star*, August 3, 1861.

9. In WoR1: John B. Mills to William Nelson, September 21, 1861; US Legation to W. H. Seward, August 28, 1861; and Sibley to Governor of Sonora, December 16, 1861. Deverell, *Whitewashed Adobe*, 11–22.

10. Matthew Keller to E. V. Sumner, August 10, 1861, WoR1. John Mathew Keller, "Mathew Keller: Biographical Sketch, 1959," Box 4, Folder 10, Matthew Keller Papers, HUNT.

11. In WoR1: J. H. Carleton to D. C. Buell, July 31, 1861; W. S. Ketchum to Adjutant General, September 5, 1861. Thomas K. Tate, *General Edwin Vose Sumner: A Civil War Biography* (Jefferson, NC: McFarland, 2013).

12. Robinson, *Los Angeles in Civil War Days*, 78–81.

13. In WoR1: E. V. Summer to Montgomery C. Meigs, September 20, 1861; Carleton to R. W. Kirkham, October 29, 1861, and April 1, 1862.

14. In WoR1: Carleton to Drum, December 21, 1861; J. McAllister to Carleton, September 10, 1861; Edward E. Eyre to Carleton, October 21, 1861.

15. *Alta*, February 8, 1861, and February 17, 1862. *Star*, March 2, 1861. Wright to Drum, October 5, 1861, WoR1.

16. Because of environmental limits, the Pacific Department hoped to bypass the desert by landing soldiers on the Mexican mainland, either at San Blas or Mazatlán, and to march on Texas from there. In WoR1: E. F. Beale to Sumner, September 5, 1861; Sumner to Townsend, September 7, 1861; Wright to Townsend, October 31, 1861; Carleton to Drum, December 21, 1861.

17. In WoR1: J. W. Davidson to Buell, August 13, 1861; Carleton to Wright, October 9, 1861; J. R. West to Benjamin C. Cutler, September 19, 1861. *Alta*, September 29, 1861, October 5, 1861, October 8, 1861, October 13, 1861, and October 24, 1861.

18. Carleton to Wright, October 9, 1861, WoR1.

19. Drum, Report, May 30, 1862, WoR1.

20. *Alta*, May 11, 1858. *Star*, July 16, 1859, and September 3, 1859. In SR, HUNT: Division . . . of New San Pedro, October 11, 1862, Box 24, Folder 4; Wilmington Partition, 1862, October 31, Box 24, Folder 5; *Banning, et al. v. Downey, et al.*, No. 877, Partition Suit, October 11, 1862, Box 23, Folder 15.

21. S. H. Parker to Seward, October 22, 1861, WoR2. Wright to L. Thomas, December 28, 1861, WoR1. The *Orizaba*'s passenger list included "M. Bakonia." *Alta*, October 20, 1861. Bakunin knew Gwin's one-time clerk, Joseph Heco (Hikozō Hamada, famous as the first Japanese national to gain US citizenship), with whom he recently had voyaged from Japan. See Philip Billingsley, "Bakunin in Yokohama: The Dawning of the Pacific Era," *International History Review* 20, no. 3 (September 1998), 547–50, 555.

CHAPTER 19: WEAPONIZING THE LANDSCAPE

1. In *WoR1*: Report of James McNulty, October 1863; West to Cutler, November 7, 1861; West to Drum, January 28, 1862; Carleton to West, March 16, 1862; Eyre, July 6, 1862; T. L. Roberts, July 19, 1862. *Alta*, May 14, 1862. William A. Bell, *New Tracks in North America: A Journal of Travel and Adventure* (London: Chapman & Hall, 1869), 78–83.

2. In *WoR1*: Wright to Drum, October 7, 1861; Carleton to Drum, December 21, 1861.

3. Carleton to Eyre, February 8, 1862, vol. 40, E.3656, RG 393, NARA. In *WoR1*: Rigg to Carleton, February 14, 1862, and February 15, 1862. *Alta*, January 18, 1862, February 17, 1862, and March 15, 1862.

4. In *WoR1*: Rigg to Carleton, April 2, 1862; Carleton to Fergusson, April 21, 1862. *Alta*, February 17, 1862, and April 26, 1862. See Andrew E. Masich, *The Civil War in Arizona: The Story of the California Volunteers, 1861–1865* (New York: Oxford University Press, 2012). Aurora Hunt, *Army of the Pacific: 1860–1866* (1951; Mechanicsburg, PA: Stackpole Books, 2004).

5. In *WoR1*: West to Cutler, November 12, 1861; Carleton to Drum, December 21, 1861; Rigg to Cutler, March 1, 1862; Rigg to Carleton, April 2, 1862.

6. In *WoR1*: Carleton to Drum, December 21, 1861; Rigg to Carleton, April 2, 1862. *Alta*, June 23, 1862. Carleton found that village leaders preferred to trade for non-essential, nonedible items such as tobacco, looking glasses, beads, paint, and metal objects. This exacerbated the tribes' dependence on the US.

7. In *WoR1*: Carleton to West, November 7, 1861; Drum to Carleton, December 10, 1861; Carleton to Rigg, November 4, 1861, and February 12, 1862; West to Cutler, November 21, 1861; Carleton to Kirkham, April 11, 1862; Cutler, Special Orders, No. 53, May 1, 1862; Carleton to Y. Pesqueria, May 2, 1862; General Orders, No. 20, September 5, 1862. *Alta*, July 9, 1862, July 10, 1862, and July 12, 1862. Report of D. Fergusson, December 6, 1862, S. Exec. Doc. No. 1, 37th Cong., Special Session.

8. Carleton to Drum, May 25, 1862, *WoR1*.

9. In *WoR1*: Rigg to Carleton, February 14, 1862; Drum, Inspection Report, May 30, 1862; Carleton to Drum, April 1, 1862. M. A. McLaughlin to Drum, July 22, 1863, *WoR2*. J. Kellogg to Drum, March 17, 1864, E.171, Box 5, RG 77, NARA. *Alta*, August 16, 1862.

10. Megan Kate Nelson, *The Three-Cornered War: The Union, the Confederacy, and Native Peoples in the Fight for the West* (New York: Scribner, 2020), 101–10.

11. In *WoR1*: Henry Connelly to Canby, June 15, 1862; Eyre, July 8, 1862.

12. In *WoR1*: Report of McNulty, October 1863; Carleton to Drum, July 22, 1862. *Alta*, August 16, 1862.

13. Cutler to Rigg, March 12, 1862, *WoR1*.

14. In *WoR1*: "Skirmish . . . Fort Bowie, April 25, 1863"; "Skirmish . . . Chiricahua Mountains, September 8–9, 1863."

15. In *WoR1*: Carleton, Order, December 23, 1861; Carleton to McCleave, March 15, 1862; Rigg to Carleton, March 30, 1862; Cutler to Rigg, March 15, 1862; E. D. Shirland to West, August 10, 1862.

16. In *WoR1*: E. E. Eyre, July 6, 1862, and July 8, 1862; T. L. Roberts, July 19, 1862; Carleton to Drum, July 22, 1862.

17. Carleton to West, March 13, 1863; Carleton to Carson, October 12, 1862; Carleton to Thomas, November 9, 1862; *Condition of the Tribes*, S. Rep. No. 156, 39th Cong., 2nd Sess. (1867). Special Orders, No. 49, February 1, 1864, *WoR2*. Josephy, *Civil War in the American West*, 278–80. Nelson, *Three-Cornered War*, 166–74.

18. Carleton to Carson, August 18, 1863; Carleton to Thomas, June 17, 1863; Carle-

ton to Thomas, March 12, 1864; *Condition of the Tribes* (1867). Nelson, *Three-Cornered War*, 188–99. Sides, *Blood and Thunder*, 419–42.
19. Rigg to Cutler, "Expedition . . . to Fort Goodwin, May 16 to August 2, 1864." WoR1. Bell, *New Tracks in North America*, 51.

CHAPTER 20: STARS FALL AND RISE AT WAR'S END

1. Lincoln's Second Inaugural Address, March 4, 1865, sermonized on the Almighty's "own purposes" in war and emancipation. Prior to the war, Republicans and others spoke of slavery's "ultimate extinction." Reply Speech at Jonesboro, September 15, 1858, *Abraham Lincoln, Speeches and Writings, 1832–1858* (New York: Library of America, 1989), 603.
2. Smith, *Freedom's Frontier*, 176–92, 209–13. Waite, *West of Slavery*, 210–28. On the persistence of Indian slavery during the war years, see Madley, *An American Genocide*, 290–309.
3. In WoR1: Fergusson to Cutler, September 21, 1862; Evans, July 1, 1862, "Expedition . . . to Owen's River, June 11 to October 8, 1862." *Alta*, June 26, 1863. Sitton, *Grand Ventures*, 80–81.
4. James F. Curtis to Secretary of War, September 20, 1863, WoR2.
5. *Alta*, December 3, 1863. *Star*, December 12, 1863, July 2, 1864, and May 7, 1864. On LA mining investments at Kern, Tulare, Cuyama, Panamint, and parts of Arizona, see at HUNT: Boxes 1 and 3, Matthew Keller Papers, and Box 1, Papers of John Dustin Bicknell.
6. *Alta*, December 3, 1863, and August 22, 1864.
7. B. D. Wilson to J. L. Brent, September 21, 1865, Box 2, Joseph Lancaster Brent Papers, HUNT. R. W. Kirkham, Report to J. F. Rusling, January 8, 1867, E.171, Box 5, RG 77, NARA. *Los Angeles News*, March 25, 1865. *Wilmington Journal*, September 15, 1866.
8. John W. Shore to Brent, July 9, 1865, Box 2, Joseph Lancaster Brent Papers, HUNT. Maurice H. Newmark and Marco R. Newmark, *Sixty Years in Southern California* (New York: Newmark & Newmark, 1916), 296. *Star*, July 25, 1863.
9. *Wilmington Journal*, April 15, 1865, July 22, 1865, August 19, 1865, and August 26, 1865, RB 274677, HUNT. Desert war promised good business. Arizona's legislature requested $250,000 from Congress to continue the fight. H. Mis. Doc. No. 18, 38th Congr., 2nd Sess. (1865).
10. *Alta*, February 17, 1862, May 15, 1862, and May 29, 1862.
11. Hunt, *Carleton*, 286–88, 297–303. Nelson, *Three-Cornered War*, 200–225, 235–46.
12. Statement of Robert Bent, *Condition of the Tribes* (1867), 96.
13. Carleton, Testimony to J. R. Doolittle, et al., July 3, 1865, and Reply, July 25, 1865, *Condition of the Tribes* (1867), 323–25, 433–35.
14. Paul Andrew Hutton, *The Apache Wars: The Hunt for Geronimo, the Apache Kid, and the Captive Boy Who Started the Longest War in American History* (New York: Crown, 2016), 307–10.
15. "General E. O. C. Ord," [Mss, n.d.], Box 1, Folder 41, ORDF2, BOOTH. Bernarr Cresap, *Appomattox Commander: The Story of General E. O. C. Ord* (New York: Barnes, 1981).
16. Edward Ord to James Placidus Ord, May 23, 1864, Box 1, Folder 48, ORDF1, BOOTH. Ron Chernow, *Grant* (New York: Penguin Press, 2017), 482–83.
17. Telegrams, e.g., Lincoln to Grant, January 31, 1865, in Lincoln, *Speeches and Writings, 1859–1865* (New York: Library of America, 1989), 678. E. M. Stanton to Ord,

January 30, 1865; Lincoln to T. T. Eckert, January 30, 1865, and Davis to R. E. Lee, February 28, 1865, in *Jefferson Davis, Constitutionalist*, vol. 6.
18. XXV Corps, led by Bavarian-born Gottfried (Godfrey) Weitzel, was the largest unit of African American soldiers. On occupied Richmond, see Caroline E. Janney, *Ends of War: The Unfinished Fight of Lee's Army after Appomattox* (Chapel Hill: University of North Carolina Press, 2021), 112–23.
19. Elizabeth R. Varon, *Appomattox: Victory, Defeat, and Freedom at the End of the Civil War* (New York: Oxford University Press, 2014), 89, 92. Mary A. Benjamin, "Tale of a Table," *American Heritage* 16, no. 3 (April 1965).
20. Cresap, *Appomattox Commander*, 118.

EXTRAVAGANT COSTS AND HUGE OPPROBRIUM

1. G. M. Wheeler to A. A. Humphreys, June 2, 1869, and Wheeler to John P. Sherburne, March 31, 1870, E.171, Box 6, RG 77, NARA. William H. Goetzmann, *Exploration and Empire: The Explorer and the Scientist in the Winning of the American West* (New York: Alfred A. Knopf, 1966), 469–77.
2. Richard White, *The Republic for Which It Stands: The United States during Reconstruction and the Gilded Age, 1865–1896* (New York: Oxford University Press, 2017), 26–34, 55–59.
3. Jonathan Levy, *Ages of American Capitalism: A History of the United States* (New York: Random House, 2022), 200. Eric Foner, *Reconstruction: America's Unfinished Revolution* (New York: Harper & Row, 1988).
4. In E.171, Box 5, RG 77, NARA: Division of the Pacific, General Orders, No. 8, September 28, 1865; General Orders, No. 30, November 19, 1866; E. B. Babbitt to Chief Quartermaster, November 22, 1866; R. W. Kirkham to J. F. Rusling, January 14, 1867; Kirkham, "Report on Expenditures, January 8, 1867." Roger Jones to James B. Fry, July 15, 1867, in *Report of the Secretary of War, Part I* (Washington, DC: GPO, 1867), 82–85.
5. James F. Rusling, "National Cemeteries," *Harper's Monthly Magazine* 33 (August 1866), 310–22. James F. Rusling, *The Great West and Pacific Coast* (New York: Sheldon, 1877).
6. Report of Inspection . . . Drum Barrracks, by Rusling, May 6, 1867, and McFerran to Ekin, June 18, 1867, E.225, Box 527, RG 92, NARA. "Report of . . . Halleck, September 18, 1867" and "Report of . . . McDowell, September 14, 1867," in *Report of the Secretary of War, Part I* (1867), 72, 124–28.

CHAPTER 21: BANNING RIDES THE TURNING TIDE

1. US Department of Interior, Census Office, *Report on Transportation Business of the United States at the Eleventh Census* (Washington, DC: GPO, 1894), 6.
2. John W. Robinson, *Southern California's First Railroad: The Los Angeles & San Pedro Railroad, 1869–1873* (Los Angeles: Dawson's Book Shop, 1978). Richard Rayner, *The Associates: Four Capitalists Who Created California* (New York, W. W. Norton, 2008), 65–83.
3. White, *Railroaded*, 26–38. *New York Times* (NYT), May 21, 1866.
4. *Alta*, March 23, 1868, June 23, 1868, and July 23, 1868.
5. "Memorial of Henry Baldwin Tichenor, [n.p.] [1883]," F860 .T55, BANC. "Swamp and Overflowed Lands Claim, February 15, 1866," Box 24, Folder 3, SR, HUNT.
6. "San Gabriel," No. 840, General Photograph File, SEAV. Banning, "Settlement of

Wilmington" (1883), BANC. *Star*, January 23, 1869, and March 20, 1869. Sitton, *Grand Ventures*, 110.

7. Deverell, *Railroad Crossing*, 23.

8. California, Legislature. Senate and Assembly, "Report of the Board of Tide Land Commissioners and the State Board," in *Appendix to Journals of Senate and Assembly* (Sacramento: D. W. Gelwicks, 1870). The California Pacific and Western Pacific were two other terminal upstarts.

9. In Box 2, WBH, STANF: Various items, Folder 6, and Hyde to Boller, February 25, 1877, Folder 14. See also White, *Railroaded*, 88–92. By 1864, Olmsted had code-signed New York's Central Park. He went on to design Brooklyn's Prospect Park, the 1893 World's Exposition grounds in Chicago, and a proposal for San Francisco's Golden Gate Park.

10. In Box 2, Folder 6, WBH, STANF: Hyde to B. S. Alexander, August 26, 1867, and Hyde to John Wilson, May 5, 1867.

11. In Box 2, Folder 6, WBH, STANF: Hyde to "wife," December 8, 1867; Hyde to H. D. Bacon, August 9, 1867; November 9, 1867; and November 30, 1867.

12. *Alta*, April 15, 1868, and January 11, 1873. In Box 2, WBH, STANF: John Wilson to Hyde, February 8, 1868, Folder 7; Hyde to Bacon, March 24, 1872, Folder 10; Memorandum of Sale of Terminal Pacific Railroad, April 10, 1871, Folder 11. Davidson to Benjamin Peirce, October 1, 1870, E.22A, Box 504B, RG 23, NARA. With Hyde on board, the SPRR continued its Goat Island scheme.

13. In Box 2, WBH, STANF: Hyde to Bacon, December 2, 1869, and December 9, 1869, Folder 8; Received of E. E. Hewitt, September 26, 1872, Folder 10; Hyde to Banning & Co., August 30, 1873, Folder 11. Agreement with Banning would be finalized after the 1872 subsidy election. Robinson, *Southern California's First Railroad*, 90–96.

14. Hyde to Marietta Hyde, December 13, 1869, Box 2, Folder 8, WBH, STANF. See also *William Birelie Hyde: Letters of a California Engineer and Lobbyist*, ed. William Hyde Irwin and Charles A. Chapin (Stockton: Augusta [Hyde] Bixler Farms, 1988).

CHAPTER 22: PAST PLANS AND UNDYING HABITS

1. Cooper, *Jefferson Davis*, 620–29.

2. *Report of the Chief of Engineers . . . 1867* (Washington, DC: GPO, 1867), 2, 53–54. *Report of the Chief of Engineers . . . 1868* (Washington, DC: GPO, 1868), 4–5.

3. Williamson's pattern of illness appears in M505 Registers of Letters Received, Roll No. 4, vol. 6, NARA, and E.726, Box 1 and Box 2, RG 48, NARA. *Cullum's Register*, vol. 2, 1117 (117–21), 1373 (346–47), https://penelope.uchicago.edu. *Annual Report of the Secretary of War for 1870*, vol. 1 (Washington, DC: GPO, 1870), 49–52.

4. Tide Land Reclamation Co., *Fresh Water Tide Lands of California* (San Francisco: M. D. Carr, 1869), 10–11. Williamson to Chief of Engineers, July 29, 1867, W652, E.25, Box 34, RG 77, NARA.

5. J. B. Weller, M. S. Latham, W. M. Gwin, et al., to the Secretary of War, July 12, 1855; Davis [reply], September 14, 1855, Roll 125, ADB, NARA. Banning to Williamson, January 2, 1869, *Report of the Chief of Engineers . . . 1869* (Washington, DC: GPO, 1869), 482. The press talked up a government breakwater: *Star*, July 4, 1868, and August 15, 1868; *Alta*, September 29, 1868.

6. *Alta*, August 2, 1865, and August 4, 1865. *Brother Jonathan* wrecked off Point St. George.

7. *Ex Parte McCardle*, 74 US 506 (1869).

8. Williamson, "Report, February 13, 1869," W1317, E.25, Box 37, RG 77, NARA. Banning proposed further federal improvements. *Star*, April 29, 1869.
9. Davidson to Mateo Keller, April 26, 1870, and "Notes on San Pedro Breakwater," Carton 7, GD, BANC. California, "Senate Concurrent Resolution Relative to the Erection of a Breakwater at Wilmington Harbor," in *Appendix to Journals of Senate and Assembly* (Sacramento: D. W. Gelwicks, 1870).
10. B. S. Alexander, "Report on Breakwater at Wilmington," in *Appendix to Journals of Senate and Assembly* (1870), 11. G. H. Mendell to B. D. Wilson, February 5, 1872, Box 21, Benjamin D. Wilson Collection, HUNT.
11. *Star*, November 6, 1858. *Reports of Explorations and Surveys*, vol. 3, 136.
12. "Dear General", n.d. [1868–1871], Box 1, Folder 5, ORDL, STANF. The *Wilmington Journal* (August 19, 1865) called on the Army to make "quick though it may be an expensive job of the Apaches." Ord to William Kelly, March 10, 1869, Box 1, ORDC, BANC.
13. US Army Corps of Engineers, *Analytical and Topical Index to the Reports of the Chief of Engineers, 1866–1900*, vol. 2 (Washington, DC: GPO, 1903), 1303–4. Banning, "Settlement of Wilmington" (1883), BANC. Hyde to Huntington, December 29, 1872, Box 1, Folder 3, WBH, STANF.
14. Edward Ord to Molly, December 11, 1871, and Edward Ord to Lucy, n.d. [January 1872], Box 1, Folder 5, ORDL, STANF. The historical database for Prospect Hill Cemetery lists Albert E. Ord (age: one year) buried in July 1872. A. A. Humphreys to Ord, October 21, 1873, Box 3, ORDC, BANC. In 1871, F. V. Hayden completed a geological study of Yellowstone canyon, including paintings by Thomas Moran, which led Congress to create the first national park. A military survey focused on gold and coincided with a similar study of the Black Hills. The Yellowstone River, and tributaries like the Little Bighorn, soon reignited with Indian war.

CHAPTER 23: RAILROAD SWINDLES

1. George Davidson to Tom Davidson, February 3, 1869, Box 60, GD, BANC. Thomas Davidson Jr. was in the navy's construction bureau and designed armored vessels and torpedo boats.
2. US Coast Survey and George Davidson, *Pacific Coast: Coast Pilot of California, Oregon, and Washington Territory* (Washington, DC: GPO, 1869), 15–16.
3. In Carton 19, GD, BANC: *What the Coast Survey Has Done for the War* (New York: Charles B. Richardson, 1865); Report on the Defences of the Delaware River, 1863; and Bache to Thaddeus Stevens, January 9, 1862. Odgers, *Bache*, 168–74.
4. Nancy Clarke (Fowler) Bache to Ellie Davidson, September 7, 1864, Box 2, GD, BANC. Lawson, *Autobiography* (1879), NOAA. Edmonds, *History of the Central High School*, 78. *Evening Star* (Washington, DC), February 18, 1867, and February 21, 1867.
5. Davidson to Peirce, September 28, 1870, E.22A, Box 504B, RG 23, NARA. *Alta*, July 14, 1869.
6. In E.22A, Box 504A, RG 23, NARA: Davidson to Peirce, October 14, 1869, November 30, 1869, and December 19, 1869.
7. Davidson to Peirce, November 17, 1870, and December 9, 1870, E.22A, Box 504B, RG 23, NARA.
8. At NARA: Davidson to Peirce, April 21, 1871, June 8, 1871, and October 18, 1871, E.22A, Box 504B, RG 23.
9. Davidson to Mateo Keller, April 26, 1870, Carton 7, and Alexander to Davidson,

December 30, 1869, Box 1, GD, BANC. Davidson to Peirce, October 1, 1870, E.22A, Box 504B, RG 23, NARA.

10. Davidson to Peirce, February 22, 1871, Letterbook 15, Carton 1, GD, BANC. Davidson to Peirce, March 27, 1871, E.22A, Box 504B, RG 23, NARA. Davidson made modest investments for himself, family, and friends. See Davidson to Belle, July 31, 1853, and Davidson to Tom, December 31, 1853, Box 59, GD, BANC.

11. Davidson to Peirce, March 27, 1871, and November 1, 1871, E.22A, Box 504B, RG 23. NARA.

12. Hyde to C. P. Huntington, June 22, 1872, and August 27, 1872, Box 2, Folder 10, WBH, STANF. Hyde to W. L. Jones, November 4, 1872, Ten Letters, California Historical Society.

13. Hyde to Huntington, November 8, 1872, Ten Letters, California Historical Society. R. M. Widney, "Los Angeles County Subsidy: Which Subsidy Shall I Vote for; or, Shall I Vote against Both?" (1872), RB 376778, HUNT.

14. Hyde to E. E. Hewitt, January 29, 1873, and Hyde to Huntington, January 5, 1873, Box 1, Folder 3, WBH, STANF.

15. Sitton, *Grand Ventures*, 111–17. White, *Railroaded*, 172–76. For a later example of the arrangements, see in Box 2, Folder 1, RBC, HUNT: Agreement, Southern Pacific . . . and Banning, December 11, 1883, and A. N. Towne to Banning, January 30, 1884.

16. In Box 2, WBH, STANF: Obituaries, Folder 6; Hyde to Bacon, August 22, 1873, Folder 11; Hyde to Huntington, September 25, 1873, Folder 12.

17. Ord to E. D. Townsend, March 21, 1869, and Nathan Porter to Ord, June 19, 1869, E.225, Box 527, RG 92, NARA. Drum Barracks Aqueduct [map], Folder 196A, SR, HUNT. *Wilmington Canal and Reservoir Co. v. Manuel Domínguez*, 50 Cal. 505 (1875).

18. Montgomery C. Meigs letters, January 2, 1872, January 9, 1872, and January 27, 1872, Meigs Family Papers, Library of Congress.

19. Meigs to Louisa, January 9, 1872, Meigs Family Papers, Library of Congress. *Alta*, January 24, 1872. The Iwakura Mission would travel on to Chicago, Washington, DC, and Britain. They failed to renegotiate the existing US treaty, signed in 1860.

20. On Alexander's visit to Hawaii, see *Alta*, January 5, 1873, April 7, 1873, and May 18, 1873. Seward continued to predict US annexation of British Columbia. *NYT*, September 21, 1869. Richard W. Meade of the Navy negotiated an 1872 Samoa harbor treaty, but the Senate refused it. Congress ratified a subsequent treaty in 1878.

21. Meigs to Louisa, January 29, 1872, Meigs Family Papers, Library of Congress. *Arizona Citizen*, June 15, 1872.

22. Meigs to Secretary of War, August 19, 1872, E.225, Box 527, RG 92, NARA.

23. Mendell to Chief of Engineers, November 30, 1875, George H. Mendell Letterbooks, vol. 1, C-B 303, BANC. Mendell to Wilson, February 5, 1872, Box 21, Benjamin D. Wilson Collection, HUNT.

Chapter 24: An Old Ranchero Defies the Times

1. John W. Shore to Brent, July 9, 1865, Box 2, Joseph Lancaster Brent Papers, HUNT.

2. *Alta*, March 7, 1869.

3. See Gillingham, *Rancho San Pedro*, 225–26, 309–10. Partitions, easement, and roads in Box 23, Folder 14 and Folder 15, SR, HUNT.

4. *Statistics of the Population of the United States at the Tenth Census*, vol. 1 (Wash-

ington, DC: GPO, 1880), 51. *Population of the United States at the Eleventh Census*, vol. 1, part 1, (Washington, DC: GPO, 1895), 11. Glenn S. Dumke, *The Boom of the Eighties in Southern California* (San Marino, CA: Huntington Library, 1944), 46–47, 277.

5. Two Domínguez sons were deceased by the 1880s.

6. H. D. Barrows, "Two Notable Pioneers," *Publications of the Historical Society of Southern California* 4 (1897–1899): 56–64. George Hansen to Davidson, June 4, 1871, Box 13, GD, BANC. In Japan, Davidson observed the transit of Venus. He continued to India in 1875, where he met with British geodesists engaged in the Great Trigonometrical Survey.

7. On the Los Angeles Canal & Reservoir Co., see Box 2, Folder 42, Matthew Keller Papers, HUNT.

8. In Box 24, Folder 7, SR, HUNT: "Survey 73"; "Domínguez Heirs . . . Report of Referees, ca. June 27, 1884," *Los Angeles Herald (LAH)*, June 25, 1885. *Los Angeles Times (LAT)*, June 15, 1885. *Los Angeles Evening Express*, June 15, 1885.

9. Vol. 8, Hansen Diary 1884, July 31, Box 19, SR, HUNT. Regarding Juan Toro: In an unrelated land case, the city paid $275 to a man with this name for gathering evidence from local Spanish records. *LAT*, October 1, 1882.

10. Frank D. Carpenter, *Geographical Surveying, Its Uses, Methods and Results* (New York: D. Van Nostrand, 1878), 83–87. *Manual of Instructions for the Survey of Public Lands* (1855), 17–18.

11. Hansen's Comments on Ord's Survey—November 19, 1869, reprinted in W. W. Robinson, "Story of Ord's Survey," *The Quarterly: Historical Society of Southern California* 19, no. 3 (1937), 128.

12. On the "Ord order" of June 1, 1877, see Alice L. Baumgartner, "The Line of Positive Safety: Borders and Boundaries in the Rio Grande Valley, 1848–1880," *Journal of American History* 101 (March 2015), 1119. Sherman attempted to dissuade President Hayes by describing Ord as highly capable, yet less wealthy than other generals of retirement age. Cresap, *Appomattox Commander*, 331.

13. Molly Ord to "Edd," December 21, [n.d.], Box 16, Folder 9, ORDF2, BOOTH. In Box 2, ORDP, BANC: William M. Evarts to Ord, February 5, 1881; Consulate General, Havana, to John Davis, July 21, 1883. Ord to C. P. Huntington, April 26, 1881, HM 40685, HUNT.

14. In Box 24, Folder 7, SR, HUNT: "Petition to Partition in Superior Court, No. 3284, Report of Referees, ca. June 27, 1884"; and "South Line of Wilmington."

15. *LAT*, May 7, 1882. In Box 2, Folder 1, RBC, HUNT: Indenture, March 20, 1879; Deed, February 25, 1880; Agreement between Southern Pacific Railroad Company and P. Banning, December 11, 1883. *McFadden v. Banning*, #4434 (1879), LAACR, HUNT.

16. In Box 2, Folder 1, RBC, HUNT: Banning to Robert Robbinson, October 17, 1879, Letterbooks, vol. 1, and Banning to Robbinson, February 18, 1880. McFadden's lawyers were the SP-allied firm of Bicknell & White. Hence, he was not the San Francisco "shark" that Banning suspected behind the plot. Following the verdict, the Bannings added to their tideland claims. McFadden did likewise at the Newport estuary, thirty miles south.

17. Banning, "Settlement of Wilmington" (1883), BANC.

18. *Alta*, November 22, 1883, and July 10, 1884. Matthew Josephson, *The Robber Barons* (New York: Harcourt, Brace, 1934), 222. *Santa Clara County v. Southern Pacific Railroad Co.*, 118 US 394 (1886).

19. *LAT*, March 10, 1885. *LAH*, June 18, 1884. Sitton, *Grand Ventures*, 129, 148. May-

mie Krythe, *Port Admiral: Phineas Banning, 1830–1885* (San Francisco: California Historical Society, 1957), 228.

CHAPTER 25: CONQUEST OVERTURNED, THEN EXPANDED

1. In Box 19, SR, HUNT: Vol. 192, Field Books, Box 11, and Hansen Diary, 1884, vol. 8.
2. Testimony of William Banning, Box 1, Folder 3, and W. H. Norway, Field Notes of Mormon Island, July 1880, Box 1, Folder 2, DEH, WRCA. *De Guyer v. Banning*, No. 4991, LAACR2, HUNT. *LAH*, August 3, 1881. P. Banning to [illeg?], March 5, 1880, Letterbooks, vol. 2, RBC, HUNT.
3. Hansen Testimony, Motion for New Trial, January 4, 1889, in *De Guyer*, No. 4991, LAACR2, HUNT. "Descriptions of Tide Land locations wholly or partly within the boundaries of the San Pedro Rancho, August 19, 1886," Box 24, Folder 7, SR, HUNT.
4. *De Guyer v. Banning*, 25 Pac. 252 (1890).
5. *United Land Association v. Knight*, 24 Pac. 818 (1890). *Knight v. United Land Association*, 142 US 161 (1891).
6. *De Guyer v. Banning*, 91 Cal. 400 (1891). *LAT*, April 21, 1891, and September 27, 1891. *LAH*, September 27, 1891.
7. *LAH*, May 5, 1892. *LAT*, June 30, 1891, and May 5, 1892. Gillingham, *Rancho San Pedro*, 277–79.
8. Deverell, *Railroad Crossing*, 93–122. Charles D. Willard, *The Free Harbor Contest at Los Angeles* (Los Angeles: Kingsley-Barnes & Neuner, 1899).
9. US Board, *Deep Water Harbor at San Pedro Bay* (Los Angeles: Times-Mirror, 1894). *Deep Water Harbor at Port Los Angeles . . . February 28, 1896* (Washington, DC: GPO, 1896). *LAT*, March 7, 1895.
10. In GD, BANC: "Memoranda in the Matter of [Disorganizing] the Coast Survey," [n.d.] and "Considerations against the Proposed Transfer of the US Coast and Geodetic Survey to the US Navy," (1883), Carton 19; "Notes, n.d., on San Pedro Harbor Commission," Carton 7.
11. In GD, BANC: Mendell to Davidson, May 10, 1895, and June 26, 1895, Box 16; Obituaries, Memorial Resolutions, Carton 19. Mendell and Davidson knew each other at least since 1872, when they served (with B. S. Alexander) on a federal commission to study irrigation for California's Central Valley. Mendell directed the San Pedro jetty project until 1886.
12. Charles T. Healey to Davidson, February 6, 1888, Box 13, GD, BANC. *Report of the Board of Railroad Commissioners, 1893–1894* (Sacramento: State Office, 1894), 37. In Box 48, Folder 778, Meyer Lissner Papers, STANF: Joseph H. Call, "Waterways and Railways," *Los Angeles Financier*, September 5, 1908, and "Business at San Pedro," *Los Angeles Financier*, September 5, 1908.
13. At HUNT: H. H. Markham to J. D. Bicknell, May 6, 1886, Box 6, and J. E. Foulds to S. M. White, March 28, 1888, Box 7, Papers of John Dustin Bicknell. Bicknell, Tidelands Application, March 1888, Box 2, Folder 2, RBC. *LAT*, February 22, 1901. Oscar T. Shuck, *History of the Bench and Bar of California* (Los Angeles: Commercial Printing House, 1901), 642–46, 1137. Deverell, *Railroad Crossing*, 44, 49.
14. US Board of Engineers, *Deep-Water Harbor at Port Los Angeles or at San Pedro* (Washington, DC: GPO, 1897).
15. *White, De Guyer v. Banning . . . Brief of Defendants* (Washington, DC: McGill & Wallace, [1895?]). White, De Guyer v. Banning, Supplemental Brief (Chas. W. Palm, 1895), 24. Patton Sr. (1856–1927), father of the future World War II general, married

Benjamin D. Wilson's daughter Ruth. Patton's half-sister, Anne O. Smith, married Phineas Banning's son Hancock.

16. In Box 14, Papers of John Dustin Bicknell, HUNT: White to Bicknell, December 5, 1895, Folder 2; White to Bicknell, March 24, 1896, Folder 5. White did speak about the issue in Congress. White, "Deep Sea Harbor . . . Speech of Hon. Stephen M. White, May 8, 9, and 12, 1896" ([n.p.], 1896), SEAV. *LAT*, May 26, 1897.

17. Brief reprinted in Holmes Conrad, "Title to a Harbor," *LAT*, March 30, 1897.

18. *De Guyer v. Banning*, 167 US 723 (1897). *LAT*, March 16, 1897, and May 26, 1897. The court cited Henry Hancock's 1857 map as the definitive "final survey," unaware that lines of the Wilmington waterfront were Ord's measures of 1855. The Court wrongly assumed the work had been done by Hancock, a legitimate US surveyor, who had died in 1883. "Field notes of Final Survey by Hancock, December 11, 1857, Box 2, Folder 54, DEH, WRCA.

19. Container 52, Reel 34, Papers of John Marshall Harlan, Library of Congress. Robert Harlan, James's mixed-race ward (son, too, perhaps) and former slave, lived briefly in Gold Rush California before returning east. Peter S. Canellos, *The Great Dissenter* (New York: Simon & Schuster, 2021), 67–68, 371–92. On Harlan's legacy, see also Eric Foner, *The Second Founding: How the Civil War and Reconstruction Remade the Constitution* (New York: W. W. Norton, 2019), 152–67.

20. *LAT*, May 26, 1897. George H. Smith to James A. Gibson, September 14, 1909, Box 3, Folder 1, RBC, HUNT.

21. *LAH*, November 11, 1897; *LAT*, November 13, 1897.

Still Teeming with Life

1. Saltwater wetlands rank only behind freshwater wetlands and tropical rainforests in the primary production of organic compounds.

Conquests by Another Name

1. Stanley Karnow, *In Our Image: America's Empire in the Philippines* (New York: Random House, 1989), 78–80, 87–100. David J. Silbey, *A War of Frontier and Empire: The Philippine-American War, 1899–1902* (New York: Hill & Wang, 2007), 35–40. Walter T. K. Nugent, *Habits of Empire: A History of American Expansion* (New York: Alfred A. Knopf, 2008), 266–72. Hahn, *A Nation without Borders*, 491–97. *LAH*, April 26, 1898, and May 4, 1898.

2. Silbey, *A War of Frontier and Empire*, 31–34. George C. Herring, *From Colony to Superpower: US Foreign Relations since 1776* (New York: Oxford University Press, 2008), 309–18, 320–22.

3. Herring, *From Colony to Superpower*, 321, 326–29. Silbey, *A War of Frontier and Empire*, 60–66.

4. Hiltzik, *Iron Empires*, 131–37, 141–45.

5. Hiltzik, *Iron Empires*, 150–54, 190–203.

6. White, *Railroaded*, 404–9. Hiltzik, *Iron Empires*, 203–6. Josephson, *The Robber Barons*, 222–27.

7. *LAH*, July 16, 1898, and July 24, 1898. McKinley replaced Alger with Elihu Root in August 1899.

8. *LAH*, March 23, 1898, March 26, 1898, April 26, 1898, and May 3, 1898. See also Harrison Gray Otis Philippine Islands Album, 1899, P-080, SEAV.

9. White to John T. Gaffey, June 24, 1898, Box 92, Papers of Abel Stearns, HUNT.

LAH, July 7, 1898. *San Francisco Examiner*, January 26, 1899, in Stephen Mallory White Scrapbooks, Box 1, SEAV. Robert F. Troy, "Stephen Mallory White," *Journal of the American-Irish Historical Society* 10 (New York: American-Irish Historical Society, 1911), 193.

CHAPTER 26: ARRIVALS OF THE SEASON

1. *LAT*, May 7, 1904, and May 4, 1904. Notes in "Rancho San Pedro . . . Partition 1884/85: Estuary" list several squatters by name. Box 24, Folder 7, SR, HUNT. Charles T. Healey, Map, June 1887, ZNE Binder 7-402-420, California State Lands Commission.

2. Daniel Carpenter, *The Forging of Bureaucratic Autonomy: Reputations, Networks, and Policy Innovation in Executive Agencies, 1862–1928* (Princeton: Princeton University Press, 2001); William J. Novak, *The People's Welfare: Law and Regulation in Nineteenth-Century America* (Chapel Hill: University of North Carolina Press, 1996). Stephen Skowronek, *Building a New American State: The Expansion of National Administrative Capacities, 1877–1920* (New York: Cambridge University Press, 1982).

3. *LAH*, November 17, 1898. *Report of the Chief of Engineers . . . 1898* (Washington, DC: GPO, 1898), Appendix RR2 and RR3. *Report of the Chief of Engineers . . . 1899* (Washington, DC: GPO, 1899), 549, 553, 975, Appendix QQ2 and QQ3.

4. H. C. Corbin to J. J. Meyler, November 10, 1898, Box 1 (ov), Folder 2, James J. Meyler and Robert G. Meyler Papers, Collection No. 1015, SEAV. *National Cyclopaedia of American Biography*, vol. 35 (New York: James T. White, 1949), 260–61.

5. H. St. L. Coppee, "Evolution of Dredging Machinery" *Engineering News* 35 (April 30, 1896), 291–92. Charles Prelini, *Dredges and Dredging* (New York: Van Nostrand, 1911).

6. Frank E. Leonard, "Pioneers of Centrifugal Pump Dredging." *Engineering News* 37 (June 17, 1897), 378–80. J. J. Peatfield, "Dredging on the Pacific Coast," *Overland Monthly* 24 (September 1894): 323–27. *Bowers v. Von Schmidt* 63 Fed. 572 (1894). *Von Schmidt v. Bowers* 80 Fed. 121 (1897). Mendell to Chief of Engineers, June 7, 1884, E.37, Box 122, RG 77, NARA.

7. Hydraulic dredges could work between eighteen and twenty hours per day. W. R. Bassick to C. T. Leeds, March 21, 1910, W-7, Box 120, RG 77, NARA. Allen Boyer McDaniel, *Excavating Machinery* (New York: McGraw Hill, 1913).

8. *Report of the Chief of Engineers . . . 1898*, Appendix XX and 1F; Plate No. 5. H. St. L. Coppee, "Improvement of the Mississippi River by Dredging," *Engineering Magazine* 15 (April 1898), 463–66. "Hydraulic Suction Dredge for the Navigation Improvements of the Mississippi River," *Engineering News and American Railway Journal* 35 (April 23, 1896), 277–79. "Mining Debris in California Rivers" H. Exec. Doc. No. 98, 47th Cong., 1st Sess. (1882).

9. Meyler, "Survey of Inner Harbor, January 6, 1900," *Report of the Chief of Engineers . . . 1900* (Washington, DC: GPO, 1900), Appendix SS5, 4200–4205. Meyler, "Map of San Pedro Harbor," Collection No. 1015, SEAV. Brysson Cunningham, *A Treatise on the Principles and Practice of Harbour Engineering* (Philadelphia: J. B. Lippincott, 1908), 222–38.

10. Meyler, "Survey of Inner Harbor, January 6, 1900," 4202.

11. Obituaries: *LAH*, December 13, 1901. *NYT*, December 13, 1901. Appointed to the canal project in July 1907, Jadwin served under General George W. Goethals. In W-6, Box 120, RG 77, NARA: Fries to Jadwin, February 24, 1908, and Jadwin to

Fries, February 5, 1908. *Annual Report of the Isthmian Canal Commission* (Washington, DC: GPO, 1908), 47.

12. In W-7, Box 120, RG 77, NARA: C. E. Ellicott to J. H. Willard, September 16, 1903, and Llewellyn Iron Works to Willard, September 16, 1903. *Report of the Chief of Engineers . . . 1903* (Washington, DC: GPO, 1903), Appendix UU and 2175–77. *LAH*, February 28, 1905. *LAT*, September 2, 1906. EKB, "Terminal Island, April 1941," Box 2, Folder 54, DEH, WRCA. "Our West Door: Greatest Nation, Largest Ocean, Only Orient," *Los Angeles Examiner*, December 16, 1906, Box 48, Folder 778, Lissner Papers, STANF. The US government spent $100,000 (several millions in 2020 dollars) to build the *San Pedro*.

13. In W-6, Box 120, RG 77, NARA: Jadwin to A. A. Fries and John C. Oakes, February 5, 1908; Leeds to Ellicott Machine, May 7, 1910. *LAH*, July 15, 1908; *LAT*, December 25, 1911, and April 7, 1912.

14. *NYT*, August 15, 1900. George Kennan, *E. H. Harriman: A Biography*, vol. 1 (Boston: Houghton Mifflin, 1922), 233. White, *Railroaded*, 404–9. Hiltzik, *Iron Empires*, 245–49.

15. On the jiujitsu performers, see *NYT*, February 4, 1906.

16. Kennan, *Harriman*, vol. 1, 241. Kennan, *Harriman*, vol. 2, 2.

CHAPTER 27: LEVIATHANS DIG IN FOR A FIGHT

1. *LAH*, February 28, 1905.

2. Davidson, *Coast Pilot of California, Oregon, and Washington* (Washington, DC: GPO, 1889), 40–41. In E.171, Box 6, RG 77, NARA: Mendell to Chief of Engineers, July 1, 1890, and "Mendell asks . . . ," March 19, 1890. Hughes, "Notes as to Location and Width of Channel, July 2, 1924," Box 1, Folder 22, DEH, WRCA.

3. Hiltzik, *Iron Empires*, 246–55, 336–38. Josephson, *Robber Barons*, 450.

4. C. T. Healey to Davidson, February 6, 1888, Box 13, GD, BANC. Kennan, *Harriman*, vol. 1, 258–60.

5. "On Navigability of San Pedro Inner Harbor, June 11, 1908," Box 1, Folder 15, DEH, WRCA. Southern Pacific Co., et al., to W. H. Taft, February 9, 1907, Box 36, TEG, HUNT. In Box 2, Folder 3, RBC, HUNT: George J. Smart to US Harbor Board, May 29, 1903; W. Banning to Chief of Engineers, August 18, 1905.

6. R. H. Ingram to L. C. Easton, August 24, 1905, W-9, Box 120, RG 77, NARA. Ingram announced the arrival of Julius Kruttschnitt, Harriman's chief of operations and special assistant, and Edgar E. Calvin, SP vice president and general manager.

7. D. E. Hughes, "The Sickle or Perfect Railway Curve," *Transactions of the Technical Society of the Pacific Coast* 9, no. 3 (1892), 63–87.

8. Diagram, from interview between Capt. C. H. McKinstry and H. Banning, June 7, 1905, Box 3, Folder 1, RBC, HUNT. "On Navigability . . . Inner Harbor, June 11, 1908," Box 1, Folder 15, DEH, WRCA. In RG 77, NARA: Fries to W. H. Heuer, October 12, 1906, W-6, Box 120, and Hughes to Leeds, January 11, 1919, W-1a, Box 119.

9. *LAT*, December 12, 1906, and December 29, 1906. *Los Angeles Examiner*, December 16, 1906, Box 48, Folder 778, Lissner Papers, STANF.

10. *LAH*, July 8, 1905, May 27, 1906, and July 30, 1905. Henry E. Huntington co-owned the Pacific Electric interurban railway, which built a branch to Naples. He remained an ally of Harriman's SP.

11. *LAH*, October 29, 1905, September 7, 1907, December 12, 1905, and May 19, 1906. *LAT*, June 5, 1908. Long Beach later acquired Cerritos Slough for its municipal port. *Bolsa Land Co. v. Burdick*, 15 Cal. 254 (1907).

12. Hughes to "Dear Friend," March 7, 1938, Box 2, Folder 54, DEH, WRCA.

CHAPTER *28:* THE PEOPLE'S PORT

1. Josephson, *Robber Barons*, 277–80, 418–31, 445–50. *US v. E. C. Knight Co.*, 156 US 1 (1895).
2. Hiltzik, *Iron Empires*, 339–41. On Harlan's opinion in *Northern Securities*, see Canellos, *The Great Dissenter.*
3. Hancock Banning to Banning Co., November 14, 1907, Box 2, Folder 4, RBC, HUNT.
4. "A Story of Growth" (Los Angeles: Security Trust & Savings Bank, 1913), General Collection #1299, Box 54, "Harbors," SEAV. City of Los Angeles, Annexation Map, www.navigatela.lacity.org. Paul Soifer, "Water and Power for Los Angeles," in *Development of Los Angeles City Government: An Institutional History, 1850-2000*, vol. 1, ed. Hynda L. Rudd (Los Angeles: Los Angeles City Historical Society, 2007), 219–20. *LAH*, February 27, 1906, and March 2, 1906.
5. *LAH*, December 24, 1907.
6. "Major General Amos A. Fries" (biographical manuscript), and "Book of Services," Box 1, Folder 2, Amos A. Fries Papers, University of Oregon Special Collections.
7. In Fries Papers, University of Oregon: *Washington (DC) Herald*, September [?], 1937, and *Evening Star* (Washington, DC), November 7, 1935, Box 1, Folder 2; in Box 2, Folder 15: Fries to Herbert Hoover, March 16, 1931; Elizabeth C. Fries to [?], [n.d.]; A. A. Fries to Marion T. Hughes, April 14, 1950. See Fries, *Sugar Coating Communism for Protestant Churches: Chart Showing Interlocking Membership of Churchmen, Socialists, Pacifists, Internationalists, and Communists* (Washington, DC: [n.p.], [1932?]), Box 20, Folder 8, Radical Pamphlet Collection, Library of Congress.
8. "Attorneys for Applicants . . . In the Matter of the Application of Ruth W. Patton . . . and Hancock Banning for leave to wharf out beyond pier-head line and to dredge a channel in Wilmington or Inner San Pedro Harbor," Box 3, Folder 1, RBC, HUNT. Lawrence M. Friedman, *A History of American Law* (New York: Simon & Schuster, 1973), 207–9.
9. In Box 2, Folder 4, RBC, HUNT: James A. Gibson to Hancock Banning, January 6, 1907; Fries to William Banning, June 27, 1907; R. H. Ingram to Banning Co., et al., January 3, 1907; Resolutions No. 98, December 20, 1906, and No. 126, July 5, 1907, Wilmington Municipal Records, Los Angeles City Archives (LACA).
10. James W. Ingram III, "Charters: A History of the City's Constitution," in *Development of Los Angeles City Government*, 7–11. In Box 2, Folder 4, RBC, HUNT: Banning Co. to W. D. Stephens, January 2, 1907, and W. J. Washburn to Fries, January 23, 1907.
11. Petition to W. H. Taft, February 9, 1907, and Petition, Chas. Nelson Co., et al., to Taft, January 24, 1907, Box 36, TEG, HUNT. In Box 2, Folder 4, RBC, HUNT: Gibbon to Southern Pacific, San Pedro, Los Angeles & Salt Lake Railroad, and Banning Co., March 27, 1907; and Gibbon to Banning Co., March 29, 1907.
12. Deverell, *Railroad Crossing*, 154–66.
13. Ron Chernow, *The House of Morgan: An American Banking Dynasty and the Rise of Modern Finance* (New York: Grove Press, 2001), 21–22, 121–30.
14. In RBC, HUNT: A. Mackenzie to Secretary of War, April 15, 1908, Box 2, Folder 5; Banning Co. to Taft, December 9, 1907, and W. Banning to Taft, January 2, 1908, Box 2, Folder 4. *Los Angeles Express*, June 15, 1908, Box 48, Folder 778, Lissner Papers, STANF. War Secretary Taft pledged to overrule the chief of engineers, but resigned before doing so. His successor, Luke E. Wright, later withdrew the offer of quit-claim deeds. *LAH*, June 2, 1908, and August 6, 1908. *LAT*, June 2, 1908.
15. "On Navigability of . . . Inner Harbor, June 11, 1908," Box 1, Folder 15, DEH, WRCA. Gibbon to Southern Pacific Co.; San Pedro, Los Angeles & Salt Lake Rail-

road Co.; and Banning Co., March 27, 1907; Gibbon to Banning Co., March 29, 1907, Box 2, Folder 4, RBC, HUNT. Petition, Chas. Nelson Co., et al., to W. H. Taft, January 24, 1907, Box 36, TEG, HUNT.

16. Fries to Otwell, January 5, 1908, W-10, Box 120, RG 77, NARA. Amos A. Fries, "San Pedro Harbor," *Out West* (October 1907), 314, 331–32. Fries, "A Free Harbor Essential," *Los Angeles Financier*, September 5, 1908, Box 48, Folder 778, Lissner Papers, STANF. The Engineers were not alone. The Chamber of Commerce announced it would oppose the Peninsula Plan. W. J. Washburn to Fries, January 23, 1907, Box 2, Folder 4, RBC, HUNT.

17. Hughes to Fries, July 8, 1909, and "Respecting the instructions that quit claim deeds," May 24, 1908, W-10, Box 120, RG 77, NARA. History of the Inner Bay of San Pedro, 1908, Box 1, Folder 14, DEH, WRCA.

18. Fries to Heuer, October 27, 1908, W-9, Box 120, RG 77, NARA. Plan of Development of San Pedro Harbor, February 1907, Box 2, Folder 54, DEH, WRCA.

19. "The Determination of the Plane of Ordinary High Tide for Pacific Coast Harbors . . . A Discussion by D. E. Hughes and Otto Von Geldern," *Journal of the Association of Engineering Societies*, 44, no. 4 (1910), in W-10, Box 120, RG 77, NARA.

20. In Box 1, DEH, WRCA: "Random Notes on Lands and Tides, July 15, 1927," Folder 1; "On Navigability of . . . Inner Harbor, June 11, 1908," Folder 15; Memorandum: Harbor Line Surveys, November 30, 1915, Folder 17. In same, Box 2, Folder 54: Hughes to Fries, January 19, 1908; Hughes to Fries, May 18, 1908; Hughes to [J. A.] Anderson, May 14, 1920. Hughes argued that Ord's lines were a valid approximation of acreage, though never intended as an official property boundary. *LAT*, June 2, 1908.

21. US Harbor Lines, July 29, 1908, Box 1, Folder 17, DEH, WRCA. In Box 48, Folder 778, Lissner Papers, STANF: "Proposed Lines for Inner Harbor," *Los Angeles Express*, February 21, 1907, and "Plans of Corporations Completely Upset," *Los Angeles Express*, February 21, 1907. In P-9, Box 107, RG 77, NARA: "Description of Proposed US Harbor Lines"; and Chief of Engineers to W. H. Taft, December 16, 1908.

CHAPTER 29: TIDELANDS IMPERIALISM

1. Mendell to Wilson, February 5, 1872, Box 21, Benjamin D. Wilson Collection, HUNT. In E.171, Box 6, RG 77, NARA: Mendell to Chief of Engineers, July 1, 1890, and "G. H. Mendell asks . . . ," March 19, 1890. Mendell requested assistance from the state legislature and tideland commissioners. Hughes, "Notes as to Location and Width of Channel, July 2, 1924," Box 1, Folder 22, DEH, WRCA. Geraldine Knatz, *Port of Los Angeles: Conflict, Commerce, and the Fight for Control* (Santa Monica: Angel City Press, 2019), 16–19.

2. In Box 2, RBC, HUNT: [Unsigned] to E. E. Calvin, December 10, 1907, Folder 4; W. Banning to Banning Co. and R. H. Ingram, May 7, 1908, Folder 5.

3. *LAH*, January 26, 1908. Theodore Roosevelt, Eighth Annual Message, December 9, 1908, millercenter.org. Hiltzik, *Iron Empires*, 345–50.

4. In Box 2, Folder 5, RBC, HUNT: W. Banning to Banning Co., May 14, 1908, and Memorandum of Agreement. The SP paid $250,000.

5. *LAH*, August 17, 1907. *People ex rel. McConnell v. City of Wilmington*, 151 Cal. 649 (1907), LA No. 2000.

6. In Box 2, Folder 5, RBC, HUNT: W. Banning to R. H. Ingram, June 3, 1908, and various coded messages.

7. J. A. Anderson to W. B. Mathews, June 5, 1908, Box 1, Folder 15, DEH, WRCA. *LAH*, June 4, 1908, and June 9, 1908.
8. Harbor Commission minutes, in Clarence Matson to T. E. Gibbon, June 22, 1912, Box 36, TEG, HUNT. City Council Minutes, vols. 75 and 76, LACA.
9. Charles D. Willard, "The Inner Harbor at San Pedro," *Land of Sunshine* 14 (1901), 76. *LAT*, March 7, 1895, and November 11, 1906. *LAH*, August 19, 1906, January 12, 1907, April 3, 1909, and April 4, 1909. "Report of the Consolidation Committee, 1909," Box 36, TEG, HUNT.
10. *LAH*, July 25, 1909, August 26, 1908, November 7, 1908, November 15, 1908, and April 21, 1909.
11. Fries to Hughes, August 21, 1911, Box 2, Folder 15, Fries Papers, University of Oregon. City Council Minutes, June 22, 1909, vol. 78, LACA. Fries became internationally known as the founder of the US chemical warfare program. Later, he would advocate for the nonmilitary use of tear gas, against labor strikes and civil unrest especially.
12. Kennan, *Harriman*, vol. 2, 326. *LAH*, March 19, 1909, and March 21, 1909. City Council Minutes, August 24, 1909, vol. 78, LACA.
13. Gibbon, "Los Angeles Harbor and Its Importance to the Realty Board, October 19, 1911," Box 36, TEG, HUNT. Gibbon, "Important European and Oriental Commerce," *LAH*, August 1, 1909.
14. *LAH*, June 21, 1908.
15. "Copy. Letter . . . to the Directors of the Merchants' Association, as Published in the San Francisco *Call*, January 17, 1904," Box 3, Folder 30, Reuben Brooks Hale Papers, ms 912, California Historical Society. *LAH*, June 21, 1908. The *Lusitania* never crossed the canal. A German U-boat sank it on May 7, 1915, killing 1,200 persons.
16. *LAH*, April 25, 1909.
17. *LAH*, October 9, 1909. President Taft's sister Frances (and her husband, Dr. W. A. Edwards) resided near Los Angeles.
18. *LAH*, March 15, 1908, and April 19, 1908.
19. William Harding Carter, *The Life of Lieutenant General Chaffee* (Chicago: University of Chicago Press, 1917). Hutton, *Apache Wars*, 307–10.
20. *LAH*, October 12, 1909, March 15, 1908, and October 18, 1910.
21. *LAH*, May 28, 1912.

CHAPTER 30: A CITY RETURNS TO CIVIL WAR
1. *LAH*, August 26, 1911. Grand Army of the Republic and Kindred Societies, National Encampments, Library of Congress, www.loc.gov.
2. *LAT*, October 21, 1906, May 26, 1909, and June 13, 1909.
3. *LAH*, October 6, 1909, and *LAT*, August 30, 1911. In RBC, HUNT: W. Banning to E. E. Calvin, June 12, 1909, Box 5, Folder 6; various items, Box 3, Folder 1; and H. Banning to Heuer, September 15, 1909, Box 2, vol. 1, (Addendum, 1985). California's supreme court later grouped the complaints into three appellate cases.
4. *LAT*, October 6, 1909, and October 21, 1909.
5. *LAH*, October 3, 1908, January 10, 1909, January 31, 1909, and March 5, 1909.
6. J. J. McNamara was a leader of the ironworkers' union. LA sought an advantage over San Francisco, where industries were unionized. *San Francisco Call*, April 24, 1911. Library of Congress, Chronicling America, *Los Angeles Times* Bombing, loc.gov.
7. "Los Angeles Disaster Reviewed by Gene Debs," *Labor Journal* (Everett, WA), October 21, 1910.
8. *LAT*, March 29, 1911. *LAH*, December 3, 1910.

9. "Ownership of Tide Land" and "Opinion of the Court, 1911; *People of the State of California v. Southern Pacific, et al.*, No. 64,535," Box 5, Folder 4, Walter Bordwell Papers, STANF.

10. *Christian Science Monitor*, March 4, 1911. H. Banning to Frank Karr and A. M. Jamison, January 27, 1910. Box 5. Folder 6, RBC, HUNT.

11. *LAH*, October 25, 1911, and October 22, 1913. *Los Angeles Record*, March 13, 1912. *People v. Southern Pacific Railroad Co., et al.*, 166 Cal. 614 (1913). *Banning Co. v. California*, 240 US 142 (1916).

12. *LAH*, September 16, 1911, and June 24, 1912. *LAT*, September 16, 1911, and October 21, 1911. Gibbon to C. H. McKinstry, April 3, 1912, Box 36, TEG, HUNT. Resolution, August 9, 1910, vol. 82, City Council Minutes, LACA.

13. *LAH*, October 17, 1911.

14. *LAH*, October 25, 1911. "Plain Facts About the Aqueduct, Power Development, and the Harbor," *Los Angeles Record*, March 13, 1912, Box 48, Folder 778, Lissner Papers, STANF.

15. *LAH*, January 14, 1913. *LAT*, January 22, 1913. "Red tape" included four commissions (Public Works, Harbor, Art, and the Harbor Advisory Board); the city council and its harbor committee; the city engineer's office and its harbor engineer; the mayor; a Harbor Bureau within Public Works; and the Harbor Commission's hired consulting engineer. Horace B. Ferris to Harbor Commissioners, January 5, 1912, Box 36, TEG, HUNT.

16. Chaffee's contributions to LA harbor, for better or worse, are absent from his one published biography, Carter's *Life of . . . Chaffee*, 280–87.

17. *LAT*, January 7, 1912, and January 16, 1912. Knatz, *Port of Los Angeles*, 98–103.

18. In Box 36, TEG, HUNT: [unsigned] to E. P. Goodrich, June 14, 1912; "Information whose source must be treated as confidential"; "Gibbon Writes Open Letter to Mayor," *Los Angeles Record,* June 27, 1912. *LAH*, June 24, 1912, and June 26, 1912. *LAT*, July 6, 1912, and July 24, 1912.

19. Ingram, "Charters: A History of the City's Constitution," in *Development of Los Angeles City Government*, 13–16.

20. *LAH*, May 2, 1913.

21. *LAH*, September 7, 1912, and September 9, 1912.

22. *LAT*, October 20, 1912.

PEOPLE OF THE EARTH

1. I am grateful to Chief Red Blood Anthony Morales, Tribal Chairman of the Gabrieleño San Gabriel Band of Mission Indians, for his generous assistance. I also draw upon the work of tribal educators Julia Bogany, Kimberly Morales-Johnson, Cindi Alvitre, and Craig Torres.

2. On collecting of and trade in indigenous relics, see *LAH*, April 17, 1891, November 15, 1893, and June 30, 1894.

3. Vol. 3.5, Reel 105, Images 575–578, John P. Harrington Papers, National Museum of Natural History, Smithsonian Institution. Thank you to Professor Justin D. Spence of the University of California, Davis, for locating the original transcript of Juncos's story.

CONCLUSION

1. At San Pedro Historical Society: City of Los Angeles, *Annual Report: Board of Harbor Commissioners* (n.p., 1924), 88.22.01, and "Los Angeles: The Great Seaport of

the Southwest" (n.p., 1924), 92.48 Pe 155. US Corps of Engineers, *The Ports of Los Angeles and Long Beach* (Washington, DC: GPO, 1931), 8–9.

2. Knatz, *Port of Los Angeles*, 189–92, 195–201, 257–58. See also, Richard Webster Barsness, "The Maritime Development of San Pedro Bay, California" (PhD diss., University of Minnesota, 1963).

Selected Bibliography

MANUSCRIPT COLLECTIONS

The Bancroft Library, University of California, Berkeley

Hubert Howe Bancroft, Miscellaneous California Dictations
Hubert Howe Bancroft: Records of the Library and Publishing Companies
George Davidson Correspondence and Papers
Frémont Family Papers
John Tracy Gaffey Correspondence and Papers
William McKendree Gwin Papers
George H. Mendell Letterbooks
Edward Otho Cresap Ord Correspondence and Papers
Edward Otho Cresap Ord Papers
United States District Court (California: Southern District) Land Case Files
United States General Land Office, Reports by California Deputy Surveyors
Von Schmidt Family Papers
R. S. Williamson Correspondence
Benjamin Davis Wilson Papers

Booth Family Center for Special Collections, Georgetown University

Ord Family Papers

Gerth Archives and Special Collections, California State University, Dominguez Hills

Robert C. Gillingham Working Papers
Rancho San Pedro Collection

The Huntington Library

Records of the Banning Company
Collection Related to Edward Fitzgerald Beale
Papers of John Dustin Bicknell
Joseph Lancaster Brent Papers
Thomas E. Gibbon Papers
Henry Edwards Huntington Collection

Matthew Keller Papers
Los Angeles Area Court Records
James Mandeville Papers
Correspondence of George H. Mendell
William Jones Rhees Papers
Solano-Reeve Papers.
Papers of Abel Stearns
Benjamin Davis Wilson Collection

Library of Congress

Grand Army of the Republic & Kindred Societies, National Encampments
Los Angeles Times Bombing, Chronicling America
Papers of John Marshall Harlan
Meigs Family Papers

Seaver Center for Western History Research, Los Angeles County Museum of Natural History

Ephemera Collection
General Photograph File
James J. Meyler and Robert G. Meyler Papers
Harrison Gray Otis Philippine Islands Album
Stephen Mallory White Scrapbooks

National Archives & Records Administration

RG 23—Records of the Coast and Geodetic Survey
RG 23—Correspondence of Alexander Dallas Bache
RG 48—Records of the Secretary of the Interior—Office of Exploration and Surveys
RG 60—General Records of the Department of Justice—Office of the Attorney General—California Land Claims
RG 77—Records of the Office of the Chief of Engineers—United States Army
RG 92—Records of the Office of the Quartermaster General—United States Army
RG 107—Records of the Office of the Secretary of War
RG 393—Records of United States Army Continental Commands—Division and Department of the Pacific

Department of Special Collections, Stanford University

Walter Bordwell Papers
William Birelie Hyde Papers
Meyer Lissner Papers
Edward Otho Cresap Ord Letters
Stephen Mallory White Papers

Miscellaneous

Amos A. Fries Papers, Special Collections and University Archives, University of Oregon
John Peabody Harrington Papers, National Museum of Natural History, Smithsonian Institution
William Birelie Hyde, Ten Letters, California Historical Society

Wyman Field Notes, Department of Ornithology and Mammalogy, Los Angeles County Museum of Natural History

Joseph Lane Correspondence Collection, Oregon Historical Society

Rancho San Pedro: Documents, UCLA Library Special Collections, University of California, Los Angeles

David E. Hughes Papers, Special Collections and University Archives, University of California, Riverside

ADDITIONAL SELECTED SOURCES

Banner, Stuart. *American Property: A History of How, Why, and What We Own.* Cambridge: Harvard University Press, 2011.

Barr, Juliana. *Peace Came in the Form of a Woman: Indians and Spaniards in the Texas Borderlands.* Chapel Hill: University of North Carolina Press, 2007.

Barrow, Mark V., Jr. *Nature's Ghosts: Confronting Extinction from the Age of Jefferson to the Age of Ecology.* Chicago: University of Chicago Press, 2009.

Batzer, Darold P., and Rebecca R. Sharitz. *Ecology of Freshwater and Estuarine Wetlands.* Berkeley: University of California Press, 2006.

Baumgartner, Alice L. *South to Freedom: Runaway Slaves to Mexico and the Road to Civil War.* New York: Basic Books, 2020.

Beckert, Sven. *The Monied Metropolis: New York City and the Consolidation of the American Bourgeoisie, 1850–1896.* New York: Cambridge University Press, 2001.

Bensel, Richard Franklin. *Yankee Leviathan: The Origins of Central State Authority in America, 1859–1877.* New York: Cambridge University Press, 1990.

Blight, David W. *Race and Reunion: The Civil War in American Memory.* Cambridge: Belknap Press of Harvard University Press, 2001.

Brands, H. W. *The Last Campaign: Sherman, Geronimo and the War for America.* New York: Doubleday, 2022.

Brooks, James F. *Captives and Cousins: Slavery, Kinship, and Community in the Southwest Borderlands.* Chapel Hill: University of North Carolina Press, 2002.

Capozzola, Christopher. *Bound by War: How the United States and the Philippines Built America's First Pacific Century.* New York: Basic Books, 2020.

Carpenter, Daniel. *The Forging of Bureaucratic Autonomy: Reputations, Networks, and Policy Innovation in Executive Agencies, 1862–1928.* Princeton: Princeton University Press, 2001.

Casas, María Raquél. *Married to a Daughter of the Land: Spanish-Mexican Women and Interethnic Marriage in California, 1820–1880.* Reno: University of Nevada Press, 2007.

Chaffin, Tom. *Pathfinder: John Charles Frémont and the Course of American Empire.* New York: Hill & Wang, 2002.

Chernow, Ron. *Grant.* New York: Penguin Press, 2017.

Cooper, William J., Jr. *Jefferson Davis, American.* New York: Vintage, 2000.

Cozzens, Peter. *The Earth Is Weeping: The Epic Story of the Indian Wars for the American West.* New York: Alfred A. Knopf, 2016.

Cronon, William. *Nature's Metropolis: Chicago and the Great West.* New York: W. W. Norton, 1992.

Cumings, Bruce. *Dominion from Sea to Sea: Pacific Ascendancy and American Power.* New Haven: Yale University Press, 2009.

Davis, Jack E. *The Gulf: The Making of an American Sea.* New York: Liveright, 2017.

Delay, Brian. *War of a Thousand Deserts: Indian Raids and the U.S.-Mexican War.* New Haven: Yale University Press, 2008.

Deverell, William F. *Whitewashed Adobe: The Rise of Los Angeles and the Remaking of Its Mexican Past*. Berkeley: University of California Press, 2004.

————. *Railroad Crossing: Californians and the Railroad, 1850–1910*. Berkeley: University of California Press, 1996.

De Wolk, Roland. *American Disruptor: The Scandalous Life of Leland Stanford*. Berkeley: University of California Press, 2019.

Downs, Gregory P. *After Appomattox: Military Occupation and the Ends of War*. Cambridge, MA: Harvard University Press, 2015.

Faragher, John Mack. *Eternity Street: Violence and Justice in Frontier Los Angeles*. New York: W. W. Norton, 2016.

Foner, Eric. *The Second Founding: How the Civil War and Reconstruction Remade the Constitution*. New York: W. W. Norton, 2019.

————. *Reconstruction: America's Unfinished Revolution, 1863–1877*. New York: Harper & Row, 1988.

Frazier, Donald S. *Blood and Treasure: Confederate Empire in the Southwest*. College Station: Texas A&M University Press, 1995.

Garcia, Matt. *A World of Its Own: Race, Labor, and Citrus in the Making of Greater Los Angeles, 1900–1970*. Chapel Hill: University of North Carolina Press, 2001.

Go, Julian, and Anne L. Foster, eds. *The American Colonial State in the Philippines: Global Perspectives*. Durham, NC: Duke University Press, 2003.

Greenberg, Amy S. *A Wicked War: Polk, Clay, Lincoln and the 1846 U.S. Invasion of Mexico*. New York: Alfred A. Knopf, 2012.

Groom, Winston. *Kearny's March: The Epic Creation of the American West, 1846–1847*. New York: Alfred A. Knopf, 2011.

Guelzo, Allen C. *Fateful Lightning: A New History of the Civil War and Reconstruction*. New York: Oxford University Press, 2012.

Hahn, Steven. *A Nation without Borders: The United States and Its World in the Age of Civil Wars*. New York: Viking, 2016.

Hämäläinen, Pekka. *Indigenous Continent: The Epic Contest for North America*. New York: Liveright, 2022.

Henderson, Timothy J. *A Glorious Defeat: Mexico and Its War with the United States*. New York: Hill & Wang, 2007.

Hiltzik, Michael. *Iron Empires: Robber Barons, Railroads, and the Making of Modern America*. New York: Mariner Books, 2021.

Hutton, Paul Andrew. The *Apache Wars: The Hunt for Geronimo, the Apache Kid, and the Captive Boy Who Started the Longest War in American History*. New York: Crown, 2016.

Hurtado, Albert. *Indian Survival on the California Frontier*. New Haven: Yale University Press, 1988.

Hyde, Anne F. *Empires, Nations, and Families: A New History of the North American West, 1800–1860*. Lincoln: University of Nebraska Press, 2011.

Hyslop, Stephen G. *Contest for California: From Spanish Colonization to the American Conquest*. Norman, OK: Arthur H. Clark, 2012.

Igler, David. *The Great Ocean: Pacific Worlds from Captain Cook to the Gold Rush*. New York: Oxford University Press, 2013.

Isenberg, Andrew C. *Mining California: An Environmental History*. New York: Hill & Wang, 2005.

Jacobson, Matthew Frye. *Barbarian Virtues: The United States Encounters Foreign Peoples at Home and Abroad, 1876–1917*. New York: Hill & Wang, 2000.

Jacoby, Karl. *Shadows at Dawn: A Borderlands Massacre and the Violence of History*. New York: Penguin Press, 2008.

Janney, Caroline E. *Ends of War: The Unfinished Fight of Lee's Army after Appomattox.* Chapel Hill: University of North Carolina Press, 2021.

Johnson, Susan Lee. *Roaring Camp: The Social World of the California Gold Rush.* New York: W. W. Norton, 2000.

Johnson, Walter. *River of Dark Dreams: Slavery and Empire in the Cotton Kingdom.* Cambridge: Belknap Press of Harvard University Press, 2013.

Karnow, Stanley. *In Our Image: America's Empire in the Philippines.* New York: Random House, 1989.

Karp, Matthew. *This Vast Southern Empire: Slaveholders at the Helm of American Foreign Policy.* Cambridge, MA: Harvard University Press, 2016.

Kim, Jessica. *Imperial Metropolis: Los Angeles, Mexico, and the Borderlands of American Empire, 1865–1941.* Chapel Hill: University of North Carolina Press, 2019.

Knatz, Geraldine. *Port of Los Angeles: Conflict, Commerce, and the Fight for Control.* Santa Monica: Angel City Press, 2019.

Kropp, Phoebe S. *California Vieja: Culture and Memory in a Modern American Place.* Berkeley: University of California Press, 2006.

Langum, David J. *Law and Community on the Mexican California Frontier: Anglo-American Expatriates and the Clash of Legal Traditions.* Norman: University of Oklahoma Press, 1987.

Lavender, David. *The Great Persuader: The Biography of Collis P. Huntington.* Niwot: University Press of Colorado, 1969.

Levy, Jonathan. *Ages of American Capitalism: A History of the United States.* New York: Random House, 2022.

Lytle-Hernández, Kelly. *City of Inmates: Conquest, Rebellion, and the Rise of Human Caging in Los Angeles, 1771–1965.* Chapel Hill: University of North Carolina Press, 2017.

Madley, Benjamin. *An American Genocide: The United States and the California Indian Catastrophe.* New Haven: Yale University Press, 2016.

McCurry, Stephanie. *Women's War: Fighting and Surviving the American Civil War.* Cambridge, MA: Belknap Press of Harvard University Press, 2019.

McDonough, James Lee. *William Tecumseh Sherman: In the Service of My Country: A Life.* New York: W. W. Norton, 2016.

McPherson, James. *Battle Cry of Freedom: The Civil War Era.* New York: Oxford University Press, 1988.

Miller, Lillian. *The Lazzaroni: Science and Scientists in Mid-Nineteenth Century America.* Washington, DC: Smithsonian Institution Press, 1972.

Monroy, Douglas. *Thrown among Strangers: The Making of Mexican Culture in Frontier California.* Berkeley: University of California Press, 1993.

Nelson, Megan Kate. *The Three-Cornered War: The Union, the Confederacy, and Native Peoples in the Fight for the West.* New York: Scribner, 2020.

Ngai, Mae. *The Chinese Question: The Gold Rushes, Chinese Migration, and Global Politics.* New York: W. W. Norton, 2021.

Nugent, Walter T. K. *Habits of Empire: A History of American Expansion.* New York: Alfred A. Knopf, 2008.

Orsi, Richard. *Sunset Limited: The Southern Pacific Railroad and the Development of the American West.* Berkeley: University of California Press, 2005.

Pérez, Erika. *Colonial Intimacies: Inter-Ethnic Kinship, Sexuality, and Marriage in Southern California, 1769–1885.* Norman: University of Oklahoma Press, 2018.

Quinn, Arthur. *The Rivals: William Gwin, David Broderick, and the Birth of California.* New York, Crown, 1994.

Rayner, Richard. *The Associates: Four Capitalists Who Created California.* New York, W. W. Norton, 2008.

Reid, Joshua. *The Sea Is My Country: The Maritime World of the Makahs*. New Haven: Yale University Press, 2015.

Reséndez, Andrés. *The Other Slavery: The Uncovered Story of Indian Enslavement in America*. Boston: Houghton Mifflin Harcourt, 2016.

Richards, Leonard L. *The California Gold Rush and the Coming of the Civil War*. New York: Alfred A. Knopf, 2007.

Richardson, Heather Cox. *West from Appomattox: The Reconstruction of America after the Civil War*. New Haven: Yale University Press, 2007.

Sandos, James A. *Converting California: Indians and Franciscans in the Missions*. New Haven: Yale University Press, 2004.

Shelton, Tamara Venit. *A Squatter's Republic: Land and the Politics of Monopoly in California, 1850–1900*. Berkeley: University of California Press, 2013.

Silbey, David J. *A War of Frontier and Empire: The Philippine-American War, 1899–1902*. New York: Hill & Wang, 2007.

Sitton, Tom. *Grand Ventures: The Banning Family and the Shaping of Southern California*. Berkeley: University of California Press, 2010.

Smith, Stacey. *Freedom's Frontier: California and the Struggle over Unfree Labor, Emancipation, and Reconstruction*. Chapel Hill: University of North Carolina Press, 2013.

Sobel, Dava. *Longitude: The True Story of a Lone Genius Who Solved the Greatest Scientific Problem of His Time*. New York: Walker & Company, 1993.

St. John, Rachel. *Line in the Sand: A History of the Western U.S.-Mexico Border*. Princeton: Princeton University Press, 2011.

Torres-Rouff, David. *Before L.A.: Race, Space, and Municipal Power in Los Angeles*. New Haven: Yale University Press, 2013.

Varon, Elizabeth R. *Appomattox: Victory, Defeat, and Freedom at the End of the Civil War*. New York: Oxford University Press, 2014.

Vileisis, Ann. *Discovering the Unknown Landscape: A History of America's Wetlands*. Washington, DC: Island Press, 1997.

Waite, Kevin. *West of Slavery: The Southern Dream of a Transcontinental Empire*. Chapel Hill: University of North Carolina Press, 2021.

Weber, David J. *The Mexican Frontier, 1821–1846*. Albuquerque: University of New Mexico Press, 1982.

West, Elliott. *Continental Reckoning: The American West in the Age of Expansion*. Lincoln: University of Nebraska Press, 2023.

White, Richard. *The Republic for Which It Stands: The United States during Reconstruction and the Gilded Age, 1865–1896*. New York: Oxford University Press, 2017.

———. *Railroaded: The Transcontinentals and the Making of Modern America*. New York: W. W. Norton, 2011.

Wilcove, David S. *No Way Home: The Decline of the World's Great Animal Migrations*. Washington, DC: Island Press, 2008.

Woods, Michael E. *Arguing until Doomsday: Stephen A. Douglas, Jefferson Davis, and the Struggle for American Democracy*. Chapel Hill: University of North Carolina Press, 2021.

Zappia, Natale A. *Traders and Raiders: The Indigenous World of the Colorado Basin, 1540–1859*. Chapel Hill: University of North Carolina Press, 2014.

Zedler, J. B. *The Ecology of Southern California Coastal Salt Marshes: A Community Profile*. FWS/OBS-81/54. Washington, DC: US Fish & Wildlife Service, Biological Services Program, 1982.

Zesch, Scott. *The Chinatown War: Chinese Los Angeles and the Massacre of 1871*. New York: Oxford University Press, 2012.

Illustration Credits

page 87: Library of Congress, Prints & Photographs Division, LC-DIG-pga-09964

page 93: NPG.69.68, National Portrait Gallery, Smithsonian Institution

page 97: Robert B. Honeyman, Jr. Collection of Early Californian and Western American Pictorial Material, BANC PIC 1963.002:0478:08—A, The Bancroft Library, University of California, Berkeley

page 101: *Reports of Explorations and Surveys to Ascertain the Most Practicable and Economical Route for a Railroad from the Mississippi River to the Pacific Ocean*, Volume V, Part II, Geological Report, x F593 .U58, The Bancroft Library, University of California, Berkeley

page 107: *Pacific Coast: Coast Pilot of California, Oregon, and Washington.* Fourth edition. Washington, DC: Government Printing Office, 1889. Rare Books: 491603, The Huntington Library, San Marino, California

page 109: Library of Congress, Prints and Photographs Division, LC-DIG-ppmsca-32191

page 111, left: NPG.77.260, National Portrait Gallery, Smithsonian Institution; partial gift of Dr. Roland A. Bill

page 111, right: NPG.80.113, National Portrait Gallery, Smithsonian Institution; gift of Varina Webb Stewart, conserved with funds from the Smithsonian Women's Committee

page 115: Ord Family Papers 2, Box 20, Folder 17, Georgetown University Library Booth Family Center for Special Collections, Washington, DC

page 122: Library of Congress, Prints and Photographs Division, LC-DIG-pga-03316

page 127: Library of Congress, Prints and Photographs Division, LC-DIG-cwpb-06522

page 133: *Memorial and Biographical History of the Counties of Fresno, Tulare, and Kern* (Lewis Publishing Co., 1892), F 868. A15 M5, The Huntington Library, San Marino, California

page 136: Robert B. Honeyman, Jr. Collection of Early Californian and Western American Pictorial Material, BANC PIC 1963.002:0478:14—A, The Bancroft Library, University of California, Berkeley

page 144, left: Ord Family Papers 2, Box 1, Folder 49, Georgetown University Library Booth Family Center for Special Collections, Washington, DC

page 144, right: Rancho San Pedro Collection rsp_73, The Gerth Archives and Special Collections, California State University, Dominguez Hills

page 152: Workman and Temple Family Homestead Museum, City of Industry, California

page 156: SR_Box_23(14).02, Solano-Reeve Collection, The Huntington Library, San Marino, California

page 167: Robert B. Honeyman, Jr. Collection of Early Californian and Western American Pictorial Material, BANC PIC 1963.002:0478:10—A, The Bancroft Library, University of California, Berkeley

page 171: NPG.66.37, National Portrait Gallery, Smithsonian Institution

page 175: Portraits from the George Davidson Papers, BANC PIC 1946.010—PIC, The Bancroft Library, University of California, Berkeley

page 185: Library of Congress, Prints and Photographs Division, LC-USZ62-29061

page 192: Box 22, Volume 15, Item 1, Banning Family Collection of Photographs, Part II, photCL 451, The Huntington Library, San Marino, California

page 196: Rare Books: 492435, The Huntington Library, San Marino, California

page 200: Library of Congress, Prints and Photographs Division, LC-DIG-pga-10050

page 205: J. Ross Browne, *Adventures in the Apache Country* (Harper & Bros., 1869), Rare Books: 33995, The Huntington Library, San Marino, California

page 213: Library of Congress, Prints and Photographs Division, LC-DIG-cwpb-07382

page 217: SR_Map_0194, Solano-Reeve Collection, The Huntington Library, San Marino, California *page 224*: J. Ross Browne, *Adventures in the Apache Country* (Harper & Bros., 1869), Rare Books: 33995, The Huntington Library, San Marino, California

page 228: J. Ross Browne, *Adventures in the Apache Country* (Harper & Bros., 1869), Rare Books: 33995, The Huntington Library, San Marino, California

page 234: J. Ross Browne, *Adventures in the Apache Country* (Harper & Bros., 1869), Rare Books: 33995, The Huntington Library, San Marino, California

page 237: Library of Congress, Prints and Photographs Division, LC-DIG-stereo-1s00348

page 242, top left: Library of Congress, Prints and Photographs Division, LC-DIG-cwpb-06221

page 242, top right: Library of Congress, Prints and Photographs Division, LC-DIG-cwpbh-01212

page 242, bottom: Library of Congress, Prints and Photographs Division, LC-DIG-cwpb-02928

page 243: CXB 417, California State Lands Commission, Sacramento

page 247: Library of Congress, Prints and Photographs Division, LC-DIG-ppmsca-28282

page 252: J. Ross Browne, *Adventures in the Apache Country* (Harper & Bros., 1869), Rare Books: 33995, The Huntington Library, San Marino, California

page 258: photCL Pierce 05636, The Huntington Library, San Marino, California

page 264: CHS-43320, University of Southern California Digital Library, California Historical Society Collection

page 273: Robert B. Honeyman, Jr. Collection of Early Californian and Western American Pictorial Material, BANC PIC 1963.002:0535—C, The Bancroft Library, University of California, Berkeley

page 281: *Harper's Weekly* (March 16, 1872), Rare Books: 499752, The Huntington Library, San Marino, California

page 284: No. 20389, Security Pacific National Bank Collection, Los Angeles Public Library

page 290: Box 3, Item 420, Banning Family Collection of Photographs, Part II, photCL 451, The Huntington Library, San Marino, California

page 294, left: Rancho San Pedro Colleection, rsp_74, The Gerth Archives and Special Collections, California State University, Dominguez Hills

page 294, right: P457, Anaheim Public Library

page 301: GPF.0513, Seaver Center for Western History Research, Los Angeles County Museum of Natural History

page 309: Library of Congress, Prints and Photographs Division, LC-DIG-pga-07734

page 318: ZNE Binder 7-402-420, California State Lands Commission, Sacramento

page 321: Robert B. Honeyman, Jr. Collection of Early Californian and Western American Pictorial Material, BANC PIC 1963.002:0828—C, The Bancroft Library, University of California, Berkeley

page 327: George Kennan, *E.H. Harriman: A Biography* (Houghton Mifflin, 1922), HE2754.H2 K45, The Huntington Library, San Marino, California

page 330, top: *Out West* 27:4 (October 1907): 312–13, F851 .O9, The Huntington Library, San Marino, California

page 330, bottom: *Out West* 27:4 (October 1907): 312–13, F851 .O9, The Huntington Library, San Marino, California

page 343: *Los Angeles Express*, June 15, 1908, Folder 778, Box 48, Meyer Lissner Papers, m0070, Department of Special Collections and University Archives, Stanford University Libraries

page 346: Library of Congress, Manuscript Division, vol. 4, p. 198, Series 15, Box 86, Theodore Roosevelt Papers

page 353: Box 3, Banning Family Collection of Photographs, Part I, photCL 180, The Huntington Library, San Marino, California

Index

Page numbers in *italics* indicate illustrations.